Visual Prosthesis and Ophthalmic Devices

OPHTHALMOLOGY RESEARCH

JOYCE TOMBRAN-TINK, PhD, AND COLIN J. BARNSTABLE, DPhil

SERIES EDITORS

Visual Prosthesis and Ophthalmic Devices: *New Hope in Sight,* edited by *Joyce Tombran-Tink, PhD, Colin J. Barnstable, DPhil, and Joseph F. Rizzo, MD, 2007*

Retinal Degenerations: Biology, Diagnostics, and Therapeutics, edited by *Joyce Tombran-Tink, PhD, and Colin J. Barnstable, DPhil, 2007*

Ocular Angiogenesis: *Diseases, Mechanisms, and Therapeutics,* edited by *Joyce Tombran-Tink, PhD, and Colin J. Barnstable, DPhil, 2006*

Visual Prosthesis and Ophthalmic Devices

New Hope in Sight

Edited by

Joyce Tombran-Tink, PhD

Department of Ophthalmology and Visual Science
Yale University School of Medicine, New Haven, CT

Colin J. Barnstable, DPhil

Professor and Chair, Department of Neural and Behavioral Sciences
Director, Penn State Hershey Neuroscience Research Institute
Co-Director, Penn State Neuroscience Institute
Penn State University College of Medicine, Hershey, PA

Joseph F. Rizzo III, MD

Ophthalmology Department, Harvard Medical School
Department of Ophthalmology, Massachusetts Eye and Ear Infirmary
Director, Center of Innovative Visual Rehabilitation
Veteran's Administration Hospital, Boston, MA

Humana Press
Totowa, New Jersey

Softcover re-print of the Hardcover 1st edition 2007

999 Riverview Drive, Suite 208
Totowa, New Jersey 07512
humanapress.com

Cover design by Karen Schulz

Cover illustrations: *(Foreground)* Graphics provided by The Intraocular Retinal Prosthesis Lab at the Doheny Eye Institute at University of Southern California. *(Background)* Courtesy of Tun Min Soe (NeoMedix) and Don Minckler, MD, MSc (UC Irvine).

For additional copies, pricing for bulk purchases, and/or information about other Humana titles, contact Humana at the above address or at any of the following numbers: Tel.: 973-256-1699; Fax: 973-256-8341; E-mail: orders@humanapr.com or visit our website at www.humanapress.com

This publication is printed on acid-free paper. ∞

ANSI Z39.48-1984 (American National Standards Institute) Permanence of Paper for Printed Library Materials.

Additional material to this book can be downloaded from http://extras.springer.com.

ISBN 978-1-59745-449-0 (eBook)

Preface

The history of medicine has been substantially defined by a small number of monumental discoveries. Most of these breakthroughs have emerged from the biological sciences. One of the first great breakthroughs was the recognition by Koch in 1884 that pathogens could be transmitted from one living organism to another to cause disease. This profound concept led to a revolution in the approach to patient care that ultimately led to introduction of "sterile" techniques that greatly improved survivals of patients. This knowledge promoted the discovery of antibiotics 40 years later, which dramatically increased life expectancy throughout the more developed parts of the world.

Another great milestone that has influenced medical care was the use of anesthesia for surgery, which was first introduced in 1846. Collectively, these three discoveries armed physicians with the knowledge and means to substantially reduce the prevalence of infectious disease, which was and still remains the leading cause of death throughout the world, and to perform a much wider range of surgeries with greatly improved survivals. The improved life-expectancies enabled the medical community to focus on a wider range of medical problems and solutions to disease.

Today's modern age of medicine is being defined mostly by the benefits of the discovery of the structure of DNA by Watson and Crick in 1953. This landmark discovery enabled the revolution in the diagnosis and treatment of heritable diseases through the use of molecular genetic techniques, including the development of the polymerase chain reaction, which, like the work on DNA itself, was ultimately honored by a Nobel prize. This knowledge spawned the field of proteomics research that is now providing insights into disease mechanisms and new biological therapies.

The physical sciences have also played important roles in the development of the field of Medicine. The most significant contributions have perhaps come from the field of Physics, which provided basic X-ray technology, and roughly 80 years later the giant leap forward with the introduction of computer-assisted imaging in 1972. Since that time, there has been a natural evolution toward more detailed and elaborate imaging methods, especially magnetic resonance imaging, functional magnetic resonance imaging, and positron emission tomography. Not surprisingly, that seminal discovery of computed tomography was also awarded a Nobel prize.

In the last decade, we have seen a fusion of the biological and physical sciences in the development of the field of Ophthalmology. Translational research in these fields has resulted in the design of inorganic materials that can aid, improve, or replace biological functions of the eye. High-quality stereoscopic microscopes enabled the modern age of ophthalmic surgery, which has been defined mostly by the introduction of artificial intraocular lenses. Since Sir Harold Ridley implanted the first artificial lens in 1949, vast improvements in materials and designs over the ensuing two decades have enabled Ophthalmologists to dramatically improve the quality of life for their patients by inserting flexible artificial lenses with minimally-invasive surgical methods. Patients no longer have to endure prolonged "recovery" times, and the visual outcomes are routinely very good. These and other advances, including the very early use of LASER technology, have led to an era of exploding technological ingenuity and have earned the field of Ophthalmology the reputation of being one of the most *avant-garde* fields of medicine.

This textbook, entitled *Visual Prosthesis and Ophthalmic Devices: New Hope in Sight*, provides the most comprehensive overview of the new technologies that are defining the modern age of Ophthalmology. Two themes are presented in this volume. In the first, we explore the interface between the eye and inorganic materials that are being used either to improve the drainage of aqueous fluid in the treatment of glaucoma or to improve the quality of vision by enhancing the ability of the eye to focus images. In the second, we explore the interface between electronics and the nervous system, in which dramatic progress has been made in improving vision with microelectronic devices that bypass irreparable ocular tissue malfunction. The technologies presented in this textbook have been developed through collaborations with many types of scientists, most notably optical engineers, materials scientists, electrical engineers, microfabrication specialists and circuit designers. This collective body of work provides a perspective on a stunning array of new technologies that redefines the limits of medical mechanics and that will influence the delivery of care to patients with many types of ophthalmic problems. Some of these innovations are "platform technologies" that with some modification could likely be used to treat other types of medical problems.

Most of the technologies discussed in this book have been developed only within a couple of decades. This pace of development and implementation is a very impressive accomplishment, given the extremely challenging obstacles that are always in the path of innovative technologies that are intended for implantation into humans.

The field of Ophthalmology is immersed in a new, exciting, and seemingly endless age that holds the promise for new diagnostic and therapeutic options to treat visual disorders. The multidisciplinary scientific approach that was required to develop the devices described in this book will likely inspire others to challenge current boundaries that limit integration of microelectronics and medicine. The authors of this textbook serve as role models for a new generation of students who have become energized by this type of applied biomedical research. Today's students will be the torchbearers in the future for biomedical innovations that restore vision to the visually-impaired and that protect vision for those patients who are under the threat of blindness.

Joyce Tombran-Tink, PhD
Colin J. Barnstable, DPhil
Joseph F. Rizzo III, MD

Contents

Contributors

JORGE L. ALIÓ • *Instituto Oftalmológico de Alicante and Division of Ophthalmology, Medical School, Miguel Hernández University, Alicante, Spain*
ALBERTO ARTOLA • *Instituto Oftalmológico de Alicante and Division of Ophthalmology, Medical School, Miguel Hernández University, Alicante, Spain*
KWABENA BOAHEN • *Department of Bioengineering, Stanford University, Palo Alto, CA*
GERALD J. CHADER • *Doheny Retina Institute, University of Southern California Medical School, Los Angeles, CA*
PAUL CHEW • *Department of Ophthalmology, National University Hospital, Singapore*
DONG-IL DAN CHO • *Nano Bioelectronics and Systems Research Center, Nano Artificial Vision Research Center, School of Electrical Engineering and Computer Science, Seoul National University, Republic of Korea*
ALAN Y. CHOW • *Optobionics Corporation, Naperville, IL*
VINCENT Y. CHOW • *Optobionics Corporation, Naperville, IL*
VIVEK CHOWDHURY • *Department of Ophthalmology, Prince of Wales Clinical School, University of New South Wales, Sydney, Australia*
HUM CHUNG • *Nano Bioelectronics and Systems Research Center, Nano Artificial Vision Research Center, Department of Ophthalmology, Seoul National University School of Medicine, Republic of Korea*
MINAS T. CORONEO • *Department of Ophthalmology, Prince of Wales Clinical School, University of New South Wales, Sydney, Australia*
CHERYL L. CULLEN • *Faculty of Veterinary Medicine, University of Calgary, Calgary, AB, Canada*
REINHARD ECKHORN • *Department of Physics, Philips University, Marburg, Germany*
HANY EL SAFTAWY • *Research Institute of Ophthalmology, Giza, Egypt*
HEINRICH GERDING • *Augenzentrum, Klinik Pallas, Olten, Switzerland*
YONG SOOK GOO • *Nano Bioelectronics and Systems Research Center, Nano Artificial Vision Research Center, Department of Physiology, Chungbuk National University School of Medicine, Republic of Korea*
HENRY L. HUDSON • *Retina Centers PC, Tucson, AZ*
MARK S. HUMAYUN • *Doheny Eye Institute, Keck School of Medicine, University of Southern California, Los Angeles, CA*
LEE J. JOHNSON • *SFA, Inc., Naval Research Laboratory, Washington, DC*
MARIA I. KALYVIANAKI • *Department of Ophthalmology, University Hospital of Heraklion, Heraklion, Crete, Greece*
MONICA KENNEY • *Department of Ophthalmology, Massachusetts Eye and Ear Infirmary, Boston, MA*
EUITAE KIM • *Nano Bioelectronics and Systems Research Center, Nano Artificial Vision Research Center, School of Electrical Engineering and Computer Science, Seoul National University, Republic of Korea*

SUNG JUNE KIM • *Nano Bioelectronics and Systems Research Center, Nano Artificial Vision Research Center, School of Electrical Engineering and Computer Science, Seoul National University, Republic of Korea*
KYO-IN KOO • *Nano Bioelectronics and Systems Research Center, Nano Artificial Vision Research Center, School of Electrical Engineering and Computer Science, Seoul National University, Republic of Korea*
SÉRGIO KWITKO • *Department of Ophthalmology, Paulista School of Medicine, Sao Paulo, Brazil; Fellow in Cornea and External Diseases at the Doheny Eye Institute, University of Southern California, Los Angeles, CA; Ophthalmologist of Cornea and External Disease Service, Hospital de Clínicas de Porto Alegre, Brazil*
GEORGE D. KYMIONIS • *Institute of Vision and Optics, University of Crete Medical School, Heraklion, Crete, Greece*
WENTAI LIU • *Department of Electrical Engineering, University of California, Santa Cruz, CA*
JOHN W. MORLEY • *School of Medical Sciences, University of New South Wales; School of Medicine, University of Western Sydney, Sydney, Australia*
RUDY M. M. A. NUIJTS • *Department of Ophthalmology, Academic Hospital Maastricht, Maastricht, The Netherlands*
IOANNIS G. PALLIKARIS • *Institute of Vision and Optics, University of Crete Medical School, Heraklion, Crete, Greece*
JOSEPH F. RIZZO III • *Department of Ophthalmology, Massachusetts Eye and Ear Infirmary; Director, Center of Innovative Visual Rehabilitation, Veteran's Administration Hospital, Boston, MA*
DEAN A. SCRIBNER • *Department of Optical Sciences, Naval Research Laboratory, Washington, DC*
JONGMO SEO • *Nano Bioelectronics and Systems Research Center, Nano Artificial Vision Research Center, Department of Ophthalmology, Seoul National University School of Medicine, Republic of Korea*
MOHAMED H. SHABAYEK • *Vissum, Instituto Oftalmológico de Alicante and Division of Ophthalmology, Medical School, Miguel Hernández University, Alicante, Spain and Research Institute of Ophthalmology, Giza, Egypt*
MOHANASANKAR SIVAPRAKASAM • *Department of Electrical Engineering, University of California, Santa Cruz, CA*
LAURA SNEBOLD • *Department of Ophthalmology, Massachusetts Eye and Ear Infirmary, Boston, MA*
JÖRG SOMMERHALDER • *Eye Clinic, Geneva University Hospitals, Geneva, Switzerland*
NAYYIRIH G. TAHZIB • *Department of Ophthalmology, Academic Hospital Maastricht, Maastricht, The Netherlands*
GUOXING WANG • *Department of Electrical Engineering, University of California, Santa Cruz, CA*
JENN-CHYUAN WANG • *Department of Ophthalmology, National University Hospital, Singapore*
JAMES D. WEILAND • *Doheny Eye Institute, Keck School of Medicine, University of Southern California, Los Angeles, CA*
JANG HEE YE • *Nano Bioelectronics and Systems Research Center, Nano Artificial Vision Research Center, Department of Physiology, Chungbuk National University School of Medicine, Republic of Korea*

Young Suk Yu • *Nano Bioelectronics and Systems Research Center, Nano Artificial Vision Research Center, Department of Ophthalmology, Seoul National University School of Medicine, Republic of Korea*

Kareem A. Zaghloul • *Department of Neurosurgery, Hospital of the University of Pennsylvania, Philadelphia, PA*

Jingai Zhou • *Nano Bioelectronics and Systems Research Center, Nano Artificial Vision Research Center, School of Electrical Engineering and Computer Science, Seoul National University, Republic of Korea*

Mingcui Zhou • *Department of Electrical Engineering, University of California, Santa Cruz, CA*

Companion CD

Color versions of illustrations listed here may be found on the Companion CD attached to the inside back cover. The image files are organized into folders by chapter number and are viewable in most Web browsers. The number following "f" at the end of the file name identifies the corresponding figure in the text. The CD is compatible with both Mac and PC operating systems.

1

Retinal Prosthetic Devices

The Needs and Future Potential

Gerald J. Chader, PhD

Inherited retinal degenerations (RDs) are a leading cause of severe vision loss and blindness throughout the world. In most cases, the vision loss is irreversible and untreatable. This leads to many problems for the affected person and for society. For the individual, difficulty in performing everyday functions increases dramatically. Also, job-related duties suffer or must be abandoned. Similarly, quality of life is degraded as face recognition and reading ability become difficult or impossible. Loss of independence, especially in elderly but otherwise healthy individuals, is a significant factor that, along with the actual loss of vision, can lead to severe depression. For society, the cost is more impersonal but no less significant. Costs of care of the visually impaired are significant and often are for an extended period of time. Loss of productivity of the individual must also be calculated into the equation. In the case of early-onset RD (e.g., Leber Congenital Amaurosis) this can be an entire lifetime.

What is "inherited RD"? Inherited RD is not a single entity but rather a heterogeneous family of retinal diseases that have the final common end point of vision loss or complete blindness. There are two main groupings within the family, the retinitis pigmentosa (RP) group and the macular degeneration (MD) group. The MD group is usually further divided into the early-onset MDs (e.g., Stargardt disease) and the late-onset age-related macular degeneration (AMD). Phenotypically, the hallmarks of the RP-group of diseases are loss of peripheral and night vision, signifying an early role of rod dysfunction in these diseases. For the MDs, loss of central (macular) vision is an early symptom, indicative of loss of cone function. Genotypically, genetic mutations (dominant, recessive, X-linked) play virtually a singular role in causing the RP-types of disease. To date, more than 100 genes have been identified whose mutations lead to a form of RP. Early-onset MDs like Stargardt disease also seem to be caused by mutations in a single gene, which lead to the disease phenotype. In contrast, AMD is a complex disease with both genetic and environmental components. It is only recently that several genes have been identified whose mutations lead to or predispose to AMD. Clearly though, factors such as smoking also greatly increase the risk of developing AMD.

How common is RD? The numbers of RP and early-onset MD patients in any one country are not very large—roughly about 100–200,000 affected individuals in the

From: *Ophthalmology Research: Visual Prosthesis and Ophthalmic Devices: New Hope in Sight*
Edited by: J. Tombran-Tink, C. Barnstable, and J. F. Rizzo © Humana Press Inc., Totowa, NJ

United States. Comparable numbers are estimated in other populations with a worldwide prevalence of 1:2000 to 1:4000 (depending on the data source quoted). However, with AMD it is estimated that there are more than 6 million affected individuals in the United States alone. It is the leading cause of severe visual impairment in people more than 55 yr of age. Comparable numbers appear to be present in Europe based on a fairly similar population. In other areas such as Africa and the Far East, less information is available but it has been believed that the numbers are fewer. In Japan, for example, it has been assumed that AMD numbers are low but it is now clear that either these numbers have been underestimated or that AMD is rising dramatically. The number of patients with AMD in industrialized countries of North America and Europe are predicted to rise to "epidemic" levels in the next two decades. Thus, the number of individuals who are losing their sight owing to RD worldwide are currently in millions with increasing millions at risk. Notably, no treatment for RP is currently available and, for patients with dry AMD (estimated at about 90% of all AMD patients), treatments are also lacking.

Good vision depends on good function of photoreceptor cells in the retina. In turn, photoreceptor cells depend on good function of choroid and retinal pigment epithelial (RPE) cells as well as secondary neurons with which they interact in the neural retina. Thus, the RDs are caused by simple mutations in important photoreceptor cell genes (e.g., the opsin gene), but can also be caused by problems in RPE and chroidal cells caused by other gene mutations (e.g., *RPE65* gene mutations). As mentioned previously, environmental effects such as smoking must also be taken into consideration in AMD. It is probable that actual photoreceptor cell death is by apoptosis—programmed cell death. Thus, therapies that delay apoptotic cell death are being considered as RD treatments along with gene or cell replacement therapies.

With these considerations, what theoretical options do patients with RD have for future treatment and how do these prospective treatments compare with implantation of a retinal prosthesis? Gene therapy, for example, has the potential of virtually "curing" an RD as is seen in canine experiments on RPE65 gene replacement therapy. However, long-term efficacy and safety of gene therapy has yet to be firmly established. Also, less than one-half of genes causing the RDs have been identified, leaving the remaining half untreated. Also, gene therapy can only be performed when viable photoreceptor cells are present. Many of the RDs progress rapidly or are caught too late in life for successful gene therapy.

In a second treatment paradigm, many neuron-survival agents are now known that can delay photoreceptor neuron death in RP animal models. However, many of these effects are short lived and probably need repeated administration. Some of these agents are known to have adverse side effects, whereas many others still need rigorous testing. Drug delivery to the retina is also a problem as pharmaceutical treatments for neurotrophic agents under current consideration involve invasive procedures. As with gene therapy, viable photoreceptor cells need to be present for such pharmaceutical therapy to be effective. As a third treatment option, photoreceptor cell transplantation offers the theoretical possibility of sight restoration in cases in which photoreceptor cell death is significant or even complete. However, studies in this area have met with limited success, although human trials in progress have shown some promise. Thus, potential therapies other than the electronic prosthetic device could ultimately result in at least partial sight restoration in specific populations of patients with RD. However, some of these are only very short-term solutions and others have significant safety issues. Some are yet to be fully validated on the basis of proof of principle. Most depend on the presence of viable photoreceptor cells, limiting the usefulness of the therapy to smaller numbers of patients at relatively early stages of disease.

Electronic prostheses thus offer an opportunity for sight restoration that other potential treatment modalities cannot now and might never be able to fulfill. As seen in the chapters in this book, approaches to chip development vary widely. However, in most, electrodes are fashioned to interact with retinal cells, delivering the visual signal captured by a video device. These electrodes may be subretinal or epiretinal and are limited in number—currently less than 100. However, a much larger number of electrodes are envisioned for future iterations of the chip that should lead to vast improvements in functional vision. Essentially, this should improve vision from restoration of light perception and simple shape discernment to reading ability and face recognition.

Despite years of development and the fact that some human trials are currently taking place, challenges in several areas of chip development are yet to be overcome. Long-term safety issues are still to be addressed, mainly as related to the biological interface between the chip and the tissue. Even with great flexibility and low weight, the implant (and any associated tack) will still have the potential of eroding underlying delicate retinal tissue along with damaging it with electrical energy. Conversely, the safety of the chip itself as well as signal strength and modulation needs to be assured over an extended time period as the ocular milieu must be considered to be hostile to implanted materials. Luckily, much evidence from both animal and human experiments already indicates the relative safety of chip implantation. The electronics need to be upgraded to such an extent that chips with 1000 or more electrodes will function reliably. As this signaling capacity rises, so do safety concerns.

Along with the immediate biological interface, problems with the inner retina and brain recognition of the transmitted signal as discernable visual patterns need to be addressed—probably on an individual patient basis. Essentially, are there enough inner retinal neurons left in the patient's eye to act as a "platform" for the implant and are they functional? Several investigators have morphologically examined RP eyes and have found everything from relatively intact retinas to almost total degeneration. On the average though, the inner retinal layers seem to remain alive and relatively intact long after photoreceptors in the outer retina degenerate. For example, some researchers have found up to 90% preservation of inner retina in donor eyes of patients with AMD. In RP, up to 80% of the inner nuclear layer can remain with 30% preservation of ganglion cells. These numbers compare favorably to the ear in which the cochlear implant requires only about 10% of surviving neurons to function.

Along with numbers of surviving inner retinal neurons, the question that remains for each individual patient is whether neuronal wiring of the remaining cells is adequate to allow for passage of a proper visual signal to the brain. In retinal regions with significant photoreceptor loss, long axonal-like processes (neurites) sprout and become associated with surfaces of inappropriate, secondary neurons or glia. These circuitry changes could secondarily contribute to the ERG abnormalities and progressive decline in vision seen in RP patients. However, presently it does not appear that these neurite abnormalities are severe enough in most cases to preclude passage of the visual signal to the brain. Although, patients are in small numbers, it is clear that some patients who had little or no vision prior to chip implantation have partial restoration of vision. Moreover, owing to central nervous system plasticity, it could very well be that proper circuits "rearrange" or are impressed on the retina once the chip is in place and relatively normal electrical signals are again received.

Finally, even if the chip were to allow for normal retina functioning, what will the brain perceive? Nothing? A garbled pattern? Something approaching a normal image? Here, everyone must contend with the "use or lose" phenomenon of neural activity: essentially, disuse for extended periods might preclude restoration of function. Again, the success of the cochlear implant gives us hope as cited earlier for similar success with the retinal implant. Also, the success of the limited 16 electrode retinal implant in some patients points to the ability of the brain ultimately to receive, synthesize, and make sense out of even crude inputs.

Despite all these caveats and problems, it is clear that the retinal prosthetic device has great potential in the future for sight restoration. The medical and social needs are great. The financial reward is also great with savings of medical costs and with the decreased need for long-term government assistance. Certainly, quality of life will be enhanced. Regarding problems, the improvement in the electronics needed to sharpen the visual image above the current level is certainly attainable. Safety issues seem to be satisfied, at least in the short-term and the biological interface and transmission issues do not seem insurmountable. Thus, no other treatments will be available nor will be as widely applicable to severely impaired patients with RD. In the final analysis, the cost of development of the chip is substantial and challenges remain, but who can put a price tag on restoring a person's sight?

2

Spatial-, Temporal-, and Contrast-Resolutions Obtainable With Retina Implants

Reinhard Eckhorn

CONTENTS

INTRODUCTION

Two Types of Retina Implants

In blind patients with degenerated photoreceptors a large percent of inner retinal neurons remain histologically intact and stay functionally alive *(1)*. Without visual input, the retinal ganglion cells continue to transmit spontaneously generated action potentials through the optic tract to the visual centers. They can also be artificially activated by trains of short electrical impulses and such stimuli were shown to elicit localized perceptions of light, called phosphenes *(2–5)*. This finding opened the possibility to restitute basic vision with retina implants by delivering electrical stimuli to the retina. Two main types of retina implants are currently developed using either sub- or epiretinal electrode arrays (*see* the other reviews in this Volume). Subretinal devices are implanted between the pigment epithelial layer and the outer layers of the retina so that they can activate the outer plexiform layer and bipolar cells *(6)*. Epiretinal implants are placed from the vitreous side onto the ganglion cell and nerve fiber layer of the retina *(7)*. Epiretinal implants can evoke action potentials in retinal ganglion cells and/or their axons *(8)*. Finally, both types of implants send visual information in the form of spike patterns through the axons of retinal ganglion cells along the optic nerve to central visual structures. One crucial question remains regarding the type of neural responses evoked in the visual cortex by stimulation with retina implants. Are the spatial, temporal, and intensity resolutions obtainable with current technologies sufficient, in order to provide useful vision for discriminating objects in a static environment and to perceive

From: *Ophthalmology Research: Visual Prosthesis and Ophthalmic Devices: New Hope in Sight*
Edited by: J. Tombran-Tink, C. Barnstable, and J. F. Rizzo © Humana Press Inc., Totowa, NJ

motion in dynamic scenes of everyday life? Data for estimates of perceptual resolutions achievable with retina implants are rare. Therefore, it was decided to combine knowledge from psychophysical experiments, simulating "pixelized" visual input in human vision, with the few data available from humans with degenerations of outer retinal cells. In addition, at the present state of retina-implant investigations, data from animal experiments with intact and degenerated visual systems are particularly valuable because extensive data were collected including aspects that could not yet satisfactorily, or not at all, be analyzed in blind patients.

Two Operating States: Object Recognition and Dynamic Perception Require Different Resolutions

Object recognition in normal vision is mainly based on high resolution of contrast and space and requires only a small range of sensitivity for slow motion because best performance exists during ocular fixation of the object. For this task the visual system is specialized in the representation of its central visual field. *Perception of dynamic scenes*, present with moving visual objects or induced by movements of the observer in a static environment, requires the processing of temporally changing inputs for a broad range of velocities but needs less resolution in the domains of space and contrast levels. For these tasks the visual system is well adapted in its peripheral visual-field representation. Between these extreme capabilities exists a smooth gradient, already present in the receptive field properties of retinal ganglion cells, which change dramatically from central to peripheral locations. This inhomogeneity in processing of the visual field has to be considered for retina implants. Hence, the stimulus locations of the electrodes have to be chosen carefully in order to obtain useful vision. In addition, the artificial activation of the retinal ganglion cells at different eccentricities of the retina require adequate coding of visual inputs into electrical stimulation sequences.

Estimates of Spatial and Temporal Resolutions From Anesthetized Cats

The most extensive investigations with respect to achievable resolutions with retina implants have been performed with animals (review in ref. *8*). Why is it possible to estimate potential perceptual resolutions in blind humans from experiments in anesthetized animals? First, there exists broad knowledge about the visual systems of cats *(9)*. Second, for identical visual stimuli, human visual perception has been shown to correlate with the receptive field properties of visual cortical neurons in lightly anaesthetized monkeys and cats (overview in ref. *10*). Third, one can conclude that perception of visual details requires the activation of neurons in this area *(11)*. Fourth, estimates of perceptual temporal resolution can be obtained from the time-course of cortical responses to focal retinal impulse stimuli *(12)*. Finally, the spatial extent of the cortical activations to such stimuli can give conservative estimates of spatial visual resolution *(8)*.

SPATIAL RESOLUTION

Ranges of Spatial Resolution

The neural mechanisms of object recognition are, in the intact peripheral visual system, mainly based on the capability of processing and representing local variations of contrast. A small retinal patch, corresponding to a small circular receptive field in visual space, primarily influences the output spike pattern of a single ganglion cell. Hence, spatial resolution is mainly determined by the retinal-processing network that has fed forward input to a single retinal ganglion cell. There is high resolution in the center of view and a continuous decrease to the periphery of the visual field. For clinical classification of spatial resolution three ranges have been defined that enable different basic performances *(13)*. A resolution that is better than 0.5° of visual angle is rated as "seeing" (visual acuity 1/35). This is the ultimate goal in the development of retina implants. A resolution of 2° provides the recognition of objects important for taking meals, for washing, and dressing. A resolution of 10° is sufficient for *mobility and orientation*, allowing subjects to perceive the presence of large objects and their movement directions. Even this low resolution would be highly valuable for blind people, gaining much improvement of performance and quality in daily life.

Limitation by Size of Electrode

The spatial resolution, obtainable with electrical stimulation is limited by the diameter of the stimulation electrodes. An electrode diameter of 0.3 mm at the human retinal surface covers about 1° visual angle. High resolution would therefore require the use of small electrodes, which provide the additional advantage of smaller required electrical charge for activating a retinal neuron, because the current densities and their local gradients, essential for activation, grow with decreasing surface area when the overall delivered charge is kept constant. However, the electrode sizes should not fall below the safety limits of current densities for safe biocompatible activation for long periods and for erosion of the electrodes' metal surface. The safe charge delivery capacities of electrodes used for neural stimulations has been determined for platinum as 100 $\mu C/cm^2$ and for iridium oxide as 1 mC/cm^2 *(14)*. The thin-film electrodes used in the cat experiments *(8,15)* had mostly diameters of 100 µm, corresponding to 0.5° of visual angle, so that the spread of retinal and cortical activation covered even larger angles of visual representation. In fact, electrodes giving rise to more localized high current densities, like the fiber-cone electrodes with approx 20 µm base diameter *(16)*, revealed highest resolutions of about 0.7° with epiretinal stimulation. On the other hand epiretinal film electrodes of 100 µm diameter reached only 1.2°, estimated from the cortical response distribution. With subretinal stimulation, and rectangular film electrodes of $100 \times 100\ \mu m^2$, similar resolutions of 0.9–1.3° were obtained.

Limitations by the Stimulation Current-Field

With electrodes smaller than 50 µm the retinal distribution of the stimulation current field is probably the most influential factor, because even from a point-tipped electrode this field expands, and becomes weaker, with distance between electrode tip and from the retinal neurons to be excited. Schanze and coworkers *(17)* found with cone-tipped microelectrodes in cat retina that the threshold current for activating retinal neurons doubles when the electrode is retracted from the inner limiting membrane by about 45 µm. Interestingly, the size of the activated patches in cat visual cortex did not significantly expand with stimulation current *(8)* and therefore it is assumed that the increase in the size of phosphenes is small with increasing current.

Limitations by the Degree of Retinal Degeneration

The earlier-discussed limitations of spatial resolution as a result of the size of the required stimulation electrode and the spread of current fields with distance were mainly based on data collected from intact visual systems. However, Rizzo and coworkers *(18)* found with RP-patients and epiretinal stimulation that the threshold stimulation current, required for evoking phosphenes, increases with the degree of retinal degeneration. It is argued that probably several intact retinal ganglion cells have to be activated simultaneously in order to elicit a percept. As the density of intact retinal ganglion cells decreases with increasing degeneration *(19)*, and the extent of the activating electrical field of an electrode increases with increasing current, this would explain the observed increase in threshold current with the degree of degeneration. However, larger currents require larger electrodes for safe and reliable stimulation, leading to lower spatial resolution with increasing degeneration. Even if no larger electrodes would be used with higher degree of degeneration, the spatial resolution is expected to decrease with the density of intact ganglion cells if their density is evenly distributed across the retina.

Multiple-Site Phosphenes With Axon and Soma Activations

With epiretinal stimulation the current fields may not only activate retinal neurons in the inner nuclear layer, but may also excite the fibers from neurons with receptive fields remote from the stimulation site, passing to the optic disk directly below the electrodes. If this is the case the activated retinal locations can consist of two parts, one localized directly below the electrode, probably with concentric shape. The other patch with receptive fields distally to the electrode position with respect to the optic disk and probably with elongated shape (e.g., *see* ref. *20*), evoking potentially two spatially segregated phosphenes by a single electrode. This problem can be reduced with high-aspect cone electrodes protruding with their tips through the inner fiber layer to the nuclear layer of the ganglion cells. Another possibility is usage of special forms and time-courses of electrical fields, optimized for selective activation of ganglion cells (review in ref. *21*).

Limitations by Spacing and Number of Electrodes

Minimal requirements of spatial resolution for useful artificial vision have been determined by psychophysical methods with normally sighted humans, by using "pixelized" vision. As this was intended to mimic phosphene vision with retina implants, the discrete pixels of a visual image were selected in their size, position, and spatial arrangement such that they correspond as near as possible to the locations of electrodes in a potential retinal implant and to the expected extent of phospenes *(13,22,23)*. Reading at central retinal projection sites requires about 300 pixels (20×15) within about 5° of retinal eccentricity *(23)*. If a physiologically achievable resolution of 0.75° is taken and each image pixel is substituted by a stimulation electrode with constant spacing the stimulation array will cover 3.9×2.8 mm^2 retina ($7.1 \times 5.3°$ visual eccentricity). If safe navigation in dynamic outdoor environments is required, stimulation should include eccentricities of at least 15° *(22,24)*. As the spatial resolution required for safe navigation outside the reading area can be chosen to be much coarser (at about 2° electrode pitch) this would add about another 200 electrodes for the outer "navigational belt."

Hence, as a minimal condition a total of about 500 well-functioning electrodes with comparable stimulation capabilities would be required for an implant usefully spanning a retinal stimulation area of about 8 mm in diameter.

Spatial Resolution Obtained From Recordings in Cat Primary Visual Cortex

Spatial resolution was estimated in the cat investigations on the basis of the retino-cortical spread of activation in response to a focal electrical stimulus in the retina *(8)*, which can be transformed to degrees of visual angle. This measure is important because it is directly related to the spatial extent of perceived phosphenes and cortical neurons outside the activation range, not coupled to the focal stimulus, and thus will not contribute to perception. With small fiber-cone electrodes *(16)* and epiretinal stimulation highest resolutions of about 0.7° of visual angle were obtained, whereas only 1.2° of flat foil-array electrodes of 100 μm diameter were reached. Subretinal stimulation with flat foil-array electrodes of 100 μm diameter reached resolutions of 0.9–1.3° of visual angle. The cortical activation spread, used for these investigations, may underestimate the actually obtainable spatial resolution, because visual focal stimuli result in comparable sizes of cortical activity distributions *(25)* and are associated with visual perceptions of much higher resolutions.

With the current technology of retinal stimulation individual ganglion cells cannot be activated singly by a single electrical impulse, instead near-synchronized responses are evoked from a small patch of cells in the retina. Each activated ganglion cell responds only with a single spike to epi- and subretinal impulse stimuli *(8)*. Quite importantly, one cannot differentiate which types and combination of types of retinal ganglion cells were activated, i.e., on- or off-center, magno-, parvo-, or konio-cellular neurons (for cell types *see* refs. *26,27*). However, it is assumed that all types contributed to the population responses, probably rendering vision transmitted by such arrays as colorless as the S-, M-, and L-cone systems are activated simultaneously. This has experimentally been confirmed with epiretinal stimulation in human subjects evoking phosphenes that were reported as *bright*, not *dark (3)*. It also implies that it is not possible to activate the retino-cortical network with electrical stimulation in the same way as with visual stimuli. Accordingly, with electrical stimulation, local contrasts will probably be processed at lower spatial and intensity resolutions than are obtained with visual stimuli in the intact system.

TIME AND CONTRAST RESOLUTIONS

Linked Time and Contrast Resolutions

Retinal ganglion cells transmit the visual information to the higher visual centers by temporal modulations of their spike rates. Thus, aspects of time and local contrast are carried by the same signal and their resolutions are intimately related. As the maximal spike rate of a ganglion cell is limited, the number of different successive contrast values that can be transmitted in the absence of noise, characterized by Shannon information, is mainly determined by the amount of time available for transmitting a single contrast value *(28–32)*. For example, if a neuron can discharge maximally 31 action potentials during 333 ms, which is the average duration of ocular fixation, it can theoretically code one out of 32 different intensity values within each of the three fixation intervals per second, or 5 bit of information in 333 ms and accordingly 3 × 5 = 15 bit/s. This resolution seems sufficient for most visual fixation tasks in which spatial details and high resolution of contrasts play a prominent role. If, on the other hand, a pedestrian wants to cross a road in heavy traffic, local contrast resolution is of minor importance and may be reduced to only four different values, corresponding to 2 bit. However, in such a situation temporal resolution is of vital importance and needs to be high. The example neuron can at best signal 2 bit by 0–3 spikes during a time window of 32.26 ms, hence, within 1 s it can transmit 31 times 2 bit equal to 62 bit/s. This example demonstrates that a reduction in intensity resolution, from 32 down to four values per sample results in a potential increase in temporal resolution from 3 to 31 samples or frames per second, in accordance with an overall increase in the rate of information from 15 to 62 bit/s. Can the retino-cortical pathway adapt flexibility in its temporal integration windows to these different situations? It can in fact control the integration time in a range from 5 to 50 ms (review in refs. *33–35*) by changing the average activation level of a neuron, which changes its membrane impedance. With fast changing scenes high average rates of activity are present causing high temporal resolution, whereas ocular fixation causes lower spike rates, and hence, longer integration windows. As a consequence, retina implants have to generate appropriate spike density coding so that the visual centers can use their flexible strategy in which the trade-off between intensity and time resolution is continuously optimized.

Estimated Resolutions From Information Transmitted to Cat Visual Cortex

The rate of information, evoked by a random-interval stimulus train, at a single retinal site, to a single cortical recording location in cat primary-visual cortex, has been determined for visual and for electrical stimuli by Eger and coworkers *(32)*. Visual stimuli were sequences of focal light flashes applied to the same location where the electrical impulses were applied having the identical interval pattern. Light flashes at rates between 2 and 10 Hz show a maximum of transmitted information of about 20–40 bit/s. At high flash rates of 20–50 Hz, transmission of information steeply declined, probably because of the retinal high frequency cut-off at more than 10 Hz (e.g., *see* ref. *36*). Striking differences between electrical and visual stimulation were observed. First, in contrast to visual stimulation with a spread of information across several recording locations, electrical stimulation delivered high rates of information only to a single focal cortical patch with very little information spread toward neighboring cortical locations. Second, using electrical stimulation, information rates increased continuously in parallel with stimulation rates up to the highest applied rate of 80 Hz (details in ref. *32*), whereas they showed a peak at 4–8 Hz flash rates when visual stimuli were applied. Rates of information transmitted by a single stimulation electrode to a single cortical recording site ranged from 20 to 100 bit/s. If 30 bit/s is chosen in order to estimate conservative values of contrast resolutions for the two extreme cases of high temporal resolution (30/s) and three fixations per second, respectively, the following values are obtained. With 30 bit/s and 30 frames/s, only 1 bit/frame is left for contrast coding, corresponding to two levels *(bright* and *dark)*. If a visual object can nearly be stabilized on the optical sensor array of a retina implant for intervals of 100 ms, then 3 bit would be available during that window, with a potential of discrimination for eight different contrast levels. However, these estimates are made under simplifying assumptions, including boxcar temporal integration windows and full extraction of the available information by the cortical neurons involved in phosphene perception. Although, these simplifications may lead to overestimated values it has to be kept in mind that the example estimation is based on a low information rate (30 bit/s) compared with what often has been measured (60–80 bit/s) *(32)*. The temporal resolution with retina implants, estimated from the time-courses of the neural responses in cat primary-visual cortex, are similar to those reported with natural vision and with electrical stimulation in blind humans. The average temporal resolution in the part of the visual system, representing the central visual field, is about 30 frames/s or 33 ms *(8)*.

Contrast Coding by High Impulse Rates

Suprathreshold stimulus pulse trains, applied to the retinae of RP-patients, evoked nonflickering perception of more than 40–50 Hz *(3)*. These patients described their visual percept as growing brighter with increasing stimulus frequency of more than 40–50 Hz. Thus, impulse trains more than the critical flicker fusion frequencies induced perception of brightness, increasing with stimulation frequency. As retinal ganglion cells can discharge up to 50 spikes within 50 ms (spike interval 2 ms; then adapting to lower rates), the intensity coding would have a dynamic range between about 3 and 25 spikes in this interval. It depends on the level of spontaneous activity (formally termed *noise* here) how many contrast levels can be coded in this interval. However, a blind RP-patient with implanted epiretinal electrodes could discriminate 10 levels of phosphene brightness after 10 wk training *(4)*. If this capability is related to the activation of retinal ganglion cells and if logarithmic scaling of brightness coding is assumed with stimulation rate then an increase in spike rate by about 45% for each step would explain the discrimination capability of 10 levels by the patient. At least with epiretinal stimulation, the spike rate in ganglion cells is directly coupled to stimulation rate, because each stimulus impulse elicits one action potential in cat optic tract *(8)*. With low-amplitude stimulation from the outer retinal side of an isolated chicken retina only single action potentials were evoked in retinal ganglion cells. However, an increase in delivered charge generates bursts of action potentials with this subretinal stimulation. As the bursts were fired at high frequencies they might probably code phosphene brightness by the intraburst frequency and the duration of the burst. Hence, with subretinal devices brightness might be coded by modulating the stimulation current, or at low current values, by modulating the stimulation frequency at more than the critical flicker fusion frequency.

ACKNOWLEDGMENTS

The author would like to acknowledge the cooperation of many colleagues involved in the cat retina implant project (for the presentation of the retina implant consortium *see* www.uni-tuebingen.de\rim; www.nero.uni-bonn.de; www.neuro.physik.uni-marburg.de/~retina-implant). I am also very grateful for the support by the German Ministry of Research and Technology (01KP0007, 01KP0008).

REFERENCES

1. Santos A, Humayun M, de Juan E, et al. Preservation of the inner retina in retinitis pigmentosa. Arch Ophthal 1997;115:511–515.
2. Humayun MS, de Juan E, Dagnelie G, Greenberg RJ, Probst RH, Phillips DH. Visual perception elicited by electrical stimulation of retina in blind humans. Arch Ophthalmol 1996; 114:40–46.
3. Humayun MS, de Juan E, Weiland JD, et al. Pattern electrical stimulation of the human retina. Vision Res 1999;39:2569–2576.
4. Humayun MS, Weiland JD, Fujii GY, et al. Visual perception in a blind subject with a chronic microelectrode retinal prosthesis. Vision Res 2003;43:2573–2581.
5. Rizzo JF, Wyatt JL. Prospects for a visual prosthesis. Neuroscientist 1997;3:251.
6. Zrenner E. Will retinal implants restore vision? Science 2002;295:1022–1025.
7. Eckmiller R. Learning retina implants with epiretinal contacts. Ophthalmic Res 1997;29: 281–289.
8. Eckhorn R, Wilms M, Schanze T, et al. Visual resolution with retinal implants estimated from recordings in cat visual cortex. Vision Res 2006;46:2675–2690.
9. Orban GA. Neuronal Operations in the Visual Cortex. Springer, Berlin, New York, 1984.
10. Tovee MJ. An introduction to the visual system. Cambridge Univ. Press, 1996.
11. Darian-Smith C, Gilbert CD. Topographic reorganization in the striate cortex of the adult cat and monkey is cortically mediated. J Neurosci 1995;15:1631–1647.
12. Bullier J, Hupe J-M, James AC, Girard P. The role of feedback connections in shaping the responses of visual cortical neurons. Progr Brain Res 2001;134:193–204.
13. Legge GE, Ahn SJ, Klitz TS, Luebker A. The visual span in normal and low vision. Vision Res 1997;37:1999–2010.
14. McCreery DB, Agnew WF, Yuen TG, Bullara L. Charge density and charge per phase as cofactors in neural injury induced by electrical stimulation. IEEE Trans Biomed Eng 1990;37:996–1001.
15. Wilms M, Eger M, Schanze T, Eckhorn R. Visual resolution with epi-retinal electrical stimulation estimated from activation profiles in cat visual cortex. Visual Neurosci 2003; 20:543–555.
16. Reitböck HJ. Fiber microelectrodes for electrophysiological recordings. J Neuosci Meth 1983;8:249–262.
17. Schanze T, Wilms M, Eger M, Hesse L, Eckhorn R. Activation zones in cat visual cortex evoked by electrical retina stimulation. Graefe's Arch Clin Exper Ophthalmol 2002;240: 947–954.
18. Rizzo JF, Wyatt J, Loewenstein J, Kelly S, Shire D. Methods and perceptual thresholds for short-term electrical stimulation of human retina with microelectrode arrays. Invest Ophthalmol Vis Sci 1992;44:5355–5361.
19. Stone JL, Barlow WE, Humayun MS, de Juan E, Milam AH. Morphometric analysis of macular photoreceptors and ganglion cells in retinas with retinitis pigmentosa. Arch Ophthalmol 1992;110:1634–1639.
20. Wilms M, Eckhorn R. Spatiotemporal receptive field properties of epiretinally recorded spikes and local electroretinograms in cats. BMC Neurosci 2005;6:50.
21. Rattay F, Resatz S. Effective electrode configuration for selective stimulation with inner eye prostheses. IEEE Trans. Biomed Engin 2004;51:1659–1664.
22. Cha K, Horch KW, Normann RA, Boman D. Reading speed with a pixelized vision system. J Optical Soc Am A 1992;9:673–677.
23. Sommerhalder J, Oueghlani E, Bagnoud M, Leonards U, Safran A, Pellizone M. Simulation of artificial vision: I. Eccentric reading of isolated words, and perceptual learning. Vision Res 2003;43:269–283.

24. Geruschat DR, Turano KA, Stahl JW. Traditional measures of mobility performance and retinitis pigmentosa. Optometry Visual Sci 1998;75:525–537.
25. Grinvald A, Lieke E, Frostig R, Hildesheim R. Cortical point spread function and long-range lateral interactions revealed by real-time optical imaging of macaque monkey primary visual cortex. J Neurosci 1994;14:2545–2568.
26. Waessle H, Boycott BB. Functional architecture of the mammalian retina. Physiol Rev 1991;71:447–480.
27. Nirenberg S, Carcieri SM, Jacobs AL, Latham PE. Retinal ganglion cells act largely as independent encoders. Nature 2001;411:698–701.
28. Shannon CE. A mathematical theory of communication. Bell Syst Tech J 1948;27:623–656.
29. Eckhorn R, Pöpel B. Rigorous and extended application of information theory to the afferent visual system of the cat. II. Experimental results. Biol Cybernetics 1975;17:7–17.
30. Eckhorn R, Pöpel B. Responses of cat retinal ganglion cells to the random motion of a spot stimulus. Vision Res 1981;21:435–443.
31. Rieke F, Warland DK, de Ruyter van Steveninck RR, Bialek W. Spikes: Exploring the Neural Code. MIT Press, Cambridge, 1998.
32. Eger M, Wilms M, Eckhorn R, Schanze T. Information transmission from a retina implant to the cat visual cortex. BioSystems 2005;79:133–142.
33. König P, Engel AK, Singer W. Integrator or coincidence detector? The role of cortical neuron revisited. Trends Neurosci 1996;19:130–137.
34. Agmon-Snir H, Segev I. Signal delay and input synchronization in passive dendritic structures. J Neurophysiol 1993;70:2066–2085.
35. Nelson ME. A mechanism for neuronal gain control by descending pathways. Neural Computation 1994;6:242–254.
36. Shapley RM, Victor JD. How the contrast gain control modifies the frequency responses of cat retinal ganglion cells. J Physiol 1981;318:161–179.

3
How to Restore Reading With Visual Prostheses

Jörg Sommerhalder, PhD

CONTENTS

INTRODUCTION

In daily life sight is used for mainly two types of tasks. Those requiring the recognition of small forms or objects as it is specifically done in *reading* and those requiring spatial orientation and localization in three-dimensional environments, such as whole body *mobility* and *visuo-motor coordination*. For the former the central part of the visual field is mainly used, whereas the latter rely for an essential part on the peripheral visual field. Both have to be seriously considered when developing useful vision aids for low vision and/or blind patients. This chapter is dedicated to the particular question: What are minimum requirements for visual prostheses to restore *useful reading abilities* to blind patients?

First, it will be explained why simulations are a useful tool to explore and predict some aspects of artificial vision and the term *useful reading abilities* will be discussed. Second, the results of some research groups working on simulations of artificial vision will be briefly reviewed and our own work on minimum requirements for useful artificial vision for reading will be summarized. The chapter will end with a discussion on the limits of the approaches presented, followed by some basic recommendations for visual, and especially retinal prostheses, as well as prospects for future simulation experiments.

From: *Ophthalmology Research: Visual Prosthesis and Ophthalmic Devices: New Hope in Sight*
Edited by: J. Tombran-Tink, C. Barnstable, and J. F. Rizzo © Humana Press Inc., Totowa, NJ

SIMULATIONS—A WAY TO EXPLORE ARTIFICIAL VISION ON NORMAL SUBJECTS

There are two major possibilities to determine the minimum requirements for vision aids to restore useful vision. Either several generations of such devices are developed and tested on target patients, or simulations are used to mimic the visual percepts elicited by these systems on normal volunteers. When noninvasive vision aids are considered, it is evident that the first method is straightforward and therefore generally the most adequate. However, when it comes to more invasive devices such as visual prostheses, simulations are a pertinent way to obtain information about the potential benefits of such devices. First, patients using such prostheses are not available at the initial development stages. Second, by using normal observers with a simulated impairment one can look at the effect of a single parameter without confounding its effect with that of others. Finally, the use of simulations makes it easy to repeat experiments on a single subject. A given parameter can also be varied over a full range. These advantages of simulations have been recognized by others and used to address specific questions *(1,2)*.

The knowledge of the minimum information that has to be transmitted to the brain to restore useful function is essential, theoretically and practically, in order to design visual prostheses. Some authors argue that in the case of cochlear implant development, psychophysical studies had limited value until actual devices were implanted and tested *(3)*. In fact, this is not true. The need for certain minimum cues in order to achieve a useful prosthesis was already pointed out as early as 1979 by Kiang et al. *(4)*. Breakthroughs such as the advent of multichannel cochlear implants, allowing for adequate speech recognition, were only possible as a result of psychophysical studies *(5–7)*. Simulations of artificial hearing on normal subjects were extremely useful for understanding cochlear function and for designing new speech coding strategies for cochlear implants *(8–10)*. In the case of visual prostheses, assuming that the perception of about N-spatially distinct phosphenes distributed in a given way within the visual field is necessary to code the information needed to perform a given visual task, a device which does *not* satisfy this requirement will be of very limited clinical value. It is thus extremely important to have such knowledge before proceeding to human trials.

It is clear that the pertinence of the results obtained through such simulation experiments depends directly on how close these experiments mimic future visual prosthesis. Some simplifications are indispensable and might even improve the pertinence of the results, whereas others might lead to wrong estimates.

Most visual prostheses aim to restore function by direct electrical stimulation of neural tissue*. They work according to the following basic principles: (1) a sensor captures the stimuli of interest; (2) a processor/stimulator transforms the stimuli into patterns of electrical currents; which are (3) transmitted by an array of microelectrodes to the target neural tissue; and (4) the brain attempts to make sense of the overall neural excitation pattern. Recently, some researchers have also begun to use neurotransmitters to stimulate surviving retinal neurons *(33–35)*. In such retinal prostheses, the microelectrode array is replaced by a microfluidic drug delivery system.

*Several sites in the visual pathway could be stimulated: the visual cortex *(11–16)*, the optic nerve *(17–19)*, or the retina (epiretinal approach *[20–26]* and subretinal approach *[27–32]*).

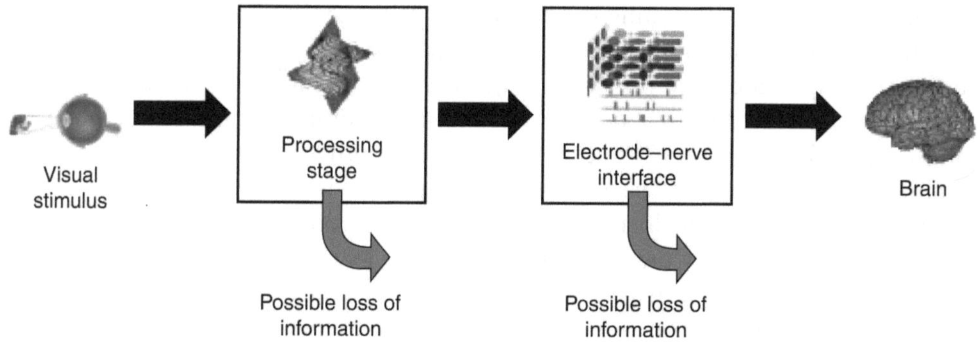

Fig. 1. The transfer of visual information to the brain when using electrical stimulation. At the processor/stimulator interface (processing stage) the stimulus is processed into discrete channels and at the electrode–nerve interface, these stimulation signals have to be transmitted to the retinal neurons. Both sites are subject of information loss.

How to determine minimum requirements for such devices to work? The brain will be capable of achieving useful function only if it is provided with sufficient information (*see* Fig. 1). The first major site of information loss is the processor/stimulator interface, because at this level visual information is split into a finite number of processing channels (e.g., corresponding to the number of available electrodes). It is mandatory that enough information be transmitted through this first interface in order to achieve useful visual function. The second major site of possible information loss is at the electrode–nerve interface. The detailed characteristics of neural activation at this boundary are largely unknown at present. They will depend on the exact nature and site of activation.

The research approaches presented in this chapter concentrate mainly on the processing stage. In other words, they are designed to assess the amount and type of visual information that is critical to perform a specific function, assuming the hypothesis that the brain can use all information transmitted through the processing stage (thus representing some sort of "absolute minimum values"). This type of approach is not new. It has already been used to better understand speech reception by patients using cochlear implants *(8,36–38)*.

WHAT ARE USEFUL READING ABILITIES?

Much attention has been directed to understand the fundamentals of reading. One of the main research centers in this field is the laboratory for low vision research at the University of Minnesota. These authors systematically studied various aspects of the psychophysics of reading on subjects with normal vision and on low-vision patients. For normal subjects, Legge et al. *(39)* reported that maximum reading rates are achieved for characters subtending 0.3–2°. Reading rates increase with field size (up to four characters, independent of character size) and with sample density when texts are matrix sampled (pixelized), but only up to a critical value depending on character size. Reading was found to be very tolerant to either luminance or color contrast reductions *(40–42)*. However, at very low (<10%) luminance contrast, reading speed drops owing to prolonged fixation times and to an increased number of saccades, presumably related to a reduced visual span *(43)*. When testing the effect of print size on reading speed in normal peripheral vision, they found that the use of larger characters improves peripheral reading to some extent up to a critical print size *(44)*. They also found that maximum possible reading speed decreased from about 808 words/min for foveal vision to about 135 words/min for peripheral vision (20° eccentricity).* Thus print size was not the only factor limiting maximum reading speed in normal peripheral vision, contradicting the "scaling hypothesis" *(45,46)*, which suggests that peripheral word recognition can be made equal to that at the fovea by increasing print size. One major reason for lower reading speeds in peripheral vision seems to be the shrinkage of the visual span from at least 10 letters in central vision to about 1.7 letters at 15° eccentricity *(47)*. It has also been demonstrated that letter recognition and reading speed in peripheral vision can benefit from perceptual learning *(48)*.

On low-vision patients, reading is similar to normal reading in several aspects *(42,43,49,50)*, but difficult to predict on the basis of routine clinical evaluations *(51)*. However, as a rule it can be stated that low-vision patients with central field deficits achieve lower reading rates than those with preserved central visual fields *(49,50)*.

The studies quoted earlier (as well as many others) have led to the identification of a series of important parameters that are critical for reading in normal and low-vision subjects. However, in practice it is quite difficult to quantify useful reading performance limits. The typical rate for Braille reading (about 50–100 words/min *see* e.g. ref. *52*) could be taken as a base value. However, in every day situations, texts are rarely available in Braille. It is thus evident that lower reading rates are already very useful to blind people. Reading rates of about 40 words/min seem to be adequate for activities of daily living, such as reading price tags or correspondence (for e.g., *see* refs. *53,54*). A recent study on eccentric reading indicates that reading accuracy (word recognition scores) is more closely correlated to qualitatively evaluated text comprehension than reading rates *(55)*. Although, "good" to "excellent" text comprehension could be achieved with reading rates as low as 10 words/min, reading scores of more than 85% were required to reach the same comprehension levels (median 96.8%). Thus, it seems adequate to define "useful reading abilities" essentially on the basis of reading accuracy (e.g., >95% of correctly recognized words in a text) and much less on the basis of reading rates (e.g., >20–40 words/min†), in spite of the fact that reading rates are often more sensitive to variations in reading conditions and that psychophysical studies on reading deal almost exclusively with reading rates *(53)*.

READING WITH SIMULATED ARTIFICIAL VISION USING THE CENTRAL VISUAL FIELD

There is only a limited number of studies on reading specifically oriented toward the development of visual prostheses. Cha et al. *(56)* used a "pixelized vision system" to

*Such high reading rates were achieved by using rapid serial visual presentation (RSVP).

†Reading rates of about 20 words/min seem to be very low compared with "normal" values of more than 100 words/min. Very few people would want to read a novel at 20 words/min, but greatly reduced reading rates may still be useful for daily living reading tasks. Clinical experience shows that low-vision patients (e.g., with central scotomata) are happy to be able to read even at very low reading rates.

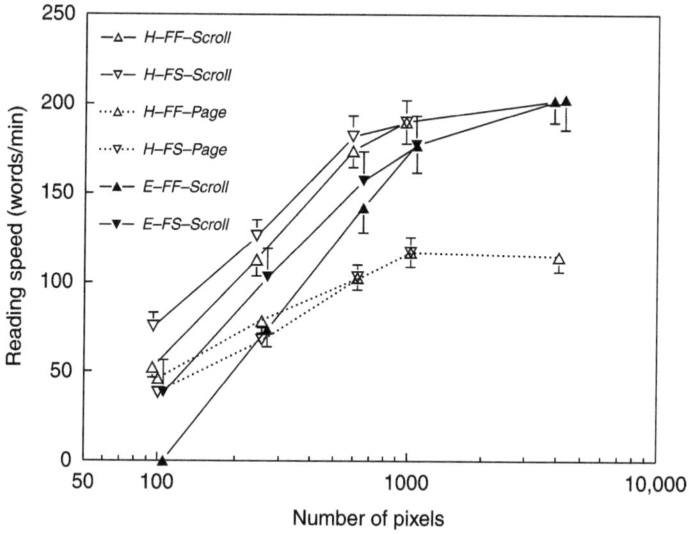

Fig. 2. Reading rates as a function of pixel number using a limited viewing area in central vision. Data published by Cha et al. *(56)*. Subjects viewed the stimuli (scrolled or full-page text) on a monitor covered with perphorated masks. These masks were used to vary pixel number and/or pixel spacing. Different viewing conditions were used: "head scanning" of the text (H) using a head-mounted display and camera, and "eye scanning" of the text (E) using a Purkinje Eye Tracker to stabilize the image on the retina. Fixed-field (FF—1.7° viewing angle) and fixed-spacing (FS—3.24' visual angle between pixels) masks were used. With fixed-field masks the sampling density decreased as the number of pixels decreased, whereas the field of view remained constant. With fixed-spacing masks the sampling density was kept constant, whereas the field of view decreased as the number of pixels decreased. Solid lines connect the scrolled-mode data and dotted lines the page-mode data. The rightmost points represent the clear-mask condition. Plotted are means ± standard errors for six subjects. For clarity, scrolled-mode data are slightly offset, and error bars are shown on one side of the symbols only (opposite the point of the triangle). Please refer to the original article for more detailed information on methods and results (reprinted from ref. *56* with permission from the Optical Society of America).

simulate artificial vision in normal subjects. Their results showed that a 25 × 25 pixel array covering four letters of text, projected on a foveal visual field of 1.7° is sufficient to provide reading rates close to 170 words/min using scrolled text, and close to 100 words/min using fixed text (*see* Fig. 2). This investigation was conducted within the framework of a project aiming at the development of a cortical visual prosthesis *(14)*.

Another research group developing an epiretinal implant *(21)* conducted similar experiments *(57)*. Prosthetic vision was simulated in the central visual field using a head mounted video display. Using free viewing conditions, subjects could inspect paragraphed text by moving a pixelizing grid on the text images with a computer mouse. The dot grid image was thus generated in real-time, but not stabilized on the retina. White text on a black background was used. Several grid parameters (dot size, grid size, dot spacing, random dropout percent, and number of gray scale levels) as well as the influence of character size and stimulus contrast were explored. Results demonstrated that perfect reading accuracy at reading rates of 30–60 words/min could be achieved in certain conditions. Reading performance deteriorated: (1) when the grid size covered less than four letters; (2) when a grid density of less than 4 pixels per character width was used; or (3) when more than 50% of the pixels were randomly turned off. Reading accuracies of about 90% at rates of 10–20 words/min could be reached with grid densities of 3 pixels per character, leading the authors to the conclusion that a 3×3 mm² prosthesis (covering a $10 \times 10°$ visual field) containing 16×16 electrodes should allow for paragraph reading. It is also interesting to mention that—with sufficient practice—similar performance levels could be achieved with low contrast stimuli (12.5%) and that only few grayscale levels (typically four) would be sufficient for reading.

A third research group, working on a biohybrid retinal implant *(58,59)* also conducted reading experiments using simulated prosthetic vision *(60)*. Their simulator consisted of a CMOS camera connected through an image-processing unit to a head-mounted display (a viewing field of $56 \times 48°$ was projected onto a visual field of $30 \times 23°$). Eye movements were not compensated. Four normal subjects were instructed to read the Japanese version of the MNREAD test *(61)*. Two parameters, pixel size (corresponding to the stimulation intensity) and number of gray levels (corresponding to the number of possible stimulation levels) were explored with constant grid density (2304 pixels). The authors suggest that for reading rates around 50 words/min, four to eight stimulation levels are necessary and that low stimulation intensities (pixel sizes) can be used. These results have to be considered with caution because it has been demonstrated that very small pixels (very small dots representing highly localized stimulation) impair reading performance *(62)*.

READING WITH SIMULATED ARTIFICIAL VISION USING AN ECCENTRIC PART OF THE VISUAL FIELD

In the experiments mentioned previously, stimuli were projected onto the central retina (the fovea). However, the anatomo-physiology of the retina does not favor a foveal location for retinal prostheses *(63)*. Retinal implants, which represent probably the most promising concept* for visual prostheses, are primarily designed to stimulate neurons of the inner retinal layers in cases of photoreceptor loss (e.g., retinitis pigmentosa). Surviving bipolar and/or ganglion cells are the targets for electrical stimulation. In the central fovea, these neurons are not present. In the parafovea, they are arranged in several superimposed layers making it difficult to activate them in predictable patterns. The best sites for retinotopic activation without major distortion are located beyond the parafoveal region at an eccentricity of 10° and more†. This means that the vision of future users of a retinal prosthesis will probably be restricted to relatively small§ peripheral areas of their visual field. However, our ability to identify objects in the periphery is poor.

---

*Retinal implants could benefit from natural processing in the still intact peripheral structures of the visual system. Besides, surgery is less invasive than for other stimulation sites (optic nerve or visual cortex), which is an important clinical advantage.

†Retinal implants transforming light to stimuli *in situ* do not allow preprocessing to prevent nonretinotopic mapping. If systems using an external camera to capture the stimuli are considered, such as those envisioned by Humayun et al. *(21)* or Rizzo and Wyatt *(25)*, the transmitting hardware could possibly include such remapping routines. Although this is technically conceivable, it might require prohibitive amounts of perceptual tests for adjustment. We are convinced that it would be optimal to try to place a retinal implant beyond 10° of eccentricity in a first instance.

§For reasons of surgical manageability, a realistic retinal implant size is situated around 3 mm of diameter.

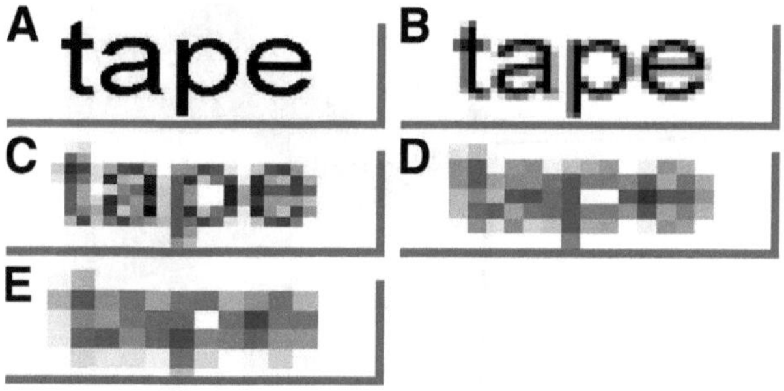

Fig. 3. The word "tape" at the five degrees of stimulus pixelization used in our experiments simulating artificial vision with a retinal implant. **(A)** Maximum screen resolution. **(B)** 875 pixels, **(C)** 286 pixels, **(D)** 140 pixels, **(E)** 83 pixels in the viewing area (reprinted from ref. *64* with permission from Elsevier).

Reading words of several letters is very difficult because of contour interaction, the so-called "crowding effect" (*see* e.g., ref. *45*). Eccentric locations as well as the fact that a retinal implant will stimulate a fixed area of the retina have apparently not been fully taken into consideration by the authors cited earlier.

In our laboratory we were interested in simulations exploring as realistically as possible the information loss occurring at the processor/stimulator interface of a retinal implant. All previously mentioned limitations, such as a fixed and eccentric implant location, the restricted size, and a finite number of discrete stimulation contacts were considered.

Stimuli with reduced information content (pixelized images of isolated words or of lines of text, *see* Fig. 3) were projected on a defined and stabilized area of the retina using a fast computer display. Gaze position data, monitored online by a video-based high-speed gaze tracking system (SMI EyeLink, Senso Motoric Instruments GmbH, Teltow/Berlin, Germany), were used to move a viewing window on the screen in front of the subject (Fig. 4). The position of the viewing window, relative to the gaze position could be offset by any value. It could be projected either onto the central retina (no offset) or onto a defined eccentric area of retina (constant nonzero offset). Please refer to the scientific articles *(55,64)* for a more detailed description of the experimental setup.

Eccentric Reading of Isolated Words and the Learning of This Task

The first experiments were done with isolated four-letter words as stimuli. Although, visual spans covering a higher number of letters might lead to better reading performances (*see* e.g., refs. *65–67*), 4 letters are considered to be the minimum letter-sequence allowing close to normal reading speeds *(39)*. Furthermore, as the aim of this study was to simulate retinal implants covering a limited area and containing a finite number of stimulation contacts, the use of a small number of letters was an advantage.

Reading Accuracy vs Stimulus Resolution and Eccentricity

Rectangular white areas were filled with black four-letter words (common French), pixelized at several degrees (*see* Fig. 3) and projected at various eccentricities in the lower visual field. Reading performance, expressed in terms of percent of fully recognized words, was measured in five normal volunteers for five stimulus pixelizations

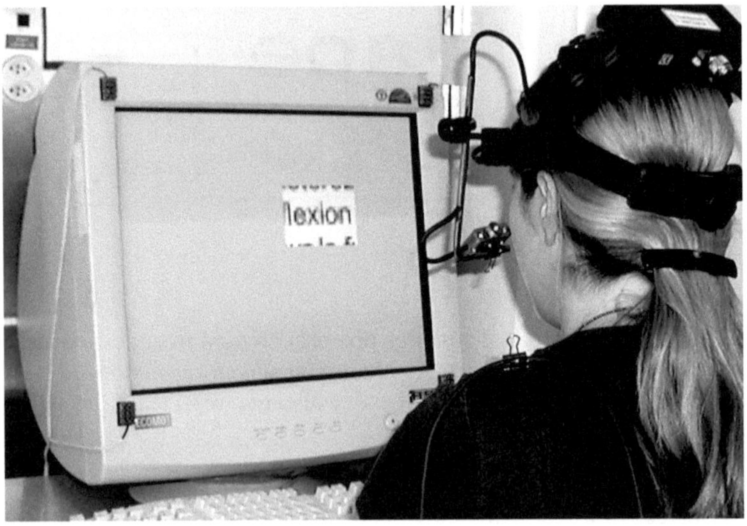

Fig. 4. The experimental setup used to simulate reading in conditions mimicking vision with a retinal prosthesis. In this case, the subject was asked to read pixelized text, using her own eye movements to move a restricted viewing window on the computer screen (reprinted from ref. *55* with permission from Elsevier).

(maximum screen resolution, 875, 286, 140, and 83 pixels) and five eccentricities (0°, 5°, 10°, 15°, and 20° in the lower visual field). Two viewing area sizes (20 × 7° and 10 × 3.5°) were used for these tests.* Please refer to Sommerhalder et al. *(64)* for a more detailed description of the experimental methods.

Reading performances for the small viewing area of 10 × 3.5° are presented in Fig. 5. For central vision, reading accuracy of four-letter words was close to perfect (>90%) down to about 300 pixels. Beyond that point, performance dropped abruptly to 50% correct or less. In peripheral vision, maximum reading performance decreased dramatically with increasing eccentricity. At eccentricities of 10°, 15°, and 20°, maximum reading performance was limited to values of 89%, 57%, and 30% correct, respectively. The large viewing area (20 × 7°) led to similar results, however, the performance decrease because of stimulus eccentricity was less pronounced (maximum reading performance was about 60% at 20° eccentricity).

According to these data, reading seems to be almost impossible at eccentricities beyond 10°, even at high stimulus resolutions. This is a very bad result if eccentric locations for retinal implants are to be considered. A comparison of the results for four-letter word reading with recognition scores of single letters *(64)* revealed that at high eccentricities isolated letters are easier to recognize than those flanked by others. An important factor, limiting eccentric reading performance is therefore probably the aforementioned "crowding effect."

---

*These viewing areas would represent surfaces of 6×2 mm^2 and of 3×1 mm^2 on the retina, respectively. The first one was chosen to optimize reading performance at large eccentricities (taking in account the critical letter size at 20°; *see* ref. *44*), the second one represents a retinal prosthesis that would be surgically manageable.

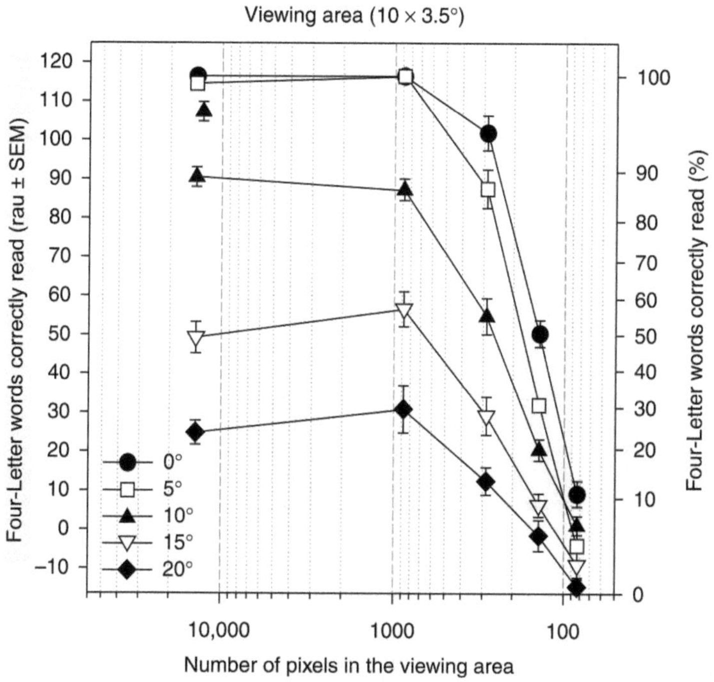

Fig. 5. Performance for single four-letter word reading vs number of pixels in a stabilized viewing area of 10 × 3.5°. Mean reading scores in rationalized arcsine units ± SEM (left scale) and in percent (%) (right scale) for five normal subjects at five eccentricities in the lower visual field. Data were statistically analyzed using scores expressed in rau (reprinted from ref. *64* with permission from Elsevier).

Learning Eccentric Reading of Isolated Words

The results from the previous experiment might underestimate performance because normal subjects are not used to eccentric reading. Future retinal implant wearers will have time to fully adapt to using their prosthesis. Therefore the effect of training on eccentric reading of isolated words was investigated *(64)*. Two subjects were trained to read four-letter words on a viewing area of 10 × 3.5° containing 286 pixels and stabilized at an eccentricity of 15° in the lower visual field. Both subjects improved their low initial reading scores impressively (subject EO from 23 to 85% and subject AR from 6 to 64%) (*see* Fig. 6). A control experiment performed at the end of the training period, using unpracticed words, indicated that the major part of the measured improvements were not due to the fact that the stimuli were chosen out of a limited word library. This confirms that the learning effect did not concern the recognition of certain (well known) words but definitely the ability to read with an eccentric part of the retina. Furthermore control experiments revealed that learning acqired with one eye transferred to the untrained eye and that learning of acquired reading persisted after a significant period of nonpractice.

These first important and promising results demonstrate that about 300 pixels are sufficient to code four-letter words in our experimental conditions. If this information is projected onto the central visual field, subjects reach close to perfect reading accuracy

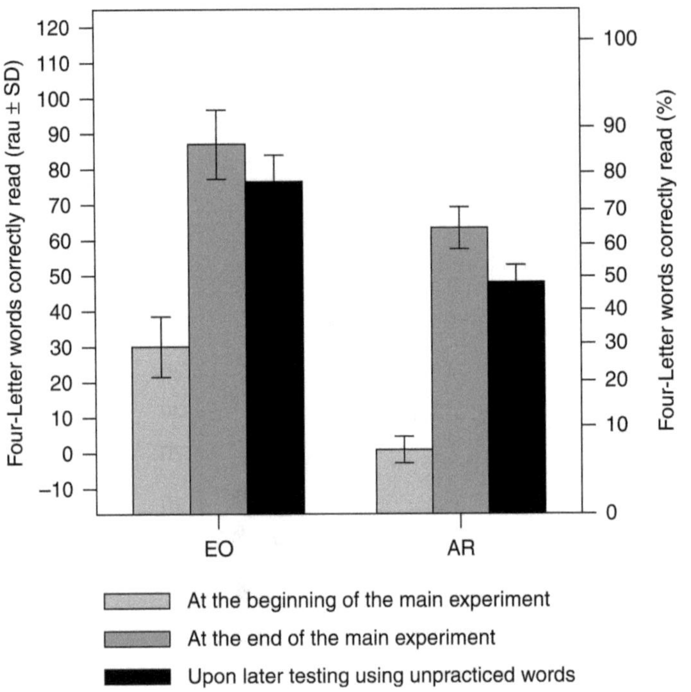

Fig. 6. Performances for four-letter word reading at 15° eccentricity and using a 10 × 3.5° viewing area containing 286 pixels. Mean reading performance with unpracticed four-letter words compared with results at the beginning and at the end of the (main) training experiment. Bars indicate mean values ± SD (reprinted from ref. *64* with permission from Elsevier).

immediately. If the same information is projected onto eccentric locations of the visual field, similar reading performances are possible but they require a significant adaptation process.

Eccentric Reading of Full-Page Texts and the Learning of This Task

Since the first study was conducted using isolated four-letter words, page navigation was not required during reading. Subjects did not have to move their gaze from one word to the next and from the end of one line to the beginning of the next one. This is very difficult for subjects who are restricted to using an eccentric viewing area; they have to learn to direct their attention to an eccentric retinal area, and to navigate precisely with this viewing area on a page of text. However, entire text reading is also expected to be easier than deciphering isolated words because subjects can make use of context information *(68,69)*.

Page navigation has essentially been studied in connection with the use of reading aids for low vision patients. The comparison of reading rates in two conditions: (a) reading without page navigation (horizontally drifting text or RSVP*) and (b) reading with manual page navigation (mouse-controlled or using a CCTV†) resulted

*Rapid serial visual presentation, involving no page navigation and very few eye movements.

†A CCTV (closed-circuit television magnifier) consists of a video camera equipped with a magnifying lens and connected to a TV monitor. The reader can thus only see a few characters at a time and has to move the video camera on the lines of text.

in significantly lower reading rates for manual page navigation *(65,70)*. Interestingly, these authors did not observe significant differences in reading rates across the four methods in patients with central field loss, i.e., for subjects who were forced to use eccentric fixation for reading.

Neither horizontally drifting text nor RSVP can realistically mimic text reading conditions using a retinal implant. Mouse-controlled and CCTV reading are both quite unnatural conditions. They rely on manual page navigation, eventually comparable to navigation with head movements mimicking reading conditions encountered by patients using a retinal implant connected to an external head-mounted camera, but not by those using a retinal implant with light to stimulus transformation *in situ*. For that reason a second study *(55)* exploring full-page text reading simulating conditions of artificial vision provided by such devices was initiated.

Three subjects, naïve to the task, were trained to read full-page texts using their eye movements to move a 10 × 7° viewing window over pixelized pages of text (*see* Fig. 4). A viewing window of double height (10 × 7° instead of 10 × 3.5° used in the previous study on four-letter word reading) was chosen to facilitate page navigation. It contained 572 pixels, the "minimum" pixel density producing good results for four-letter word reading. Small newspaper articles were cut into segments of about 25 words on seven lines and pixelized. Font and letter size were exactly the same as in the previous study. Reading performance was measured in terms of reading scores (percent of correctly read words) and reading rates (number of correctly read words per minute). First, subjects were habituated to the specific experimental conditions by using a central viewing window. Once their central reading performances stabilized, they began with eccentric reading.

For central vision, reading scores were close to perfect from the beginning and reading rates attained values between 72 and 122 words/min. For eccentric reading, initial reading scores were extremely low for two subjects (about 13% correctly read words), and astonishingly high for the third subject (86% correctly read words). All of them significantly improved their performances with time, reaching close to perfect reading scores (between 86% and 98%) at the end of the training period. Reading rates were as low as 1–5 words/min at the beginning of the experiment and increased significantly with practice up to 14–28 words/min. Qualitative text comprehension was also estimated, and was found to be "good" to "excellent" at the end of the training period. Analysis of gaze position recordings (Fig. 7), demonstrated that eye-movement control, especially the suppression of vertical reflexive saccades, constituted an important part of the overall adaptive learning process *(71)*. Control measurements, similar to those conducted for the previous study on four-letter words, also demonstrated interocular transfer and persistence of learning for at least several months. Please refer to Sommerhalder et al. *(55)* for more details.

The results of this second study suggest that retinal implants might restore full-page text reading abilities to blind patients. If such a device were capable of selective retinotopic stimulation, about 600 stimulation contacts, distributed on a retinal area of 3 × 2 mm^2, appear to be a minimum to achieve useful reading (slow but with good text comprehension). However, a significant learning process will be required to reach optimal performance, if the implant is placed outside the foveal area.

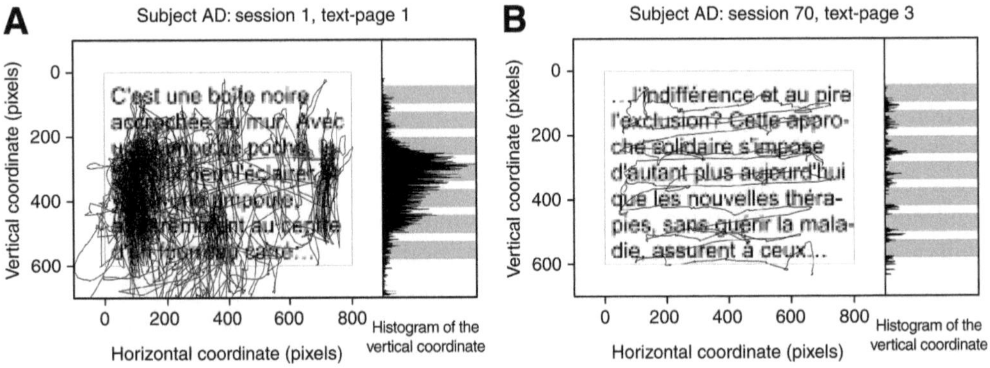

Fig. 7. Demonstration of the learning effect in full-page text reading using a $10 \times 7°$ viewing window controlled by the subject's own eye movements and stabilized at 15° eccentricity. Trajectory of the center of the viewing window (solid line) relative to the text during **(A)** the first, and **(B)** the last experimental session for subject AD. The panels on the right represent frequency histograms for the vertical coordinates of the trajectory recorded every 4 ms. Gray bars indicate the position of the lines of text. This example clearly shows that the subject was not able to scan the lines of text at the beginning of the learning process. The trajectory seems to be chaotic, with an important number of up-down eye movements (reflexive vertical saccades). At the end of the learning process, the lines of text could be scanned quite accurately. One can even observe the tracing back on difficult words and the fact that the center of the viewing window was generally placed below the text-line, a way to minimize the eccentricity of the text visualized through the viewing window (reprinted from ref. *55* with permission from Elsevier).

DISCUSSION

The simulation experiments reviewed in this chapter demonstrate that if a certain amount of information is transmitted to the visual system, reading of paragraphed text using a highly restricted visual field is possible. Good text comprehension at reading rates of about 100 words/min could be achieved when the central part of the retina was used. For reading at 15° eccentricity an important learning process was necessary to reach similar reading accuracies at much lower reading rates of about 20 words/min. Full-page text reading under conditions simulating retinal implants involves: (1) the scanning of several lines of text using an eccentric and restricted viewing window and (2) the extraction of information out of low resolution (pixelized) stimuli. If such implants are to be placed at an eccentric retinal location, the adaptation to full page text reading also requires: (3) the suppression of involuntary reflexive eye-movements; (4) the recalibration of the oculomotor system to a new eccentric fixation locus; and (5) the focussing of attention to this peripheral region of the visual field. It is impossible to analyze separately how subjects overcame all these difficulties in the previously presented studies. However, it is interesting to mention some factors influencing the learning process and others limiting final reading performance.

Factors Influencing Reading Performance

Eye movement data indicate that an important part of the overall learning process for eccentric reading can be attributed to the progressive suppression of uncontrolled reflexive eye movements. Subjects gradually learned to re-reference their eye movements to a highly eccentric retinal locus. However, the temporal course of this adaptation was not perfectly correlated with the improvement of reading performance (reading scores or reading rates). This suggests that other factors also had an influence on reading performance during the overall learning process.

Habituation to eccentric reading of isolated words (not requiring page navigation) *(64)* took a similar period of time compared with the habituation to eccentric reading of entire pages of text *(55)*. An important component of the overall learning process seems to be independent and accurate control of eye movements. Crist, Li, and Gilbert *(72)* suggested that this kind of perceptual learning is accompanied by a concomitant decrease of the "crowding effect". This decrease was found to be related to attention *(73)*, which can in turn be improved by learning (*see* e.g. ref. *74*). A significant decrease of the "crowding effect" is thus likely to be an important part of the overall learning process.

The influence of context information is difficult to assess. Final reading scores for eccentric reading of isolated four-letter words (about 75%—mean on two subjects) are low compared to final reading scores for full-page text reading (about 94%—mean on three subjects). This difference is not statistically significant for such a small number of subjects, but it supports the hypothesis that the use of context information improves reading performance, even for eccentric reading *(68,69)*.

Average reading rates for eccentric vision were 2.5–5.8 times lower than those achieved with central vision, although similar reading scores could be achieved for central and for eccentric vision. This confirms that target eccentricity is a major factor limiting the reading rate. Other authors already reported low reading rates for eccentric vision. Wensveen et al. *(75)*, for example, found that simulated 8° central scotomata produced a threefold reduction of the reading rates in young subjects. The fact that eccentric reading remains less efficient than central reading, even after adaptation to the task, might be caused by a reduction of the visual span. Legge, et al. *(47)* found that the average visual span shrinks from at least 10 letters in central vision to about 1.7 letters at 15° eccentricity; however, this low value increased with prolonged observation times. Therefore, the subjects had to either increase the number of saccades to decipher a given word, or increase fixation time to extend the visual span; both strategies result in lower reading rates.

In our experimental conditions, reading rates for central vision were found to be markedly below "normal" values (160–180 words/min for the same subjects). Thus, other factors limiting the reading rate were not due to the use of eccentric vision. First, the text was only visible through a restricted viewing window. Therefore, peripheral vision was not available to orient saccades, which probably contributed to lower reading rates. Second, the use of pixelized stimuli containing close to threshold information puts a significant load on the visual system in order to extract the relevant information. Whittaker and Lovie-Kitchin *(53)* as well as Bowers and Reid *(76)*, suggest using (for other parameters such as print size and contrast) at least several times the threshold values to achieve optimal reading rates. Finally, scanning magnified short text segments certainly takes more time than scanning normal text in a newspaper.

The difficulties and limitations discussed earlier are probably unavoidable for future implant wearers. At the present stage of development it is still technically impossible to conceive visual prostheses providing much better viewing conditions, for example, by covering large parts of the retina or by selectively stimulating individual neurons for the transmission of high-resolution images.

The Limitations of the Presented Simulation Methods

As already mentioned before (*see* "Simulations—A Way to Explore Artificial Vision on Normal Subjects"), the experiments presented here simulate the information loss at the processor/stimulator interface of retinal prostheses, in particular of retinal implants transforming light into stimulation currents *in situ*. Such simulations are useful to specify the minimum information that must be transmitted to the brain, assuming that the information transmission at the electrode/nerve interface is "perfect".

The information loss that will occur at the electrode/nerve interface is still very difficult to determine. Additional studies in retinal physiology have to be conducted to determine exactly how the surviving neurons of a degenerated retina will react to electrical (or chemical) stimulation. Research in this field will help to determine optimal electrode surface and shape, maximum electrode density, stimulation current, and so on. The quality of the electrode/nerve interface will ultimately determine whether or not retinotopic stimulation will be possible. Implantable microelectrode arrays consisting of several hundred stimulation contacts seem feasible using present technology. Zrenner et al. *(30)* as well as Peyman et al. *(28)* have already manufactured such prototypes. The results of the simulations suggest that implantable chips with an electrode-to-electrode separation of approx 100 μm might be able to generate useful percepts for reading. Multisite stimulation measurements on chicken retinae have demonstrated that such closely spaced contacts can selectively activate retinal neurons *(77)*, but these promising results on isolated retinae have yet to be confirmed by in vivo experiments and by human trials.

Other limitations in the simulation methods may be due to the oversimplification of the methods:

1. The stimuli were prepixelized fragmented texts, the gray level of these "frozen" pixels being independent of the point of gaze on the target image. However, in visual prostheses the stimulation intensity at each electrode will depend on the exact point of gaze (of the eye or the scene camera) relative to the observed image. Therefore, eye or head movements will evoke an image based on dynamic stimulation intensities. An accurate simulation of artificial vision should thus be based on "real-time" image processing. As a dynamic sequence of slightly different images contains more information than one single image, such dynamic (real-time) pixelization is likely to enhance information transmission to the visual system.
2. Stimulation images were decomposed into a finite number of pixels using a simple block-averaging algorithm. This resulted in mosaics of uniform square pixels of various gray levels. However, the patterns of neural activity elicited by electric stimulation of the retina will depend on stimulating current strength and on the distance between the electrode and the neural target *(22,77,78)*. This implies that phosphenes elicited by electrical stimulation of the retina will neither be of constant luminosity nor square-shaped. Furthermore, depending on the strength of the stimulation current, phosphenes may overlap. It could be argued that square pixelization is adequate to simulate the reduced information transmitted by a retinal implant. In a given condition, the detailed shape of each pixel does not alter the overall information content of the image. However, several studies suggest that image

detection is considerably hampered when square pixelization is used (for e.g., *see* refs. *79,80*). In more realistic simulations of artificial vision, more adequate quantization algorithms should thus replace square pixelization.

In a recent study *(62)* the influence of these simplifications on reading performance was measured by comparing prepixelized and real-time pixelsized stimuli using two pixel shapes: square and Gaussian*. Experiments were conducted in central vision and the experimental conditions were the same as those of the study on full-page text reading. The results can be summarized as: (1) Real-time pixelization significantly improves reading performance (reading scores and reading rates) and equivalent reading performances (thus also useful reading abilities) can be achieved with approx 30% lower pixel density when real-time pixelized stimuli are used; (2) For prepixelized stimuli and in optimal conditions (i.e., using optimal Gaussian function widths), Gaussian pixelization leads to slightly better reading performances than square pixelization. However, this advantage vanishes completely with real-time pixelization.

This complementary study demonstrates that real-time stimulus pixelization influences reading performance, but not sufficiently to fundamentally change the minimum requirements determined in our previous studies on the basis of off-line stimulus processing. It is likely that the advantage of real-time pixelization will be even less important for patients using visual prostheses with external head-mounted cameras as head movements in general are larger and less frequent than eye movements.

Basic Recommendations for Retinal Prostheses Concepts

Our studies *(55,62,64)* clearly indicate that 400–600 stimulation contacts covering a 3×2 mm^2 retinal area would be necessary to restore useful reading abilities. These results are in good agreement with those of other authors *(56,57)*. Extrapolated to our experimental conditions, the results of these two research groups would correspond to stimulus resolutions around 450 pixels. It seems technically conceivable to integrate about 500 electrodes onto an ultrathin externally powered 3×2 mm^2 retinal implant chip for "*in situ*" light-to-stimulation current transformation (*see* e.g., refs. *29,81,82*). It is important to target technical efforts to reach such minimum estimates, failing which the clinical value of such a device will be very limited. Reading abilities are particularly important in this context as it is strongly associated with vision-related estimates of quality of life and represents one of the main goals of low-vision patients seeking rehabilitation *(83–85)*.

A retinal implant covering a visual field of $10 \times 7°$ will only be useful for reading in every day situations when used in conjunction with adequate optics, which should typically project a highly magnified 10×7 mm^2 surface of a newspaper ($1 \times 0.7°$ visual field at a distance of 30 cm) onto the entire implant surface. For other tasks, such as mobility or visuo-motor coordination, exchangeable optics with other magnification factors will be necessary. Achieving optimal performance with such devices, especially when placed at peripheral retinal locations will require a significant adaptation process.

*Square pixels were of constant luminosity (gray level). Gaussian pixels were circular pixels following a Gaussian luminosity distribution (brighter in the center and darker toward the border). In the latter case, pixel overlapping was possible, depending on the width (σ) of the Gaussian function used.

As future users of retinal implants will wear their prosthesis permanently, it can be expected that they will benefit even more from adaptation than normal subjects in simulation experiments. The results presented are in this respect very encouraging for the future.

Finally, most of the current retinal implant concepts hinge on the possibility of retinotopic activation of surviving retinal neurons. Some of the first experiments on human subjects with very basic low-resolution retinal implants indicate that retinotopic stimulation should be possible *(23,86)*. Other authors encountered greater difficulties in obtaining reproducible percepts or the perception of simple shapes *(78,87)*. It is thus possible that extensive adaptation, cortical plasticity, and perhaps even electronic treatment* will be necessary to compensate for scarce retinotopic stimulation.

Outlook on Future Simulation Experiments of Artificial Vision

In our opinion, the "minimum requirements" for restoring useful reading abilities to blind patients have been adequately explored. Studies simulating artificial vision for other activities of daily living are and have been undertaken in our laboratory and by other research groups *(88,89)*. Some of the results have already been presented at scientific meetings *(90,91)* and will hopefully be published soon.

A crucial issue that still remains open is what and how much information can be really transmitted at the electrode/nerve interface of future visual prostheses. If more precise data on the possibilities and limitations of stimulating surviving retinal neurons become available, simulation experiments could be more closely adapted to this reality. Such simulation experiments would be designed to realistically evaluate the rehabilitation prospects with future visual prostheses. They should answer the question: What sort of vision can be restored and with which visual prostheses?

At this point it is unlikely that simulations of artificial vision will provide further breakthroughs. The next important step to validate current and future concepts of visual prostheses would be to start exhaustive psychophysical testing with passive devices and prototypes of implants in human volunteers†. Such approaches raise ethical questions, but some blind volunteers would certainly be up to the challenge, provided they are extremely well-informed and as long as major risks can be minimized.

ACKNOWLEDGMENTS

Many thanks to past and present collaborators for these studies on reading with simulated artificial vision: A. Perez Fornos, B. Rappaz, R. de Haller, E. Oueghlani, M. Bagnoud, U. Leonards, A. B. Safran, and M. Pelizzone. They are coauthors of the scientific articles summarized in this chapter. The research activity at the Geneva University Hospital is part of a larger multidisciplinary project aiming to develop a subretinal implant. The partners are: the Institute for Microsystems and the Microelectronics Laboratory at the Federal Polytechnic School of Lausanne for the development of the CMOS retinal implant chip (D. Ziegler, P. Linderholm, M. Mazza, A. M. Ionescu and P. Renaud); the Department of Physiology of the Geneva University

*Image remapping in the retinal implant.

†It has been proposed to replace extensive basic stimulation testing in human patients by measuring retino–cortical transmission in animals *(92,93)*, but such approaches can only partially replace human trials.

for "in vitro" electrophysiology (M. Lecchi and D. Bertrand); the Eye Clinic of the Geneva University Hospitals and the Laboratory of Retinal Cellular and Molecular Physiopathology (INSERM unit 592) at St-Antoine Hospitals in Paris (France) for biocompatibility and surgical techniques (J. Salzmann, S. Picaud, J. Sahel, and A.B. Safran).

Our research work is supported by the Swiss National Science Foundation (grants 3100-61956.00, 3152-63915.00, 3100A0-103918, and 3152A0-105958) and the ProVisu Foundation.

REFERENCES

1. Pelli DG. The visual requirements of mobility. In: Woo GC, ed. Low vision: Principles and applications. Springer Verlag, New York, 1986;134–146.
2. Cornelissen FW, Van den Dobbelsteen JJ. Heading detection with simulated visual field defects. Vis Impairment Res 1999;1:71–84.
3. Margalit E, Maia M, Weiland JD, et al. Retinal Prosthesis for the Blind. Survey Ophth 2002;47:335–356.
4. Kiang NY, Eddington DK, Delgutte B. Fundamental considerations in designing auditory implants. Acta Otolaryngol 1979;87:204–218.
5. Eddington DK, Dobelle WH, Brackmann DE, Mladejowsky MG, Parkin JL. Auditory prosthesis research with multiple channel intracochlear stimulation in man. Ann Otol Rhinol Laryngol 1978;87:1–39.
6. Eddington DK. Speech discrimination in deaf subjects with cochlear implants. J Acoust Soc Am 1980;68:885–891.
7. Tong YC, Dowell RC, Blamey PJ, Clark GM. Two-component hearing sensations produced by two-electrode stimulation in the cochlea of a deaf patient. Science 1983;219:993–994.
8. Shannon RV, Zeng FG, Kamath V, Wygonski J, Ekelid M. Speech recognition with primarily temporal cues. Science 1995;270:303–304.
9. Dorman MF, Loizou PC. Speech intelligibility as a function of the number of channels of stimulation for normal-hearing listeners and patients with cochlear implants. Am J Otol 1997;18 suppl:113–114.
10. Hamzavi JS, Baumgartner WD, Adunka O, Franz P, Gstoettner W. Audiological performance with cochlear reimplantation from analogue single-channel implants to digital multichannel devices. Audiology 2000;39:305–310.
11. Brindley GS, Lewin WS. The sensations produced by electrical stimulation of the visual cortex. J Physiol (Lond.) 1968;196:479–493.
12. Dobelle WH, Mladejovsky MG, Girvin JP. Artifical vision for the blind: electrical stimulation of visual cortex offers hope for a functional prosthesis. Science 1974;183:440–444.
13. Normann RA, Maynard EM, Shane Guillory K, Warren DJ. Cortical implants for the blind. IEEE Spectrum 1996;33:54–59.
14. Normann RA, Maynard EM, Rousche PJ, Warren DJ. A neural interface for a cortical vision prosthesis. Vision Res 1999;39:2577–2587.
15. Schmidt EM, Bak MJ, Hambrecht FT, Kufta CV, O'Rourke DK, Vallabhanath P. Feasibility of a visual prosthesis for the blind based on intracortical microstimulation of the visual cortex. Brain 1996;119:507–522.
16. Dobelle WH. Artificial vision for the blind by connecting a television camera to the visual cortex. ASAJO Journal 2000;46:3–9.
17. Veraart C, Raftopoulos C, Mortimer JT, et al. Visual sensations produced by optic nerve stimulation using an implanted self-sizing spiral cuff electrode. Brain Res 1998;813:181–186.

18. Delbeke J, Wanet-Delfalque MC, Gérard B, Troosters M, Michaux G, Veraart C. The Microsystems Based Visual Prosthesis for Optic Nerve Stimulation. Artif Organs 2002;26:232–234.
19. Delbeke J, Oozeer M, Veraart C. Position, size and luminosity of phosophenes generated by direct optic nerve stimulation. Vision Res 2003;43:1091–1102.
20. Humayun MS, de Juan E Jr, Dagnelie G, Greenberg RJ, Propst RH, Phillips DH. Visual perception elicited by electrical stimulation of retina in blind humans. Arch Ophthalmol 1996;114:40–46.
21. Humayun MS, de Juan E Jr, Weiland JD, et al. Pattern electrical stimulation of the human retina. Vision Res 1999;39:2569–2576.
22. Weiland JD, Humayun MS, Dagnelie G, de Juan E Jr,. Greenberg RJ, Liff NT. Understanding the origin of visual percepts elicited by electrical stimulation of the human retina. Graefes Arch Clin Exp Ophthalmol 1999;237:1007–1013.
23. Humayun MS, Weiland JD, Fujii GY, et al. Visual perception in a blind subject with a chronic microelectronic retinal prosthesis. Vision Res 2003;43:2573–2581.
24. Wyatt J, Rizzo J. Ocular implants for the blind. IEEE Spectrum 1996;33:47–53.
25. Rizzo JF, Wyatt J. Prospects for visual protesis. The Neuroscientist 1997;3:251–262.
26. Eckmiller R. Learning retina implants with epiretinal contacts. Ophthalmic Res 1997;29: 281–289.
27. Chow AY, Chow VY. Subretinal electrical stimulation of the rabbit retina. Neurosci Lett 1997;225:13–16.
28. Peyman G, Chow AJ, Chanping L, Chow VY, Perlman JI, Peachey NS. Subretinal semiconductor microphotodiode array. Ophthalmic Surg Lasers 1998;29:234–241.
29. Chow AY, Chow VY, Packo KH, Pollack JS, Peyman GA, Schuchard R. The artificial silicon retina microchip for the treatment of vision loss from retinitis pigmentosa. Arch Ophthalmol 2004;122:460–469.
30. Zrenner E, Miliczek KD, Gabel VP, et al. The development of subretinal microphotodiodes for replacement of degenerated photoreceptors. Ophthalmic Res 1997;29:269–280.
31. Zrenner E, Stett A, Weiss S, et al. Can subretinal microphotodiodes sucessfully replace degenerated photoreceptors? Vision Res 1999;39:2555–2567.
32. Zrenner E. Will retinal implants restore vision? Science 2002;295:1022–1025.
33. Safadi MR, Washko F, Lagman A, et al. Development of a microfluidic drug delivery neural stimulating device for vision. Invest Ophthalmol Vis Sci 2003;44:ARVO E-Abstract 5082.
34. Peterman MC, Mehenti NZ, Bilbao KV, et al. The Artificial Synapse Chip: a flexible retinal interface based on directed retinal cell growth and neurotransmitter stimulation. Artif Organs 2003;27:975–985.
35. Peterman MC, Noolandi J, Blumenkranz MS, Fishman HA. Localized chemical release from an artificial synapse chip. Proc Natl Acad Sci USA 2004;101:9951–9954.
36. Dorman MF, Loizou PC, Rainey D. Simulating the effect of cochlear-implant electrode insertion depth on speech understanding. J Acoust Soc Am 1997;102:2993–2996.
37. Loizou PC. Introduction to cochlear implants. IEEE Eng Med Biol Mag 1999;18:32–42.
38. De Balthasar C, Cosendai G, Pelizzone M. Simulations of the effects of electrical stimulation selectivity on speech reception with cochlear implants. Med Hyg 1999;2273:1984–1988.
39. Legge GE, Pelli DG, Rubin GS, Schleske MM. Psychophysics of reading. I. Normal vision. Vision Res 1985;25:239–252.
40. Legge GE, Rubin GS. Psychophysics of reading. IV. Wavelength effects in normal and low vision. J Optic Soc Am A 1986;3:40–51.
41. Legge GE, Rubin GS, Luebker A. Psychophysics of reading. V. The role of contrast in normal vision. Vision Res 1987;27:1165–1177.

42. Legge GE, Parish DH, Luebker A, Wurm LH. Psychophysics of reading. XI. Comparing color contrast and luminance contrast. J Optic Soc Am A 1990;7:2002–2010.
43. Legge GE, Ahn SJ, Klitz TS, Luebker A. Psychophysics of reading. XVI. The visual span in normal and low vision. Vision Res 1997;37:1999–2010.
44. Chung STL, Mansfield JS, Legge GE. Psychophysics of reading. XVIII. The effect of print size on reading speed in normal peripheral vision. Vision Res 1998;38:2949–2962.
45. Toet A, Levi DM. The two-dimensional shape of spatial interaction zones in the parafovea. Vision Res 1992;32:1349–1357.
46. Latham K, Whitaker D. A comparison of word recognition and reading performance in foveal and peripheral vision. Vision Res 1996;36:2665–2674.
47. Legge GE, Mansfield JS, Chung STL. Psychophysics of reading XX. Linking letter recognition to reading speed in central and peripheral vision. Vision Res 2001;41: 725–743.
48. Chung ST, Legge GE, Cheung SH. Letter-recognition and reading speed in peripheral vision benefit from perceptual learning. Vision Res 2004;44:695–709.
49. Legge GE, Rubin GS, Pelli DG, Schleske MM. Psychophysics of reading. II. Low vision. Vision Res 1985;25:253–265.
50. Rubin GS, Legge GE. Psychophysics of reading. VI. The role of contrast in low vision. Vision Res 1989;29:79–91.
51. Legge GE, Ross JA, Isenberg LM, La May JM. Psychophysics of reading. XII: Clinical predictors of low-vision reading speed. Invest Ophthalmol Vis Sci 1992;33:677–687.
52. Mousty P, Bertelson P. A study of braille reading: 1. Reading speed as a function of hand usage and context. Q J Exp Psychol A 1985;37:217-233.
53. Whittaker SG, Lovie-Kitchin J. Visual requirements for reading. Optom Vision Sci 1993;70:54–65.
54. Rumney NJ. Using visual thresholds to establish vision performance. Ophthalmic Physiological Optics 1995;15:S18–S24.
55. Sommerhalder J, Rappaz B, de Haller R, Perez Fornos A, Safran AB, Pelizzone M. Simulation of artificial vision: II. Eccentric reading of full-page text and the learning of this task. Vision Res 2004;44:1693–1706.
56. Cha K, Horch KW, Normann RA, Boman DK. Reading speed with a pixelised vision system. J Optic Soc Am A 1992;9:673–677.
57. Dagnielie G, Barnett D, Humayun MS, Thompson RW. Paragraph text reading using a pixelized prosthetic vision simulator: Parameter dependence and task learning in free-viewing condition. Invest Ophthalmol Vis Sci 2006;47:1241–1250.
58. Yagi T, Ito Y, Kanda H, Tanaka S, Watanabe M, Uchikawa Y. Hybrid retinal implant: fusion of engineering and neuroscience. Proc 1999 IEEE Int Conf Systems Man Cybernetics 1999;4:382–385.
59. Ito Y, Yagi T, Kanda H, Tanaka S, Watanabe M, Uchikawa Y. Cultures of neurons on microelectrode array in hybrid retinal implant. Proc 1999 IEEE Int Conf Systems Man Cybernetics 1999;4:414–417.
60. Terasawa Y, Fujikado T, Yagi T. Simulation of visual prosthesis in virtual space. International J Appl Electromagne Mech 2001/2002;15:431–436.
61. Legge GE, Ross JA, Luebker A, La May JM. Psychophysics of reading VIII. The Minnesota Low-Vision Reading Test. Optom Vision Sci 66:843–853.
62. Perez Fornos A, Sommerhalder J, Rappaz B, Safran AB, Pelizzone M. Simulation of artificial vision: III. Do the spatial or temporal characteristics of stimulus pixelization really matter? Invest Ophthalmol Vis Sci 2005;46:3906–3912.
63. Sjostrand J, Olsson V, Popovic Z, Conradi N. Quantitative estimations of foveal and extrafoveal retinal circuitry in humans. Vision Res 1999;39:2987–2998.

64. Sommerhalder J, Oueghlani E, Bagnoud M, Leonards U, Safran AB, Pelizzone M. Simulation of artificial vision: I. Eccentric reading of isolated words, and perceptual learning. Vision Res 2003;43:269–283.
65. Beckmann PJ, Legge GE. Psychophysics of reading. XIV. The page navigation problem in using magnifiers. Vision Res 1996;36:3723–3733.
66. Fine EM, Kirschen MP, Peli E. The necessary field of view to read with an optimal stand magnifier. J Am Optom Assoc 1996;67:382–389.
67. Fine EM, Peli E. Visually impaired observers require a larger window than normally sighted observers to read from a scroll display. J Am Optom Assoc 1996;67:390–396.
68. Fine, EM, Peli, E. The role of context in reading with central field loss. Optom Vis Sci 1996;73:533–539.
69. Fine EM, Hazel CA, Latham K, Rubin GS. Are benefits of sentence context different in central and peripheral vision? Optom Vis Sci 1999;76:764–769.
70. Harland S, Legge GE, Luebker A. Psychophysics of reading. XVII. Low-vision performance with four types of electronically magnified text. Optom Vis Sci 1998;75:183–190.
71. Perez Fornos A, Sommerhalder J, Rappaz B, Pelizzone M, Safran AB. Processes involved in oculomotor adaptation fo eccentric reading. Invest Ophthalmol Vis Sci 2006;47: 1439–1447.
72. Crist RE, Li W, Gilbert CD. Learning to see: experience and attention in primary visual cortex. Nature Neuroscience 2001;4:519–525.
73. Leat SJ, Li W, Epp K. Crowding in central and eccentric vision: the effects of contour interaction and attention. Invest Ophthalmol Vis Sci 1999;40:504–512.
74. Sireteanu R, Rettenbach R. Perceptual learning in visual search generalizes over tasks, locations, and eyes. Vision Res 2000;40:2925–2949.
75. Wensveen JM, Bedell HE, Loshin DS. Reading rates with artificial central scotoma with and without spatial remapping of print. Optom Vis Sci 1995;72:100–114.
76. Bowers AR, Reid VM. Eye movements and reading with simulated visual impairment. Ophthalmic Physiol Opt 1997;17:392–402.
77. Stett A, Barth W, Weiss S, Haemmerle H, Zrenner E. Electrical multisite stimulation of the isolated chicken retina. Vision Res 2000;40:1785–1795.
78. Rizzo JF, Wyatt J, Loewenstein J, Kelly S, Shire D. Perceptual efficacy of electrical stimulation of human retina with a microelectrode array during short-term surgical trials. Invest Ophthalmol Vis Sci 2003;44:5362–5369.
79. Harmon LD, Julesz B. Masking in visual recognition: effects of two-dimensional filtered noise. Science 1973;180:1194–1197.
80. Bachmann T, Kahusk N. The effects of coarseness of quantisation, exposure duration, and selective spatial attention on the perception of spatially quantised ('blocked') visual images. Perception 1997;26:1181–1196.
81. Zrenner E. The subretinal implant: can microphotodiode arrays replace degenerated retinal photoreceptors to restore vision? Ophthalmologica 2002;216:8–20.
82. Ziegler D, Linderhalm P, McCormick K, et al. An active microelectrode array of oscillating pixels for retinal stimulation. Sensors and Actuators A 2004;110:11–17.
83. Wolffsohn JS, Cochrane AL. The changing face of the visually impaired: the Kooyong low vision clinic's past, present, and future. Optom Vis Sci 1999;76:747–754.
84. Hazel CA, Petre KL, Armstrong RA, Benson MT, Frost NA. Visual function and subjective quality of life compared in subjects with acquired macular disease. Invest Ophthalmol Vis Sci 2000;41:1309–1315.
85. McClure ME, Hart PM, Jackson AJ, Stevenson MR, Chakravarthy U. Macular degeneration: do conventional measurements of impaired visual function equate with visual disability? Br J Ophthalmol 2000;84:244–250.

86. Yanai D, Weiland JD, Mahadevappa M, et al. Visual Perception in Blind Subjects with Microelectronic Retinal Prosthesis. Invest Ophthalmol Vis Sci 2003;44:ARVO E-Abstract 5056.
87. Loewenstein JI, Montezuma SR, Rizzo JF. Outer retinal degeneration: an electronic retinal prosthesis as a treatment strategy. Arch Ophthalmol 2004;122:587–596.
88. Cha K, Horch KW, Normann RA. Mobility performance with a pixelized vision system. Vision Res 1992;32:1367–1372.
89. Thompson RW, Barnett GD, Humayun MS, Dagnelie G. Facial recognition using simulated prosthetic pixelized vision. Invest Ophthalmol Vis Sci 2003;44:5035–5042.
90. Perez Fornos A, Sommerhalder J, Chanderli K, et al. Minimum requirements for mobility in known environments and perceptual learning of this task in eccentric vision. Invest Ophthalmol Vis Sci 2004;45:ARVO E-Abstract 5445.
91. Perez Fornos A, Sommerhalder J, Pittard A, Safran AB, Pelizzone M. Minimum requirements for visuomotor coordination and learning of such tasks in eccentric vision. Invest Ophthalmol Vis Sci 2005;46:ARVO E-Abstract 1533.
92. Wilms M, Eger M, Schanze T, Eckhorn R. Visual resolution with epi-retinal electrical stimulation estimated from activation profiles in cat visual cortex. Vis Neurosci 2003;20: 543–555.
93. Eger M, Wilms M, Eckhorn R, Schanze T, Hesse L. Retino-cortical information transmission achievable with a retina implant. Biosystems 2005;79:133–142.

4
Subretinal Artificial Silicon Retina Microchip Implantation in Retinitis Pigmentosa

Alan Y. Chow, MD and Vincent Y. Chow

CONTENTS

INTRODUCTION

Retinitis pigmentosa (RP) is a progressive condition that causes both central and peripheral vision loss *(1–3)*. This genetically diverse disease presents with a variable phenotypic onset, but eventually affects both eyes. No treatment is effective in restoring vision once it is lost. Although, a variety of patterns can be observed, vision loss typically occurs first in the midperiphery and progresses to involve the peripheral and finally, the central visual fields creating a funduscopic pattern of pigmented "bone spicules."

Dozens of genotypes of RP are known. Nevertheless, the common pathological damage appears to arise as a result of "outer retinal degeneration" usually associated with rhodopsin mutations. The outer portion of the inner anatomical retina (outer retina), mainly consisted of photoreceptor outer and inner segments and their cell bodies undergo apoptosis, whereas the inner portion of the inner retina (inner retina), made up of the remaining bipolar, horizontal, amacrine, and ganglion cells and nerve fiber layer may be partially spared *(4,5)*. The presence of these relatively intact remaining retinal layers prompted investigators including ourselves to study the possibility that electric stimulation of these structures may produce vision.

Electric currents applied through contacting electrodes to external structures of the eye, such as the lids, conjunctiva, and corneas are known to produce visual sensations called phosphenes in both normal *(6–8)* and blind RP patients *(9)* A similar visual evoked electrophysiological response has been demonstrated from the visual cortex

From: *Ophthalmology Research: Visual Prosthesis and Ophthalmic Devices: New Hope in Sight*
Edited by: J. Tombran-Tink, C. Barnstable, and J. F. Rizzo © Humana Press Inc., Totowa, NJ

of blind Royal College of Surgeons rats, a model of photoreceptor degeneration *(10,11)*. Direct electrical stimulation of the retinal nerve fiber layer through temporary and permanent electrodes have also been reported to evoked patterned phosphenes in RP patients *(12)* and visual evoked cortical potentials in animals *(13)*. Finally, animal studies have shown that retinal and cortical electrical activity can be elicited when the outer retina is electrically stimulated from the subretinal space with extremely low current and charge densities *(14,15)*. Based on these observations, in the year 2000 it was investigated whether a prosthesis implanted into the subretinal space could produce electrical stimulation and phosphenes in a pilot safety and feasibility study *(16–23)*.

A 2-mm diameter silicon-based semiconductor microphotodiode-array chip 25 μm in thickness called the artificial silicon retina (ASR) microchip was fabricated for implantation into the subretinal space. This chip included approx 5000 independently functioning electrode-tipped microphotodiodes, each powered solely by light falling on the retina. Individual microphotodiodes were separated from each other by an electrical channel stop and a pitch 25 μm. Each microphotodiode was 20×20 μm^2 in size with a central iridium oxide electrode of approx 9×9 μm^2. Iridium oxide was selected as the electrode material because of its high charge capacity and excellent tissue charge injection limit. All microphotodiodes on the same chip were connected to a common ground plane electrode on the back side of the chip that was also fabricated from iridium oxide. Electrical charges produced by the microphotodiodes concentrated on the front electrode and through primarily capacitive charge induction alter the membrane potentials of contacting retinal neurons. It was theorized that the result would be an electrical means of stimulating retinal cells to form retinotopically correct images in a manner similar to light. Because the implant would electrically stimulate the outer retina at an early functional stage with analog graded potentials and in a pseudobiphasic manner caused by varying ambient light conditions, subsequent visual signal processing by the remaining neuroretinal networks was thought to be possible.

In experimental cat, pig, and rat animal models, implantation of a solid ASR disk into the subretinal space blocked nourishment from the choroid and produced a model of outer retinal degeneration that histologically resembled the outer retinal degeneration of RP *(17,18,20,24,25)*. Immunohistochemistry of the overlying retina showed changes similar to changes present in patients with hereditary retinal degeneration *(20)*. Although, minimal dissolution of the ASR silicon substrate occurred over time, the ASR microchips continued to function electrically within the subretinal space *(17–19)* and demonstrated continued electrical activity for more than 3 yr after implantation *(26)*. The ASR microchips were able to induced secondary retinal and cortical responses in both normal animals and animal with retinal degeneration *(17,19)*.

On the basis of these findings, it was concluded that placement of an ASR chip into the subretinal space of a RP patient would not cause significant additional injury to the retina and may electrically stimulate remaining and partially compromised retinal cells. Implantation of the ASR chip into a midperipheral location, superior-temporal to the macular would allow safety and efficacy of the chip to be evaluated. It would also minimized surgical manipulation of the macular area, which may be functioning in many RP patients. The superior-temporal location would also lessen the possibility of macular damage should the chip gravitationally migrate inferiorly after surgical implantation.

IMPLANTATION OF THE ASR CHIP

Based on encouraging animal results, a pilot human clinical trial was designed in the late 1990's to implant a single ASR chip into the right eye of RP patients whereas retaining the left eye as a control. Between June 2000 and July 2001, Food and Drug Administration and IRB approvals were obtained and six RP patients were enrolled in an ASR safety and feasibility implantation clinical trial. Informed consent was obtained from all patients before entry. Eligible RP patients were 40 yr of age or older and were free of other significant eye diseases and medical conditions, such as substantial cataracts, uveitis, diabetes, glaucoma, or cardiac conditions. Enrollment criteria included a visual acuity of 20/800 or worse by early treatment diabetic retinopathy study (ETDRS) testing and/or a central visual field of 15°or less as determined by Humphrey automatic visual field testing (more than 10 dB loss, white III static, 31.5 apostilbs background illumination). Finally, patients had to be able to perceive electrically induced phosphenes produced by contact lens electrical stimulation. Exclusion criteria included unrealistic expectations of the study, unstable personality and significant psychiatric conditions.

For the contact lens phosphene enrollment test, electric current was provided by 1–6 photodiodes connected serially, each stimulated by a 940 nm infra-red light-emitting diode (IR LED) powered by 50 mA of current. Voltage and current produced by each photodiode was approx 0.40 V and approx 200 μA with approx 5 KΩ of measured impedance between the corneal contact lens electrode and the ipsilateral temple return electrode. Stimulation pulses were programmed by a signal generator and consisted of 50% duty cycle 5 Hz pulses with a polarity change every second and a total duration not exceeding 15 s. Thresholds for phosphene recognition in patients varied from 2 to 5 photodiodes connected electrically in series and were generally lower for patients with less severe vision loss. The initial current generated varied depending on the subject impedance and ranged from approx 200 μA for 1 photodiode to approx 600 μA for 5 photodiodes.

Upon passing vision enrollment testing, a complete medical and ophthalmic history and examination was performed including a lifestyle-quality-of-life questionnaire. Fundus photography was performed and a baseline fluorescein angiography was obtained to rule out cystoid macular edema unless contraindicated by allergy to fluorescein dye. Patients were questioned about their visual function outside of the office and asked to characterize their visual perceptions for seven aspects of visual function, comparing one eye with the other using a point rating scale. The perceptions questioned were: brightness, contrast, color, shape, resolution, movement, and visual field size. As the ASR chip would be implanted into the right eye (OD), patients were instructed to use their left eyes (OS) as a reference and to assign a fix value of 10 to the OS for comparison the OD. For example, if the brightness perception of both eyes were equal, both would be assigned a value of 10. If the brightness perceived from the OD was subjectively twice that of the OS, the OD would receive a rating of 20, and if one-half of the OS, a rating of 5. However, if a patient had no capability to see a certain percept, a value of 0 was assigned. In the latter case, if the OS could perceive a particular visual function, but the OD could not, the OS would be assigned a 10 whereas the OD was assigned a 0. Postoperatively, perceptions of the two eyes were compared with each other and to their preoperative values where possible. In the case if the OD developed subjective perception where previously it had none, it would be compared with the OS (as long as the OS had a non-0 value) as a ratio comparison with a preoperative value of 0 in the OD would not be possible (a "0" denominator would create an "infinite" improvement).

The preoperative visual acuity testing was evaluated a minimum of two times, using standard back-illuminated ETDRS charts at 0.5 m (largest letters equivalent to 20/1600) with the patient cyclopleged (cyclopentolate 1%, tropicamide 1% and phenylephrine hydrochloride 2.5%) and best corrected with their retinoscopic refraction for 0.5 m. The total number of ETDRS letters correctly identified was counted until one entire line of five letters was missed (but not counting the first line). If none of the first two lines of ETDRS letters could be identified at 0.5 m, count-fingers and hand-motions vision were evaluated and the threshold brightness for light-perception in nine visual field sectors was tested (*see* second paragraph following).

Automated visual field testing was performed with a Humphrey visual field analyzer II (HVFA) using the III and V white static spot sizes with the 30-2 and 60-4 fastpac protocols, and a custom 30° radius, 4° spot separation protocol pre- and postoperatively.

Because the HVFA instrumentation was limited in target brightness (10,000 apostilbs) and some patients saw no test targets with either the III or V static sizes, an additional visual field light threshold test was created to test nine visual field sectors in a 3 × 3 grid with <0.1 ft-candle of background room illumination. This was performed using a halogen light source coupled to a 0.5-in. diameter optical fiber that was positioned 10 cm from the patient's eye at the following nine locations from the patient's perspective: right-upper, right-middle, right-lower, middle-upper, middle-middle, middle-lower, left-upper, left-middle, and left-lower. All positions except the middle-middle position were located approx 45° from the optical axis of the eye (which is the middle-middle position). Using stacked neutral density filters in slide-holders, illuminations from 300 ft-candles down to 1e-4 ft-candles in 5 dB steps could be created for threshold testing. Threshold was established in each sector by crossing threshold at least three times in an ascending and descending staircase paradigm. The testing was continued until all 9 sectors were tested. Both the implanted and the control unimplanted eye were evaluated during each test session. In patients 1–3, this test was implemented by 4–6 mo postoperatively and in patients 4–6 by 2 m postoperatively. The test was called the Nine-Sector Test.

Electrophysiological testing was administered pre- and postoperatively with electroretinography (ERGs) and visual evoked potentials recordings using a LKC or Diagnosys Espion™ computer signal averaging system. Electrophysiological responses to both white and infrared (IR) light (940 nm) were obtained using handheld ganzfeld stimulators. IR light allowed the isolated activation of the ASR implant with subsequent electrical stimulation of the retina that was distinct from the native retina's response to visible light.

The ASR chip (Fig. 1) was implanted into the superior temporal subretinal space approx 20° from the macula in the right eye of all patients under general anesthesia. A standard 3-port vitrectomy (irrigation cannula, light-pipe, aspiration vitreous cutter) was performed with pars plana lensectomy. A retinal bleb was created using a cannula

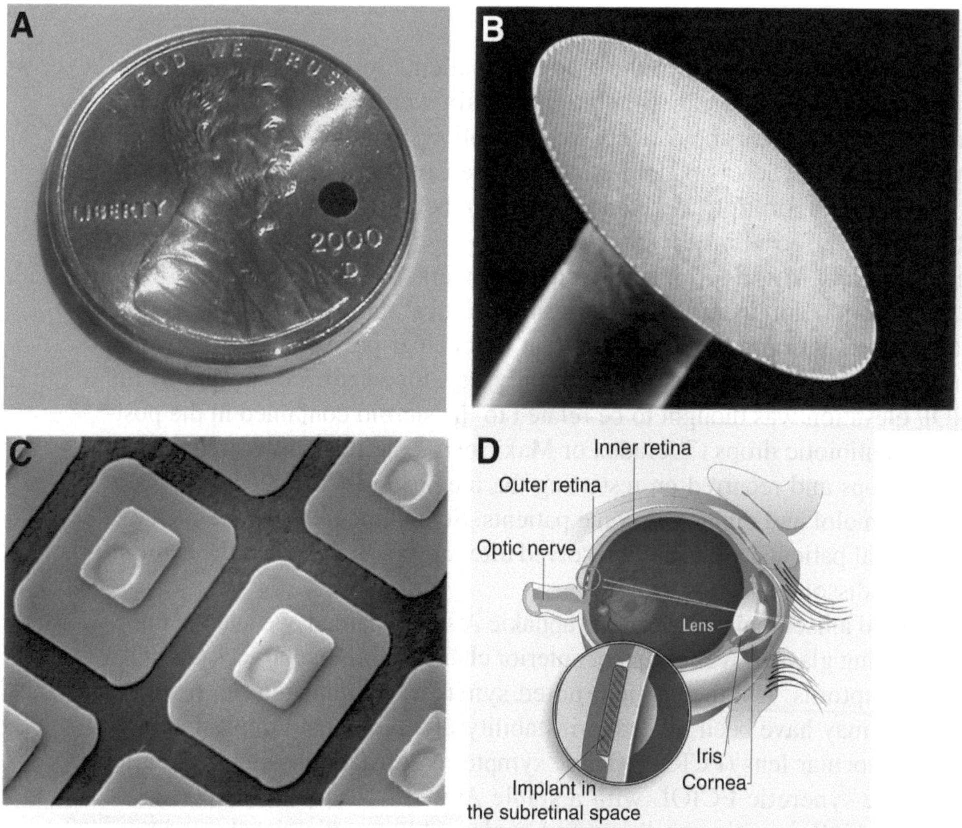

Fig. 1. Artificial silicon retina (ASR) microchip. The ASR is a silicon-chip-based NiP microphotodiode array fabricated using standard semiconductor and MEMS fabrication techniques. The device is 2 mm diameter, 25 μm thick and contains approx 5000 NiP microphotodiode pixels electrically isolated from each other by a 5 μm channel stop. Each pixel is 20 × 20 μm^2 and has a 9 × 9 μm^2 IrOx electrode. An IrOx common ground return electrode is located on the backside of the chip. No hermetic sealing was used. Pixel current was 8–12 nA under approx 800 ft-candles illumination. The ASR was placed within a Teflon® sleeve and secured to a saline-filled syringe during surgery. It was deposited into the subretinal space by the saline fluid flow pushing the ASR out of the injector sleeve. **(A)** ASR size compared with a penny. **(B)** ASR chip—×42. **(C)** ASR pixels—×1400. **(D)** Subretinal location of implanted ASR.

applied to the location of the implantation followed by hydrostatic dissection with balanced salt solution. The retinotomy was extended to 2.5 mm using vitreoretinal scissors. The ASR was inserted through the retinotomy into the subretinal space by manipulation with the cannula and a complete fluid-air exchange was performed to flatten the retina. Laser or thermal cautery was not required in most patients. The scleral incisions were closed with absorbable sutures and the eye patched with antibiotic-steroid ointment. Postoperative follow-up examinations occurred at: postoperative days 1, 2, and 4; weeks 1, 2, 4, 6, and 8; and months 3, 4, 6, 9, 12, 15, 18, 21, and 24. ERGs were performed at most follow-up visits and a fluorescein angiogram was performed at 6 mo.

SURGICAL RESULTS

Of the 15 RP patients screened for the study, 13 patients reported phosphenes on preoperative contact lens electrical stimulation and six were selected for ASR chip implantation. The genotypes of the patients were as follows: patient 1 had isolated RP without a family history, whereas patient 2 had an extensive vertical autosomal dominant family history with multiple affected family members; patient 3 had autosomal dominant RP with an affected brother and daughter; patient 4 had Usher's syndrome type II without a family history of other affected members; and patients 5 and 6 were brothers with autosomal dominant RP and a positive vertical family history.

Postoperative intervention was required to treat intraocular pressure (IOP) elevation above 25 in three out of six patients that generally developed toward the end of the 1 wk. The IOP elevation was thought to be related to the steroid contained in the postoperative steroid-antibiotic drops (Tobradex or Maxitrol) as the IOPs decreased rapidly on stopping the drops and recurred on restarting the medications. Elevated IOPs were also treated with timolol and diamox in some patients. Scratchiness of the operated eye was noted by several patients and usually resolved after approx 6 wk when the external absorbable sutures dissolved.

One patient noted aniseikonia between the aphakic ASR implanted eye and the unoperated eye when using glasses. A subsequent anterior chamber intraocular lens (ACIOL) relieved those symptoms. Another patient noted syneresis of images seen from the implanted eye that may have been related to instability of a previously implanted posterior chamber intraocular lens (PCIOL). These symptoms substantially improved after replacement of the syneretic PCIOL with a stable ACIOL. No patient experienced infection, prolonged inflammation or discomfort, undesirable entopic visual symptoms, intraocular or retinal hemorrhages, neovascularization, implant rejection, migration, or erosion through the retina.

Regarding their lens status, patients 1, 3, and 6 were pseudophakic before ASR implantation. Preoperatively, patient 2 who had bare to no light perception, had a 3+ PSC cataract (<20/200 view); patient 4 who had HM × 1 ft vision, had a 1+ anterior subcapsular opacity and a 1+NS and 1+PSC cataract (20/30 view); and patient 5 who had CF × 1–2 ft had a 1–2+ anterior subcapsular opacity, 1+ posterior cortical and 0–1+ NS cataract (20/30 view). To facilitate placement of the implant during surgery, cataracts were removed from patients 1, 3, and 6 during the ASR operation. Patients 2 and 4 were left aphakic and patient 5 eventually, underwent secondary ACIOL implantation 1 mo after ASR implantation.

Table 1 summarizes the postoperative results. At their last follow-up, no patient experienced any ASR-related complications. The retina overlying the implants appeared clear with patent vessels (Fig. 2). Fluorescein angiography showed no signs of neovascularization, vascular dropout, disruption, or leakage (Fig. 2). The anterior and posterior segments of the eye in all patients were quiet and all devices functioned electrically, demonstrated by ERG recordings to IR stimuli (Fig. 3, Top).

Preoperatively, two out of six patients (patients 5 and 6) were able to read ETDRS letters at 0.5 m. Patient 5 could read 16–25 letters with the OD and 24–28 letters with the OS preoperatively, and patient 6 could read 0 letters with the OD and 0–3 letters with the OS preoperatively. Postoperatively these two patients developed improvements in

Table 1
Visual Function Changes After ASR Implantation

Patient	*Age (yr)*	*Follow-up (mos.)*	*Lens*	*Complications*	*ETDRS visual acuity improvement*	*Subjective improvement*	*Automated visual field improvement*	*Nine sector testing improvement*
1	66	18	PCIOL	None		+		+
2	45	18	Un-corrected aphakia[a]	None		+		*
3	76	18	PCIOL	None	+	+		+
4	73	6	Un-corrected aphakia[a]	None		+		
5	59	6	ACIOL[a]	None	+	+	+	
6	59	6	ACIOL	None	+	+		

[+]Improvement measurable by the testing method.

[a]Lensectomy was performed during ASR implantation in three patients. Patient 2 had a cataract, and patients 4 and 5 had lenticular opacities that diminished visualization of the ASR during surgery. Patient 5 complained of diplopia with his aphakic glasses and received an ACIOL in a second operation. Patient 6 complained of movement of images because of instability of a previously placed PCIOL and was given an ACIOL replacement, in a second operation.

[*]Improvement compared with a muscle light of the same illumination preoperatively.

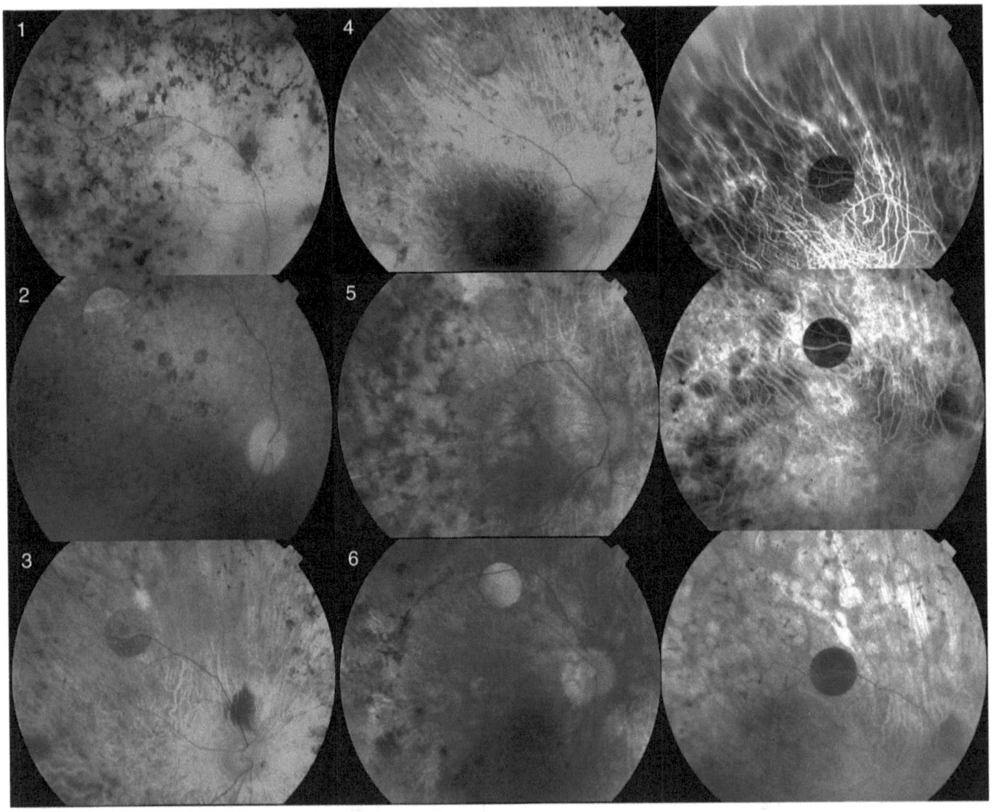

Fig. 2. Fundus photographs and fluorescein angiograms of implanted artificial silicon retina in superior-temporal retina (photo number indicates the patient number). FA is from patient 3, from top to bottom—early, mid, and late phase.

the total number of ETDRS letters read (Fig. 3, Bottom) that were consistent with their subjective impression of improved contrast, shape, and resolution perception. At 6 mo after implantation, patient 5 can read 35–41 letters with the OD and 21–28 letters with the OS, and patient 6 can read 25–29 letters with the OD and 0 letters with the OS. The smallest letters read with the OD improved from approx 20/800–20/200 for patient 5 and from <20/1600 (no letters read) to approx 20/400 for patient 6. Patient 3 who was unable to recognize any ETDRS letters preoperatively (<20/1600) with either eye was able to identify several of the largest letters consistently with the implanted OD postoperatively (~20/1280–20/1600) at 12–18 mo (Fig. 3, Bottom).

Preoperatively, positive responses from the 30–2 HVFA visual fields with the white V static target could only be obtained consistently from patients 5 and 6. Postoperatively, patient 5 demonstrated improved central and paracentral visual fields on the 30–2 protocol with the implanted OD (Fig. 4). Compared with the unoperated eyes, two implanted eyes (patients 1 and 3) showed improvement in the Nine-Sector Test at 6–12 mo after surgery. In patient 1, threshold sensitivity improved by approx 1000–1500% in all sectors and was consistent with the patient's impression that the entire visual field was brighter in the implanted eye compared with the same eye before surgery and also

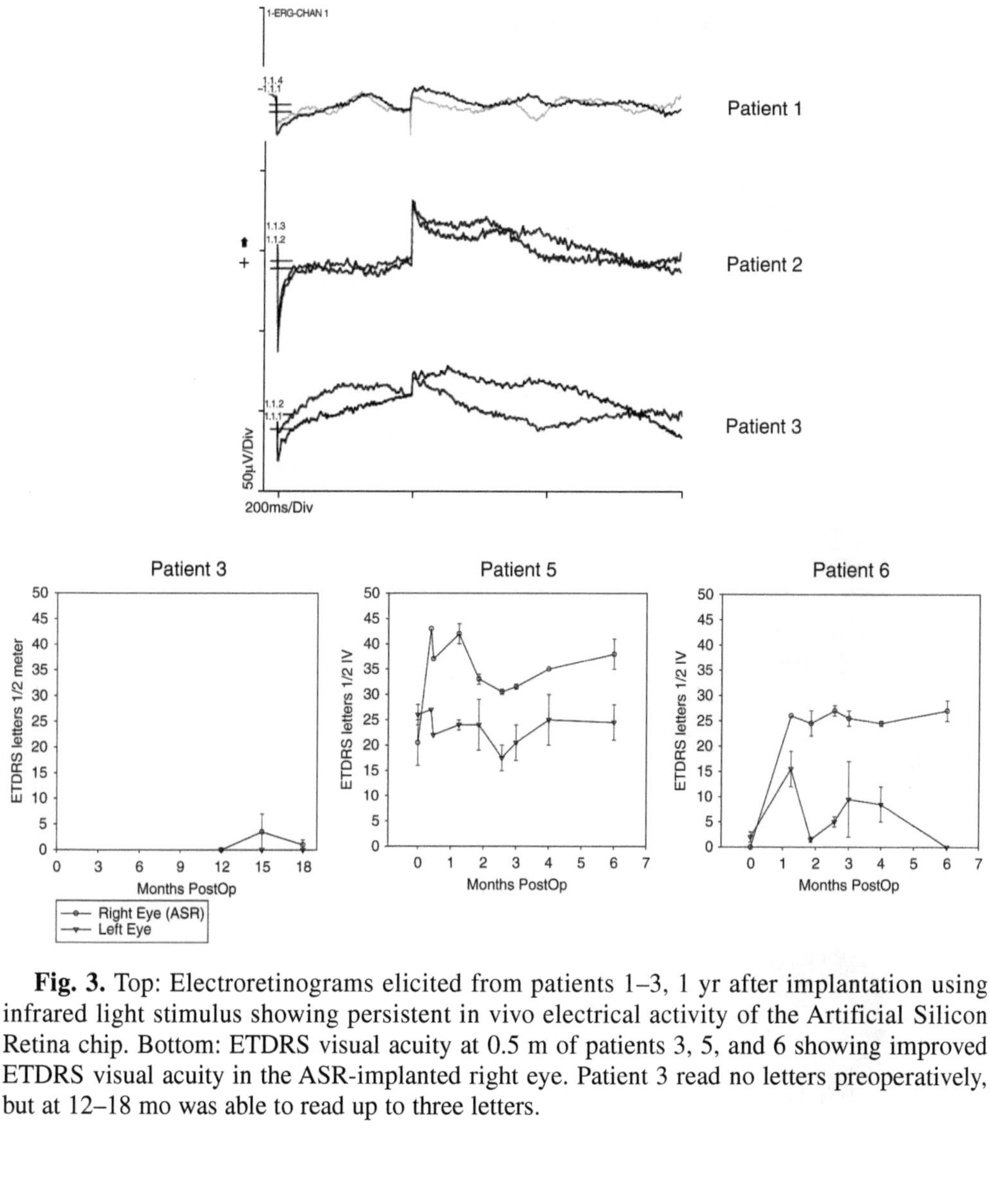

Fig. 3. Top: Electroretinograms elicited from patients 1–3, 1 yr after implantation using infrared light stimulus showing persistent in vivo electrical activity of the Artificial Silicon Retina chip. Bottom: ETDRS visual acuity at 0.5 m of patients 3, 5, and 6 showing improved ETDRS visual acuity in the ASR-implanted right eye. Patient 3 read no letters preoperatively, but at 12–18 mo was able to read up to three letters.

compared with the unoperated eye (Fig. 5). In patient 3, threshold sensitivities in the right-middle, right-lower, and middle-lower sectors of the Nine-Sector Test improved at 18 mo by approx 5000–10,000% (Fig. 5). The visual field areas of improvement in the Nine-Sector test for this patient were consistent with the patient's subjective impression that the best vision for straight ahead objects was achieved with the chin elevated to allow use of the inferior visual fields to look straight ahead. The inferior fields of improvement corresponded to the location of the implant. Patient 2 showed consistent light perception (LP) in multiple sectors of the operated eye in the Nine-Sector Test compared with the subjective bare-to-no LP in those same sectors preoperatively. The sectors of LP also corresponded to the location of the ASR chip. These perceptions

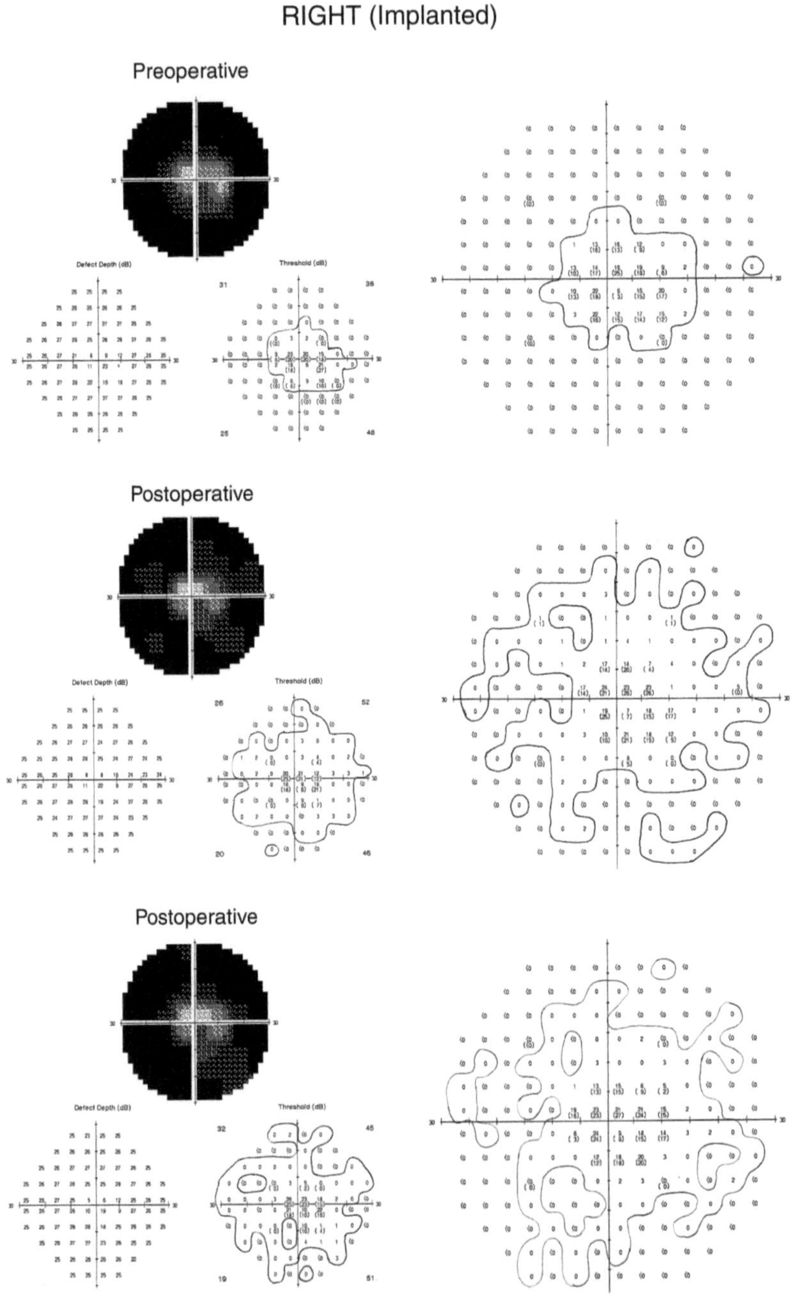

Fig. 4. *(Continued)*

were consistent with the patient's impression that postoperatively, LP developed that was present all the time in the implanted eye that improved to being able to detect shadows of people with back-lighting conditions. This patient's Nine-Sector thresholds did not improve further after 1 yr postoperatively.

Preoperatively, no patient was able to perceive color on pseudoisochromatic plate color testing. Postoperatively, patient 5 correctly identified the blue and orange dots

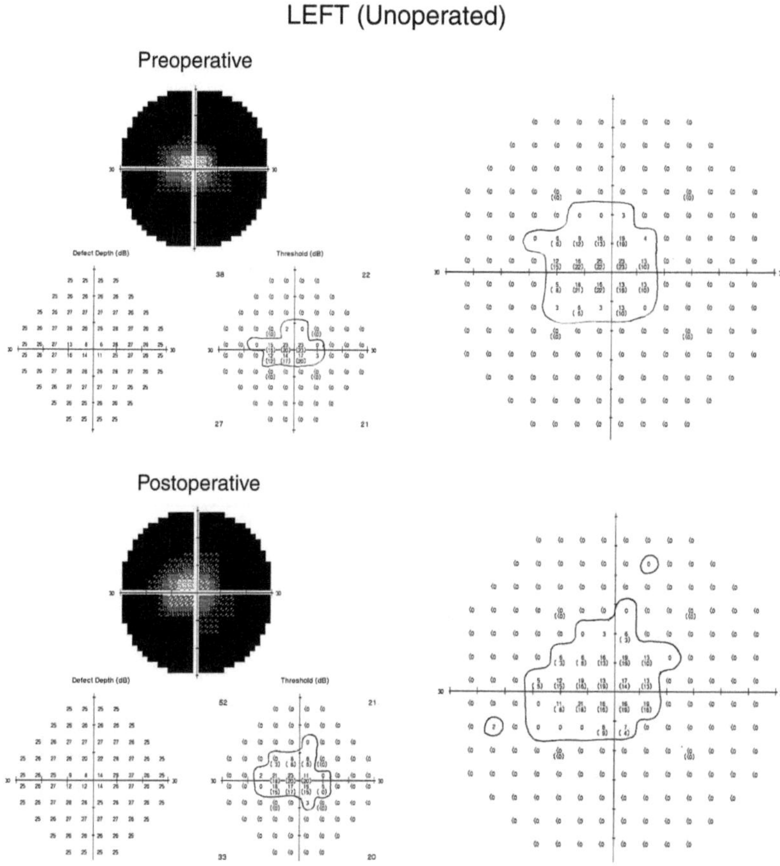

Fig. 4. Patient 5 Humphrey visual field analyzer II (HVFA). Central HVFA visual fields using the 30–2 Fastpac protocol and the white V static spot size could only be obtained consistently from patients 5 and 6 preoperatively. Postoperatively, patient 6 demonstrated no substantial changes in visual field size and no improved thresholds compared with preoperative levels. However, patient 5 demonstrated consistently improved central and paracentral visual fields in the OD postoperatively, compared with the preoperative fields. Whereas almost all of the visual field outside of 15° radius of both eyes in patient 5 was <0 dB threshold preoperatively (unrecordable and greater than 10,000 ASB threshold sensitivity—top left and right), substantial portions of the OD visual field were recordable postoperatively at 0 dB or better (left middle and left bottom). The HVFA of the unoperated OS was essentially unchanged (right middle).

of the control plate and the red and green dots of the test plates with the operated eye. The unoperated control eye remained unable to perceive any colors. Subjectively, patient 5 reported improved color perception of the environment including being able to see the green and white of highway signs, the red and white of stop signs, red and white checks on a kitchen table cloth, the green grass, and multiple colors on people's clothing.

In response to acute electrical stimulation produced by the chip using IR light stimulation, four out of six patients (patients 2–5) indicated disk-like light percepts that corresponded to the location of the implant. IR light directed into the opposite control eye did not create a visual response. However, the light response in the ASR-implanted

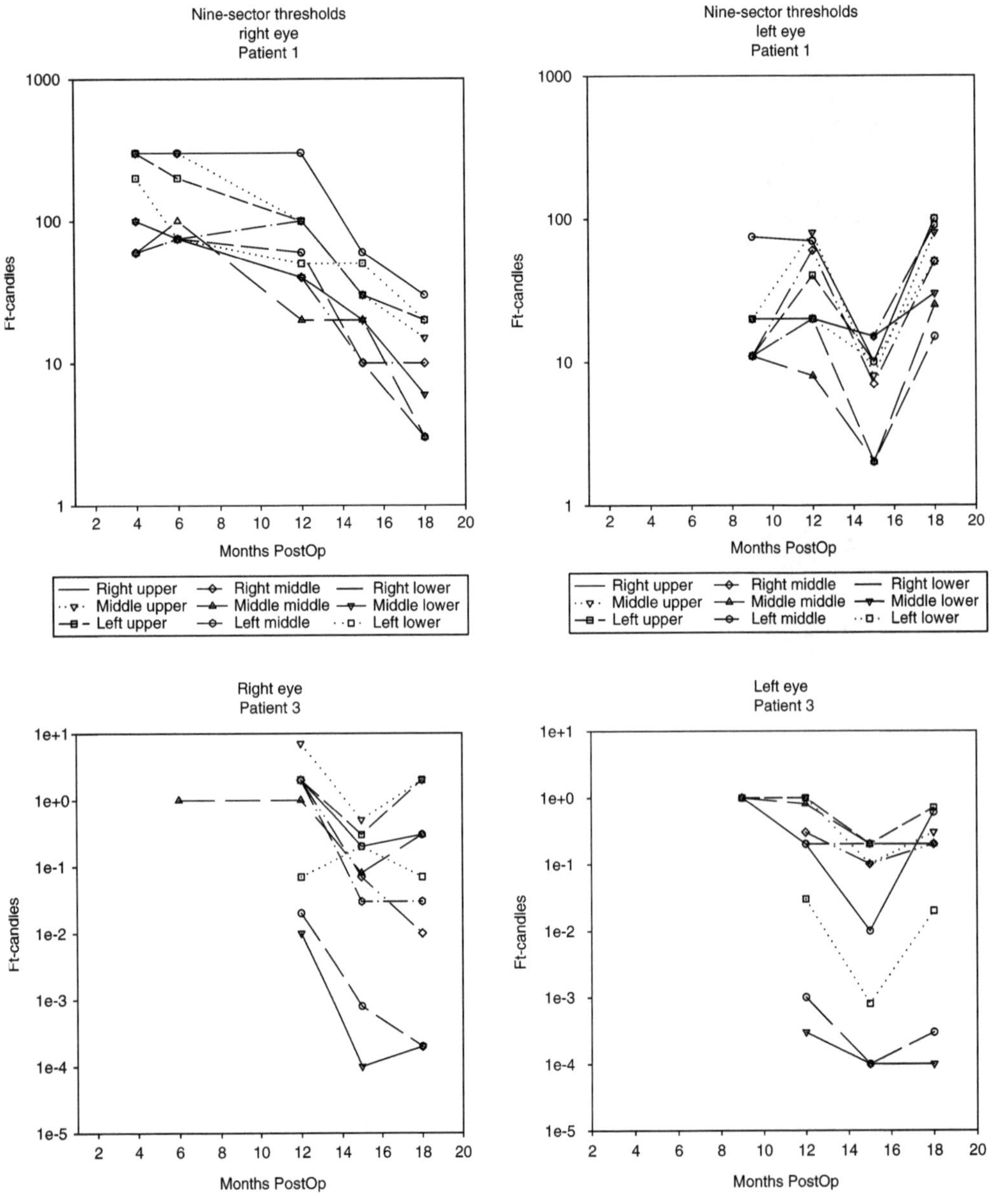

Fig. 5. Nine-sector testing in patient 1 (top) showed improved light thresholds postoperatively in the OD in all sectors from 1000 to 1500%. No persistent changes were noted in the control OS. In Patient 3 (bottom), the right-middle, right-lower, and middle-lower sectors improved 5000–10,000% postoperatively for light detection threshold. No persistent changes were observed in the control OS. The OD threshold improvements of patient 3 were in the sectors of visual field that corresponded to the implant location and were consistent with the patient's subjective impression of improved vision location.

eye was not persistent. Typically, the first test of a session resulted in the light percept, but subsequent tests did not result in light perception. The mechanism for the intermittent responses was uncertain, but may have been associated with an electrical capacitive block of the retina that resulted from a continuous direct current electrical stimulus or perhaps from a retinal fatigue effect (the repetitive light flashes observed by all patients preoperatively to external contact lens electrical stimulation was from biphasic stimulation that may have prevented capacitive block or retinal fatigue).

SUBJECTIVE COMPARISON OF VISUAL FUNCTION AFTER ASR IMPLANTATION

At 18 mo after surgery, the first group of three patients generally thought that their visual improvements had stabilized. In comparison, two of the three patients (patients 5 and 6) in the second group thought that their visual function improvements stabilized at 6 mo after surgery although patient 4 reported continuing improvement beyond that period.

Patient 1 who had LP in both eyes before surgery, reported a preoperative preimplanted to control eye comparison ratio for brightness of 5:10 and for visual field 2:10. Postoperatively, the ratios were 7:10 and 15:10, respectively, stabilizing at 18 mo. The OD implanted eye visual field size was subjectively approx 750% larger compared with the visual field before surgery. Functionally, the patient reported not having to turn the head to see lights from the right side.

Patient 2 had bare-to-no LP in the preimplanted OD and LP in the control OS before surgery and only perceived brightness, contrast, shape, and limited visual fields in the OS. Postoperatively, the patient reported substantial visual improvements in the OD characterized by LP in the inferior nasal visual field that persisted at 18 mo. The reported postoperative OD:OS ratios were for brightness 8:10, contrast 10:10, shape 10:10, and visual field size 8:10. Functionally, the patient reported being able to see shadows of people with the OD in back-light conditions.

Patient 3 had hand-motions (HM) to LP vision in both eyes before surgery. The patient compared preimplanted OD:OS control ratios as 7:10 for brightness, and 10:10 for shape, resolution, movement, and visual field size. At 18 mo postoperatively, the ratios improved for the implanted OD to 30:10 for brightness, 35:10, 50:10, 50:10, and 50:10 for the other visual functions, respectively. Functionally, the patient reported regaining the ability to use night-lights at home for navigation at night and was able to see movement on television again.

Patient 4 had HM vision OU before surgery. Preoperatively, the preimplanted OD:OS ratios were reported as 10:10 for brightness, contrast, shape, and visual field size. Postoperatively, the ratios were variable, but at maximum were improved in the OD compared with OS to: 15:10, 17:15, 17:10, and 13:10, respectively. Although, this patient could see no movement preoperatively, after surgery the perception of movement was noted to be 2/10 relative to what the patient remembered as present during his youth. Subjectively, this patient indicated that visual function was also more improved when both eyes were used together improving from 10 preoperatively to approx 25 after surgery. Functionally, the patient reported being able to navigate the yard without a cane and could readily tell which room's lights were on at night in the house.

Patient 5 had vision of CF at 1–2 ft in the OD and OS before surgery. The patient noted equal visual function OU in all perceptions (10/10) preoperatively. Postoperatively, marked vision improvement was reported in the OD without changes in the OS. The postoperative OD:OS ratios were for brightness 17:10, contrast 30:12, color 17:10, shape 15:10, resolution 35:10, movement 13:10, and visual field size 11:10. Functionally, the patient reported being able to more easily discern denominations of paper money, saw food well enough to use eating utensils, and recognized faces again, which had not been possible for approx 10 yr.

Patient 6 had HM vision OU preoperatively and reported equal visual function OU in all perceptions (10:10) before surgery. Postoperatively, the OD:OS ratios were variable between days, but maximized at brightness 20:10, contrast 25:10, color 20:10, shape 20:10, resolution 20:10, movement 20:10, and visual field size 18:10. Functionally, the patient reported being able to recognize denominations of paper money and was able to differentiate traffic lights colors. The patient reported also being able to locate cars in the street and the coffee cup at meals, both of which were not possible before surgery.

DISCUSSION

Initial results from the microphotodiode retinal prostheses implantations indicate that a silicon chip-based retinal prosthesis could be safely and consistently implanted into the subretinal space of RP patients. The ASR microchips were well tolerated and caused no discomfort, rejection, infection, inflammation, neovascularization, vessel disruption, retinal detachment, migration, or erosion of the implant through the retina. These findings were consistent with the previously reported results from animal studies showing similar biocompatibility of the implant materials (silicon, silicon oxide, titanium, and iridium oxide) *(17–20)*. The continued electrical activity of the ASR chips in human subjects was also similar to observations from the animal studies *(26)*.

Substantial and persistent visual function improvements were reported by all patients implanted with the chip. These improvements involved subjective impressions, lifestyle, and quality of life changes, tasks performance, ETDRS letter recognition, color recognition, HFVA visual fields, and the custom Nine-Sector Test of visual fields. The retinal areas and the levels of improvement however were higher than expected from a small device implanted into the superior temporal retina and stimulating only a small portion of retina. Although, phosphenes were perceived in the visual fields corresponding to the ASR chip location in four of six patients, improvements in visual function also occurred in retinal visual fields distant from the implant that included the macula region. These improvements were first noted at approx 1 wk–2 mo after surgery and continued to improve until approx 6–12 mo postoperatively.

The mechanism of visual function improvement in the retinal areas distant from the implant is unlikely to be solely from direct ASR electrical stimulation of the retina because the electrical effect of the chip decreases rapidly with distance from the chip. Also, the improved and accurate perceptions of contrast, color, resolution, movement, and visual field size are too complex to be explained by only direct electrical action of the implant. A compatible mechanism of action would include a generalized neurotrophic effect on the retina from ASR electrical stimulation.

The observation supporting this possibility was that visual function improvements did not occur immediately. Improvements began approx 1 wk–2 mo after ASR implantation and continued for up to 1 yr in some patients. Initially, patients 3 and 5 complained of worsened vision during the 1 mo after surgery before improvements were reported. Patient 2 who had no subjective LP before surgery, noted an inconsistent LP during the 1 wk after surgery, and then observed what appeared to be a "quarter size" spot of light at several feet distance in the projected visual field of the implant. In subsequent weeks, the "spot of light" increased in size to a vertical oval that eventually, expanded to cover the left hemi-field and the macula area.

Data from other studies have suggested that growth and neurotrophic-type mechanisms may be associated with electrical stimulation. The application of electric currents to a variety of organ systems has been reported to promote and maintain certain cellular functions. These cellular functions include bone growth *(27,28)*, spinal cord growth *(29)*, and cochlear spiral ganglion cell preservation *(30,31)*. Recently, deep brain electrical stimulation (DBS) of the subthalamic nucleus and the globus pallidus interna in Parkinson's patients significantly relieved tremors and spasticity *(32)*. The mechanism of improvement has been proposed to involve improved neurotransmitter balance and upregulation of growth and neurotrophic-type factors *(33,34)*.

Neurotrophic-type factors have been widely reported to have the capacity to promote and maintain retinal cellular functions. Brain-derived neurotrophic factor, neurotrophin-4, neurotrophin-5, fibroblastic growth factor, and glial cell line-derived neurotrophic factor have been shown to enhance neurite outgrowth of retinal ganglion cells and to increase their survival in cell culture *(35)*. Glial cell line-derived neurotrophic factor has been reported to preserve rod photoreceptors in the rd/rd mouse, an animal model of retinal degeneration *(36)* and ciliary neurotrophic factor has slowed photoreceptor degeneration in the rd/rd, the nervous (nr/nr) and the Q344ter mutant rhodopsin mice *(37)*. Nerve growth factor injected into the intraocular space of the C3H retinal-degenerated mouse, resulted in a temporary rescue of photoreceptor cells compared with controls *(38,39)*.

Mechanical injury stimuli, such as a penetrating wound of the sclera and retina, also, upregulates mRNA expression of basic fibroblast growth factor and ciliary neurotrophic factor and are accompanied by a transient increase of the fibroblastic growth factor receptor. These factors are believed to exert photoreceptor protective and rescue effects after injury *(40)*. It may be that chronic low-level electrical stimulation to a partially degenerated RP retina induces a similar upregulation of protective neurotrophic-type survival factors that improve the function of remaining and previously inadequately functioning photoreceptors. The early "peaking" of ETDRS letters read before a moderate decrease and plateauing of letters in patients 5 and 6 may be the result of an initial neurotrophic response to both the surgery and electric stimulation followed by a more chronic neurotrophic effect to just electrical stimulation.

Limitations of this pilot study should be mentioned and include limited controls and the necessary use of a newly developed Nine-Sector Test in some patients. Consistency of testing was attempted by using primarily one examiner (AC) to perform most of the newly developed Nine-Sector exams with assistants recording the results. A few Nine-Sector Test exams were performed by other examiners, but generally under the supervision and guidance of the main examiner. To aid in revealing potential intersession variability and placebo effects, all exams of the implanted eyes were accompanied by exams of the unoperated control eyes during each test session. Multiple preoperative evaluations of the Nine-Sector test were performed in the later enrolled patients to establish a broad preoperative baseline, but were not universally performed in the earlier patients.

Caution is appropriate in interpreting patients' subjective comparisons of visual function between their two eyes and for pre- and postoperatively times as these perceptions could be affected by the patients' impressions of whether a surgical intervention, i.e., ASR implantation, may or may not help them. Nevertheless, it is believed that careful evaluation of both subjective information along with data obtained from visual function testing is useful. It should be noted that most patients reported that their vision initially worsened, sometimes substantially for up to 1 mo after surgery before a slow improvement was occurred. Such a response would be less typical for a placebo effect.

Three of the six implanted patients had cataracts of varying degrees before ASR chip implantation that were removed during surgery. Although removal of mild cataracts may improve patient visual acuity in normally sighted individuals for high spatial frequencies, generally, it is acknowledged that removal of mild cataracts (20/30 view) would unlikely affect visual acuities in the range of patient 5 (20/200–20/800) or patient 4 (HM). Removal of a 3+ PSC cataract would also unlikely improve vision in a RP patient from no LP to LP with form recognition.

Consideration for future research will include the following: safety results and efficacy responses may be verified in a larger group of patients enrolled at multiple independent sites and with masked examiners. Multiple test sessions may be conducted preoperatively to assure that study patients are stabilized on the "learning curve" of all tests. The optimization of ASR stimulation parameters may be investigated to study the effect of varying voltage, current, duration, charge, phase, and chronicity of stimulation. Other questions that will be considered for investigation include: would implantation of multiple devices be more effective than a single device and at which retinal locations? If ASR implantation exerts a neurotrophic effect, would implantation earlier in specific types of retinal degenerative disease be more effective? Finally, would patients with other forms of retinal degenerations, such as dry age-related macular degeneration, also benefit from ASR implantation?

Silicon chip-based retinal prostheses called ASRs, containing approx 5000 electrode-tipped microphotodiodes were implanted into the right eyes of six patients in a pilot study. All ASRs functioned electrically during follow-up periods that ranged from 6 to 18 mo. No patient showed signs of implant rejection, infection, inflammation, erosion, retinal detachment, neovascularization, or implant migration. Subjective and/or objective visual function improvements occurred in all patients and included unexpected visual improvements in retinal areas distant from the implant that may involve neurotrophic mechanisms.

REFERENCES

1. National Advisory Eye Council. Report of the Retinal Diseases Panel: Vision Research: A National Plan, 1994–1998. Bethesda, Md: United States Dept of Health and Human Services; 1993. Publication NIH 93-3186.
2. Pagon RA. Retinitis pigmentosa. Surv Ophthalmol. 1988;33:137–177.
3. Berson EL, Sandberg MA, Rosner B, et al. Natural course of retinitis pigmentosa over a three-year interval. Am J Ophthalmol 1985;99:240–251.
4. Flannery JG, Farber DB, Bird AC, Bok D. Degenerative changes in a retina affected with autosomal dominant retinitis pigmentosa. Invest Ophthalmol Vis Sci 1989;30:191–211.
5. Santos A, Humayun MS, de Juan E Jr., et al. Preservation of the inner retina in retinitis pigmentosa. A morphometric analysis. Arch Ophthalmol 1997;115:511–515.

6. Brindley GS. The site of electrical excitation of the human eye. J Physiol 1955;127: 189–200.
7. Potts AM, Inoue J, Buffum D. The electrically evoked response (EER) of the visual system. Invest Ophthalmol Vis Sci 1968;7:269–278.
8. Carpenter RH. Electrical stimulation of the human eye in different adaptational states. J Physiol 1972;221:137–148.
9. Potts AM, Inoue J. The electrically evoked response (EER) of the visual system—II: Effect of adaptation and retinitis pigmentosa. Invest Ophthalmol Vis Sci 1968;8:605–612.
10. Potts AM, Inoue J. The electrically evoked response (EER) of the visual system—III: Further contribution to the origin of the EER. Invest Ophthalmol Vis Sci 1970;9:814–819.
11. Dowling JE, Sidman RL. Inherited retinal dystrophy in the rat. J Cell Biol 1962;14:73–109.
12. Humayun MS, de Juan E Jr., Dagnelie G, et al. Visual perception elicited by electrical stimulation of retina in blind humans. Arch Ophthalmol 1996;114:40–46.
13. Dawson WW, Radtke ND. The electrical stimulation of the retina by indwelling electrodes. Invest Ophthalmol Vis Sci 1977;16:249–252.
14. Knighton RW. An electrically evoked slow potential of the frog's retina—I: Properties of the response. J Neurophysiol 1975;38:185–197.
15. Chow AY, Chow VY. Subretinal electrical stimulation of the rabbit retina. Neurosci Lett 1997;225:13–16.
16. Chow AY. Electrical stimulation of the rabbit retina with subretinal electrodes and high density microphotodiode array implants. Invest Ophthalmol Vis Sci 1993;34(Suppl):835.
17. Peachey NS, Chow AY. Subretinal implantation of semiconductor-based photodiodes: Progress and challenges. J Rehabil Res Dev 1999;36:372–378.
18. Chow AY, Pardue MT, Chow VY, et al. Implantation of silicon chip microphotodiode arrays into the cat subretinal space. IEEE Trans Neural Syst Rehabil Eng 2001;9:86–95.
19. Peyman GA, Chow AY, Liang C, et al. Subretinal semiconductor microphotodiode array. Ophthalmic Surg Lasers 1998;29:234–241.
20. Pardue MT, Stubbs EB, Perlman JI, et al. Immunohistochemical studies of the retina following long-term implantation with subretinal microphotodiode arrays. Exp Eye Res 2001;73:333–343.
21. Chow AY. Artificial retina device. US Patents No. 5,016,633. 1991, No. 5,024,223. 1991.
22. Chow AY, Chow VY. Independent photoelectric artificial retina device and method of using same. US Patents No. 5,397,350. 1995, No 5,556,423. 1996.
23. Chow AY, Peachey NS. The subretinal microphotodiode array retinal prosthesis. Ophthalmic Res 1998;30:195–196.
24. Zrenner E, Miliczek KD, Gabel VP, et al. The development of subretinal microphotodiodes for replacement of degenerated photoreceptors. Ophthalmic Res 1997;29:269–280.
25. Zrenner E, Stett A, Weiss A, et al. Can subretinal microphotodiodes successfully replace degenerated photoreceptors? Vision Res 1999;39:2555–2567.
26. Pardue MT, Ball SL, Phillips MJ, et al. Status of the feline retina after subretinal implantation of an artificial silicon retina for three years. Presentation at 4th Annual VA Rehabil Res Develop Conf. Arlington, Virginia, 2002.
27. Lagey CL, Roelofs JM, Janssen LW, et al. Electrical stimulation of bone growth with direct current. Clin Orthop 1986;204:303–312.
28. Kane WJ. Direct current electrical bone growth stimulation for spinal fusion. Spine 1988;13:363–365.
29. Politis MJ, Zanakis MF. Short term efficacy of applied electric fields in the repair of the damaged rodent spinal cord: behavioral and morphological results. Neurosurgery 1988;23: 582–588.
30. Leake PA, Hradek GT, Snyder RL. Chronic electrical stimulation by a cochlear implant promotes survival of spiral ganglion neurons after neonatal deafness. J Comp Neurol 1999;412:543–562.

31. Leake PA, Hradek GT, Rebscher SJ, Snyder RL. Chronic intracochlear electrical stimulation induces selective survival of spiral ganglion neurons in neonatally deafened cats. Hear Res 1991;54:251–271.
32. The Deep-Brain Stimulation for Parkinson's Disease Study Group. Deep-brain stimulation of the subthalamic nucleus or the pars interna of the globus pallidus in Parkinson's disease. N Engl J Med 2001;345:956–963.
33. Carvalho GA, Nikkhah G. Subthalamic nucleus lesions are neuroprotective against terminal 6-OHDA-induced striatal lesions and restore postural balancing reactions. Exp Neurol 2001;171:405–417.
34. Andrews RJ. Neuroprotection for the new millennium. Matchmaking pharmacology and technology. Ann N Y Acad Sci 2001;939:114–125.
35. Bosco A, Linden R. BDNF and NT-4 differentially modulate neurite outgrowth in developing retinal ganglion cells. J Neurosci Res 1999;57:759–769.
36. Frasson M, Picaud S, Leveillard T, et al. Glial cell line-derived neurotrophic factor induces histologic and functional protection of rod photoreceptors in the rd/rd mouse. Invest Ophthalmol 1999;40:2724–2734.
37. LaVail MM, Yasumura D, Matthes MT, et al. Protection of mouse photoreceptors by survival factors in retinal degenerations. Invest Ophthalmol Vis Sci 1998;39:592–602.
38. Lambiase A, Aloe L. Nerve growth factor delays retinal degeneration in C3H mice. Graefes Arch Clin Exp Ophthalmol. 1996;234(Suppl):S96.
39. Reh TA, McCabe K, Kelley MW, Bermingham-McDonogh O. Growth factors in the treatment of degenerative retinal disorders. Ciba Found Symp. 1996;196:120–131; discussion 131–134.
40. Wen R, Song Y, Cheng T, et al. Injury-induced upregulation of bFGF and CNTF mRNAs in the rat retina. J Neurosci 1995;15:7377–7385.

5

Development of an Intraocular Retinal Prosthesis to Benefit the Visually Impaired

Wentai Liu, Mohanasankar Sivaprakasam, Guoxing Wang, Mingcui Zhou, James D. Weiland, and Mark S. Humayun

CONTENTS

INTRODUCTION

Neural prostheses have been used as treatment for a variety of neurological disorders, motivating engineers and scientists to pursue prostheses for presently incurable human diseases related to the nervous system. A retinal prosthesis is based on the principle of activating nerve cells using a device implanted on the retina. In an intraocular retinal prosthesis, the stimulation device is placed internal to the eye. Retinal prosthesis potentially targets the restoration of vision in persons affected by outer retinal degenerative diseases. The most common diseases are age-related macular degeneration (AMD) and retinitis pigmentosa (RP) *(1)*. RP is a collective name for a number of genetic defects that result in photoreceptor loss. RP affects the rods (used in night vision) first and then the cones (used in ambient daylight levels). AMD results from abnormal aging of the retinal pigment epithelium and retina. Persons with AMD will start to have distorted vision and eventually, lose most of the vision in the central 30°. In both the diseases, the vision is impaired because of the damage to the photoreceptors that convert photons to neural signals. Postmortem evaluations of retina with RP or AMD have shown that a large number of cells remain healthy in the inner retina compared with the outer retina *(2,3)*. The inner retina is made up of horizontal, bipolar, amacrine, and ganglion cells. Further, electrical stimulation of humans with RP and AMD results in the perception of light; so the neural cells can be activated, providing the hope of restoring lost vision in blind persons *(4)*. A chronic implant with 16 electrode sites on the retina

From: *Ophthalmology Research: Visual Prosthesis and Ophthalmic Devices: New Hope in Sight*
Edited by: J. Tombran-Tink, C. Barnstable, and J. F. Rizzo © Humana Press Inc., Totowa, NJ

in three blind patients has yielded promising results *(5)*. After being implanted with the prosthetic device, the patients were able to detect motion of a white bar (up, down, left, or right), detect a rectangular object, count objects, discriminate the orientation of two white bars in an "L" configuration regarding where the corner of the L was positioned, and discriminate between a dessert plate, a coffee cup, and a plastic knife. The success rate of these simple visual tasks differed between subjects and the difference is attributed possibly to the age and the number of years of blindness *(6)*. The results of acute and chronic studies have encouraged several research and development efforts for realizing chronically implantable, high-resolution retinal prostheses. However, the number of stimulations sites required for restoring vision to a useful level is not clear. Simulations of prosthetic vision suggest that 600–1000 electrodes will be required to restore visual function to a level that would allow a blind person to read, independent mobility, and facial recognition *(7–10)*. This chapter will discuss the system components required for realizing such an intraocular retinal prosthesis. The microelectronics of the system will be described in detail along with prototype examples of the electronics. Whereas this chapter will focus on retinal prosthesis, there are other approaches of visual prosthesis, some of which are listed later. Restoring lost vision through cortical prosthesis, by stimulating the visual cortex in the brain has been pursued *(11)*. This approach has the potential to treat blindness not limited to the damage of outer retina, as the stimulus is applied in the later stage of visual processing chain. A visual prosthesis through stimulation of the optic nerve to create phosphene perception is being developed *(12)*.

The advantage of retinal prosthesis on cortical and optic nerve prosthesis is that the mapping of the retina to the visual field in the space is well known. Whereas the conventional prostheses use electrical stimulation, chemical stimulation using neurotransmitters have been proposed. An artificial synapse chip that aims to direct the growth of retinal neurites to the individual stimulation sites to selectively stimulate the cells by employing microfluidic neurotransmitter delivery is described in ref. *13*.

Based on the location of the implant device with respect to the retina, retinal prostheses are categorized as epiretinal or subretinal. The epiretinal prosthesis places the implant prosthetic device that delivers the electrical stimulation to the neurons on the surface of the inner limiting membrane of the retina *(5)*. The advantages of this approach are partitioning of the microelectronics into implantable portion and wearable (external) portion, allowing for future upgrades, and control by the doctor and patient. This also uses the vitreous as a sink for heat dissipation from the microelectronics. In this method, the ganglion cells are stimulated using biphasic stimulus pulses (anodic and cathodic). The subretinal prosthesis involves implanting a microphotodiode array between the bipolar cell layer and the retinal pigment epithelium *(14)*. The advantages of this approach include closer proximity to the surviving neuron in the inner retina (bipolar cell). The retinal prosthesis described in this chapter falls into the epiretinal category.

SYSTEM COMPONENTS OF THE RETINAL PROSTHESIS

Figure 1 shows the concept of the retinal prosthesis. An external video camera captures the image in the patient's field of vision. Thus, the patient can focus on a particular object

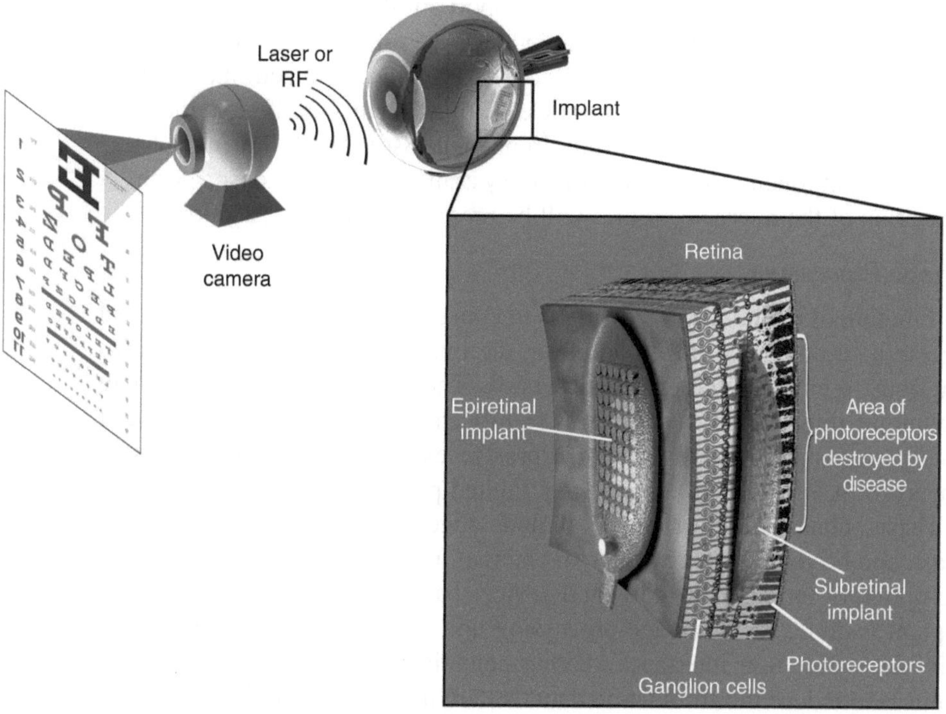

Fig. 1. Retinal prosthesis concept. (With permission, Annual Review of *Biomedical Engineering*, 2005.)

by turning his or her head (where the camera is fitted) toward the object's direction. These image data from the camera are then processed by a video signal-processing unit that converts these image data into stimulation data for the electrical stimulus. These data are then transmitted wirelessly using radio frequency waves or optical means, such as laser. These data are received inside the eye (hence, the term intraocular retinal prosthesis), converted to electrical stimulus to the retina. The electronics performing the above conversion requires electrical power, which is also wirelessly transmitted. Whereas a conductive tethering wire connecting the external unit to the implant unit is the most straightforward way to transmit power and data, there are some serious disadvantages. A wire penetrating through the skin, termed "percutaneous," may require continued medical supervision to guard against any infection. Depending on mechanical anchoring, such as to bone, percutaneous connectors may restrict movement of the tissue surrounding the wire causing friction between them. Rapid eye movement can break any penetrating wires passing through the sclera (tissue envelope covering the eyeball except the cornea) in addition to tissue damage. In addition to the technical difficulties, it also poses considerable inconvenience to the patients. Considering the difficulties of dealing with a percutaneous wire, especially, for long-term implant, wireless methods of power and data delivery are always preferred.

From the first look, it seems ideal that the prosthesis consisting of electronics would have all components fully implanted such that communication with any exterior device is minimized or even eliminated. Accordingly, for biocompatibility, more stringent power dissipation budget would be necessary to support a fully implantable prosthesis, which might be difficult to meet while trying to increase the functionality of the implant electronics. Additional benefits of external electronics include reduced internal heat dissipation, and ease of refinement, and upgrade of functionality in the external unit (which would otherwise be difficult in a fully implanted case). The following subsections list the key components of the prosthesis system with a brief description. A detailed description of the microelectronics will be presented in the next section.

External Image Processing

The main function of the image processing is to convert the image from the external video camera to a stimulus pattern. In addition, the image-processing unit may also perform functions to enhance the prosthetic vision. The human retina is not a mere receptor of photonic information, but performs significant image processing because of its layered neural network structure. As retinal prostheses can stimulate only a limited number of neurons, it is crucial to enhance this limited perception by means of image processing. These enhancement techniques include edge detection and enhancement, zoom, contrast, and brightness adjustment. An external image processing also provides the users the option of tuning their individual devices based on their visual experience after the implant is done. So the image-processing unit should be programmable and have an easily accessible user interface. Another potential functionality of the image-processing unit is to reduce the stimulation data that are transmitted wirelessly into the implant. This is done by compressing the data by encoding algorithms. In this case, the data receiver inside the eye will have to perform the reverse process, decoding, to extract the data. This type of data reduction can be inevitable in cases where the data rate of the wireless link itself is limited by other given factors. Some image-processing work similar to the ones described earlier can be found in refs. *15,16*.

Microelectronics

The microelectronics are made of three major blocks—power telemetry, data telemetry, and stimulator. The power telemetry transmits/receives the power required for all the microelectronics inside the eye. Two channels of data telemetry are required—one to transmit data to the implant (forward data telemetry) and another to transmit data from the implant to the external unit (reverse data telemetry). While the forward data telemetry transmits stimulation data, the reverse data telemetry is required for the purpose of monitoring several parameters inside the eye, including power level, temperature, pressure, pH, and electrode impedance. The stimulator is the functional replacement for the damaged photoreceptors, providing electrical stimulus to the ganglion cells, through an array of electrodes. Power and area are the key design constraints for the intraocular electronics. Reduction in power reduces the heat dissipation inside the eye, and also, results in a longer battery life. Reduction in area miniaturizes the implant and eases the packaging difficulties of the electronics and the surgery procedure to place the package inside the eye. To reduce the area, an integrated circuit (IC) approach is preferred where all the three blocks of the intraocular electronics are fabricated in one single IC. The external electronics include power transmitter, forward data transmitter, and reverse data receiver, all powered by a battery. When the area constraint on the external electronics is relaxed compared with the intraocular one, the power constraint still holds good because the power consumption directly determines the battery life and hours of continuous operation for a given battery capacity.

Stimulating Electrodes

The electrodes are the medium of interface between the stimulator electronics and the retina. At the electrode/tissue interface, the electronic current is transformed to ionic current necessary for activating the neurons. The electrode array is made up of the substrate that provides a mechanical support and the conducting material that creates the electrical connection between the stimulator and the tissue. There are several factors that decide the choice of the substrate material. The material should lend itself to micromachining as it is necessary for creating high-density electrode array with spacing in the order of micrometers between the electrodes, which is achievable by state-of-the-art micromachining processes. The substrate material should also be flexible, but not fragile, as the surface of the retina is spherical rather than planar. In order to have a good contact between the electrodes and the stimulation sites without creating too much pressure, the electrode array should follow the curvature of retina. The material should be biocompatible not to cause any infection in the eye, and also, resistant to the body fluid. The material should be a good insulator in order to isolate between the conductors that carry the stimulus current. Different materials, such as polyimide *(17)*, poly-dimethyl siloxane *(18)*, parylene *(19)* are good candidates for the substrate. The preferred conducting material is a metal for small resistivity. Platinum has been the widely used material for neural stimulation *(20)*. The conducting material should also be stable under current conduction during a long period of time. The safe stimulation limits for the electrode material is given by milliCoulomb per square centimeter, above which chemical reactions dissolve the electrode material. Other electrode materials include iridium oxide *(21)*, titanium nitride *(22)*, both of which have higher safe stimulation limits.

Packaging

The implant electronics should be housed in a hermetic package, which is a great challenge in realizing any chronically implantable device. Body fluid is very corrosive, made up of organic and inorganic materials and salts, such as chlorides of sodium, potassium. The hermetic biocompatible package should possess two main characteristics: (1) bioresistance—to shield the implantable device from the body fluids and (2) biocompatibility—to protect the tissue from any infection caused by the package material. Biocompatibility is also essential to minimize formation of any connective layer between tissue and device during the long-term *(23)*. The package should also provide hermetically sealed feed through that connect the stimulator electronics to the electrode array. Two major types of packages exist. The first one is a hard protective shell that covers the electronics platform like a cap. The materials for this type of package include titanium, ceramic, and silicon. The 16-site chronic implant used in clinical trials uses a hard case *(5)*. The second type is a conformal thin film coating around the electronics. While a hard shell increases the size of the implant, thin film coatings can produce hermetic implant almost the size of the electronics, but pose a challenge to the fabrication

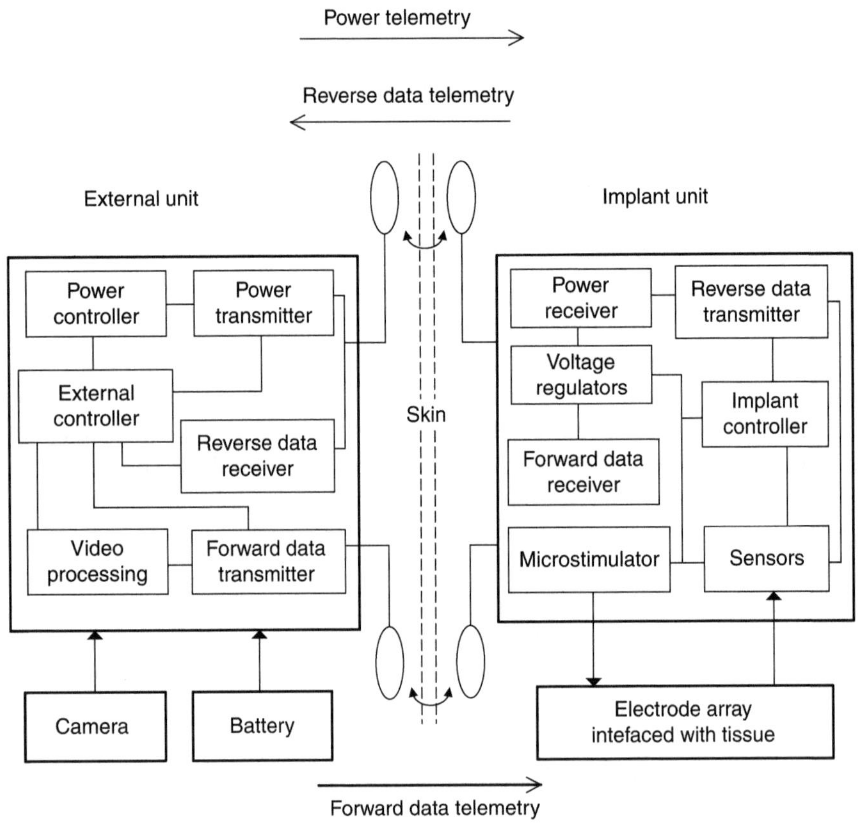

Fig. 2. Block diagram showing the microelectronics of an intraocular retinal prosthesis.

process as the coating should be compatible with the IC fabrication process *(24)*. There are two different categories of thin film coatings—organic and inorganic. The organic materials include epoxies, silicones, and polymers. Parlyene coating for a retinal prosthesis system implanted as a research device in an animal *(25)*. The inorganic materials include silicon nitride, silicon carbide, and diamond.

MICROELECTRONICS

The block diagram of the microelectronics consisting of power telemetry, data telemetry (forward and reverse), and stimulator for the intraocular retinal prosthesis is shown in Fig. 2. An inductive coupled coil pair is used to transmit power into the implant. Inductive power delivery is chosen in other schemes, such as implant battery, optical, and biothermal energy conversion means, mainly because of its ability to deliver hundreds of milliwatt of power with efficiency even up to 50%. Normally, a low frequency carrier (<10 MHz) is used to transmit power from the primary coil to the secondary coil because at low frequencies the human body is well penetrable to the magnetic field leading to higher efficiency of wireless power transfer. But, as the power carrier is chosen to be a low frequency one, a high forward data rate cannot be supported by modulating the power carrier frequency. In retinal prosthesis where the quality of the prosthetic vision increases with number of stimulus sites, large amount of information need to be transmitted continuously. Achieving large data rates is not possible with a conventional approach of a low frequency power carrier and modulating it to transmit data. It is expected that 1–2 Mbps of data will be required for 1000-site prosthesis. A novel approach is shown in Fig. 2, where the power and data are transmitted through separate pair of coils using two different carrier frequencies. This forms a hybrid dual-frequency link and allows the independent optimization of power transmission efficiency and data rate of the forward telemetry through allocation of different frequencies *(26)*. This is also referred as "*dual-band telemetry*" referring to the use of two carrier frequencies. The reverse data telemetry is used only for physiological monitoring. As these signals are low frequency signals and also do not require continuous monitoring, this data rate is low, less than 10 kbps. This allows the reverse data telemetry to reuse the power telemetry link to transmit data from the implant to the external unit.

Power Telemetry

The inductive power telemetry consists of power transmitter, inductive coil pair, and power receiver. The power transmitter operating from a battery voltage produces a time varying magnetic field by generating an alternating current (AC) through the external coil. This magnetic field is coupled to the implant coil inside the eye. The electronics in the implant converts the alternating voltage at the implant coil to the required direct current (DC) voltages. The shape and size of the coils are crucial in determining the power transfer efficiency of the wireless link. Multiple tradeoffs exist among several parameters of the coil. These include power dissipation in the coils, magnetic field generated (which should be maintained lower than the safety standards), and required battery voltage on the external unit. A formal design methodology has been developed, which starts the design from a given size and shape of the implant coil (which is determined by the surgery constraints) and proceeds to derive the remaining parameters that leads to an optimal design *(27)*.

The power transmission efficiency is mainly determined by the DC–AC conversion and the power dissipation in coils. Hence, careful design of both the power transmitter and coils is crucial for efficient wireless power telemetry. The power transmitter that converts DC voltage to AC is a high efficiency power amplifier. Of the different power amplifier topologies, Class-E amplifier is widely used for wireless powering of implants because of its high efficiency (>90%) and its ability to generate large ACs in the transmitter coil whereas operating at small voltages that can be supplied by batteries *(28)*. The Class-E amplifier is made of two inductors (including the transmitter coil), two capacitors and a transistor used as a switch. The efficiency of the Class-E amplifier depends heavily on the switching conditions of the transistor. By appropriately choosing the values of capacitors and inductors for the operating frequency of the power transmitter, the necessary low loss switching conditions can be achieved. But often, the values of the inductors and capacitors vary over time and this shifts the operation of the amplifier from its ideal conditions. To maintain optimal switching conditions and hence, the maximum efficiency at all times, a closed-loop switching scheme is used, which automatically adjusts the switching instances of the transistors by measuring the voltage (or current) of the transmitter coil or one of the capacitors. The implant unit converts the AC–DC power through rectification and regulation. A rectifier is formed by simply connecting a diode and a capacitor. The stimulator usually requires two supply voltages for generating biphasic stimulus pulses. The voltage regulation unit is made of two stages of regulation, a coarse regulation followed by a fine regulation. In the chip implementation in ref. *29*, the series regulator provides 3 mV/V of line regulation (measure of sensitivity of the output voltage to any change in input voltage) and 1 mV/mA load regulation (measure of sensitivity of output voltage to any change in the output current).

The transmitter coil is placed on the glasses of the patient. When the glasses move because of patient's motion, the physical distance between the coils changes from its nominal value. This introduces a change in the coupling between the two coils. A decrease in distance will result in more power being transmitted, which is dissipated as heat inside the eye and causes additional exposure of tissue to electromagnetic field. An increase in distance will result in less power (and voltage) being transmitted and causes malfunctioning of the electronics. An increase or decrease in the power received inside the eye can also be caused by the fluctuations in the power consumption of the implant electronics. In both cases, an ideal power transmission system would transmit the "just-needed" power under the given coupling and loading conditions, by sensing the required power by the implant. Obviously, this requires the information about the required power to be transmitted from the implant to the external unit wirelessly. The reverse data telemetry link is used for this purpose. The power required by the implant is sensed periodically, converted into digital form, and transmitted to the external unit. The external unit receives these data and on decoding the data, uses a look-up table and adjusts the power controller to vary the supply voltage of the Class-E amplifier. Figure 3 shows the voltage waveforms demonstrating the closed-loop operation of the power telemetry. When the coils are moved from their nominal position, the implant voltage measured after the rectifier increases when coils are moved closer or decreases when coils are moved apart. This information is sensed by the power measurement circuits and transmitted through the back telemetry to the external unit. The power controller increases the supply voltage of the Class-E amplifier when the rectified voltage decreases, and decreases the supply voltage when the rectified voltage increases. This restores the rectified voltage to the nominal value in less than 250 ms. It can also be seen that the voltage regulators produce a stable voltage of 14 V (which is divided into +7 V and –7 V for stimulator). Any closed-loop system should be stable to prevent the system drifting into oscillation mode. This is ensured by carefully choosing the time-interval between the power measurements inside the implant and additional circuitry to increase the stability of the closed-loop *(30)*.

Data Telemetry

Forward Data Telemetry

As mentioned earlier, the forward data telemetry operates at a higher frequency compared with the power telemetry and uses a separate coil pair for inductive link. There are several communication schemes available, such as amplitude-shift keying, frequency-shift keying, phase-shift keying (PSK), on–off keying *(31)*. In this application reducing the power consumption and bit error rate (BER) are the major design goals. BER refers to the fraction of bits received at the implant side that are different

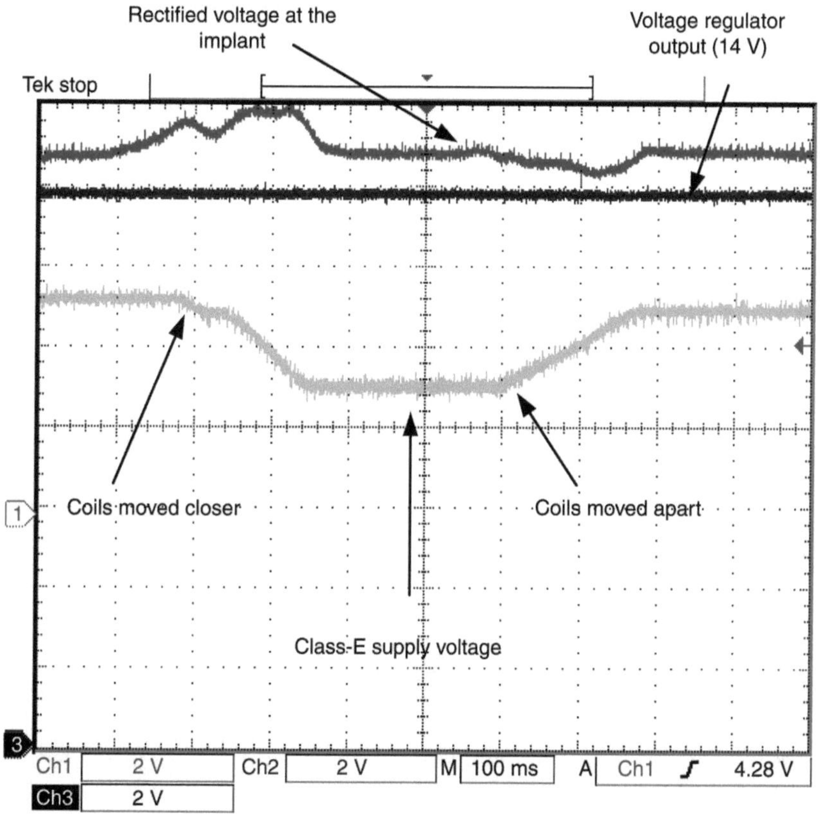

Fig. 3. Measurement waveforms demonstrating the closed-loop operation of the power telemetry.

from the ones transmitted ("1" as "0" or "0" as "1"). A low BER is desirable in retinal prosthesis where the real-time image information is sent from the external unit to the implant and the image information cannot be repeated if the image in the field of vision has changed. PSK is chosen as the modulation scheme owing to its better BER performance for a given SNR (signal-to-noise ratio) value compared with amplitude-shift keying, on–off keying, frequency-shift keying. A modification of PSK, differential PSK is used to make the receiver robust to frequency offset. It also allows for a predominantly digital implementation resulting in low-power consumption compared with an analog implementation. Figure 4 shows the sequence of the wireless data transmission. A 16-MHz carrier frequency is used to transmit a data rate of 2 Mb/s. The raw data are encoded in differentia format using digital logic. The 16 MHz carrier is then modulated with this data stream. The phase transitions can be seen in the carrier whenever there is a data transition. This modulated carrier is transmitted from the external data coil through an analog driver, which provides the necessary current to drive the coil. In the implant the data received are reduced in amplitude along with the interference from the power telemetry that operates at 1 MHz. This interference can be seen as the periodic low-frequency disturbance of the 16 MHz carrier that contains the transmitted data. This signal is then filtered to remove the power telemetry interference, and

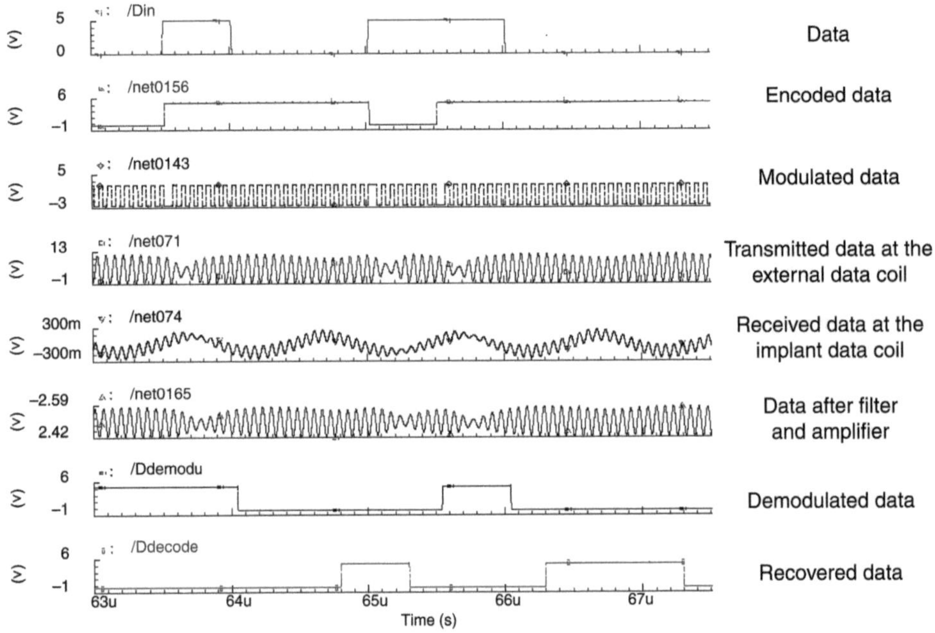

Fig. 4. Sequence of signals showing the operation of the forward data telemetry.

amplified to increase the amplitude of the data signal. The amplified signal is then processed by digital logic to recover the transmitted data. The recovered data have a small delay in the order of microseconds from the transmitted data, which is insignificant as they were used to transfer image data.

Reverse Data Telemetry

The reverse telemetry reuses the power coils. The transmission of the data is in a reverse direction to that of power telemetry, from the implant to external unit. But, unlike the forward data telemetry, which uses a dedicated data transmitter to transmit the data, the reverse telemetry link does not have a driver to drive the coil. Instead, it creates a change in the impedance viewed by the implant power coil. This method is called load-shift keying *(30)*. The impedance change is induced by a switching circuitry that disconnects the majority of the implant electronics from the power coil for a short period of time. During this short period, a storage capacitor provides the power for the disconnected implant electronics. This capacitor is charged when the data are not transmitted. This impedance variation creates a change in the reflected impedance in the transmitter power coil in the external unit. This change is detected by measuring the corresponding change in the current through the power coil in external unit. Differential mode detection is used to improve noise immunity.

Figure 5 shows the sample waveforms for the reverse telemetry. The data are transmitted in the form of pulses that operate the back telemetry switching circuitry. The width of the switching pulses is 8 μs. The change in the reflected impedance induced by the impedance variation in the implant side can be seen in the form of disturbance in

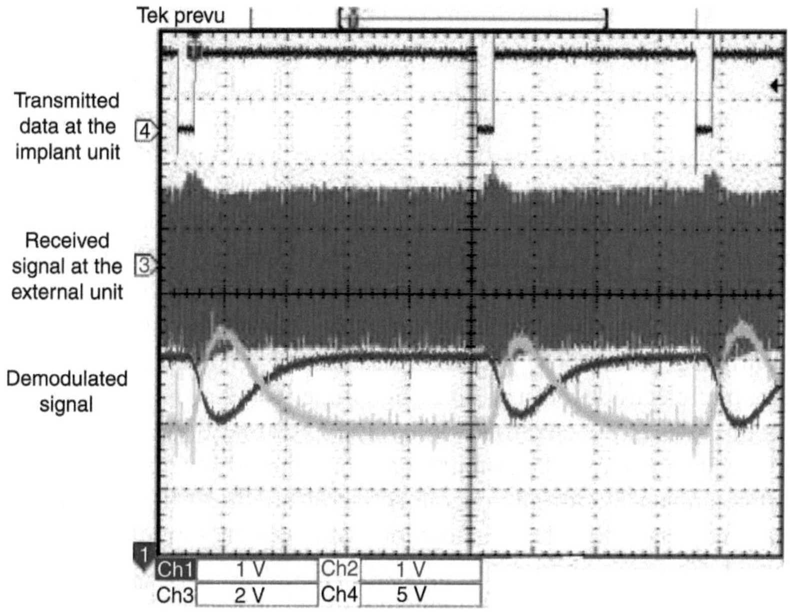

Fig. 5. Sequence of signals showing the operation of the reverse data telemetry.

the power carrier signal. This signal is measured at the output of the sensing circuitry that senses the current through the coil. It is then passed through diodes and filter, which form a demodulator, to remove the 1 MHz carrier and finally, amplified. This demodulated signal is processed using digital logic to recover the actual data. The data rate achievable for this wireless link is around 6 kbits/s.

Stimulator

The stimulator receives the stimulation data from the forward data recovery circuits along with a clock for synchronizing the communication between the external and implant units. In order to produce a biphasic-stimulation pulse, which is commonly used for electrical stimulation of tissue, the stimulator uses two supply voltages. The cathodic pulse sources current from the electrode and the anodic pulse sinks current to the electrode. The two pulses are separated by an interval to avoid the second pulse reversing the physiological effect of the first pulse. Usually, the cathodic phase is the leading one, which depolarizes the cell membrane and elicits a neural response. The amount of charge generated at the stimulation site, which should be more than the threshold for creating a response from the stimulation, is equal to the product of the amplitude of the current pulse and the pulse width. The rate at which the stimulation pulses are delivered is called the stimulation rate and is expressed in pulses per second. While stimulating the retinal sites, the stimulation rate should be more than a certain value to avoid flickering vision. Experiments have shown that nonflickering perception can be achieved with stimulation rates of 40–50 Hz *(4)*. The stimulator can be divided into two major blocks—the analog drivers that generates the stimulus pulses and the digital controller that controls the analog drivers. Each analog driver consists of biphasic-current generator operating from the positive and negative supply voltages generated by the voltage regulators of the power telemetry. The supply voltages are determined by the maximum voltage that appears at the electrode during stimulation (also referred to as compliance voltage) and the additional headroom required by the current generator on the electrode voltage *(32)*. In future high-density retinal prosthesis with more than 1000 electrodes, contrast between different pixels is necessary to effectively achieve goals, such as facial recognition. In human trials in *(5)*, tests revealed that with increasing or decreasing current the visual perception was brighter or dimmer respectively. To provide this required dynamic range, a digital-to-analog converter is required that controls the stimulus current amplitude from current generator. The output of each current generator can be connected to one or a group of electrodes. Demultiplexing of one stimulus output to more than one electrode is possible as the electrodes in the group can be time interleaved and still receive at a frequency above the stimulation rate required for nonflickering vision. This reduces the number of current generators, but also reduces the flexibility of arbitrary stimulation sequence. Besides the functional stimulation, the safety of the tissue should be ensured. Chronic accumulation of electrical charge is detrimental for biological tissue. To ensure that there is no accumulation of net charge at the electrode sites, charge balanced biphasic pulses are used for stimulation. But, two main sources of unintended charge accumulation exist, which warrants a periodical charge-removal mechanism at the electrode sites. The first source is the charge imbalance between the anodic and cathodic stimulus pulses because of circuit mismatch. The second source is the leakage from adjacent electrode sites that are stimulated. The charge removal circuit connects the electrode sites to the common ground potential, thus, discharging any residual charge. The digital controller controls the operation of the analog drivers. This includes stimulus pulse widths, stimulation rate, charge cancellation, stimulation sequence among the different drivers, storage of configuration data, error detection, and routine checks to ensure the proper functionality of the stimulator.

Figure 6 shows the die microphotograph of a stimulator chip with 60 stimulation outputs *(33)*. It consists of 60 biphasic-current generators without any demultiplexing and a digital controller. The circuit can deliver a maximum stimulus current of 200 μA, 400 μA, or 600 μA, which is programmable by configuration through the forward telemetry. For each current, 16-level resolution is provided by the anodic and cathodic digital-to-analog converter. Eight stimulation profiles, specifying the anodic and cathodic pulse widths and the interpulse interval are available, each of which can be programmed individually. The chip occupies an area of $5.5 \times 5.25\ mm^2$ and fabricated in AMI 1.2-μm complementary metal oxide semiconductor (CMOS) technology. The chip operates at +7 and –7 V with a compliance voltage of around ± 6 V (1 V headroom) and consumes 50 mW whereas delivering 600 μA into electrode impedances of 10 k at a stimulation rate of 60 Hz. Some sample stimulus waveforms generated by the chip are shown in Fig. 6. The waveforms demonstrate the flexibility of the chip to generate a wide variety of stimulation patterns. The different stimulation parameters resulting in the waveforms shown, were configured by the digital controller to the stimulus generators.

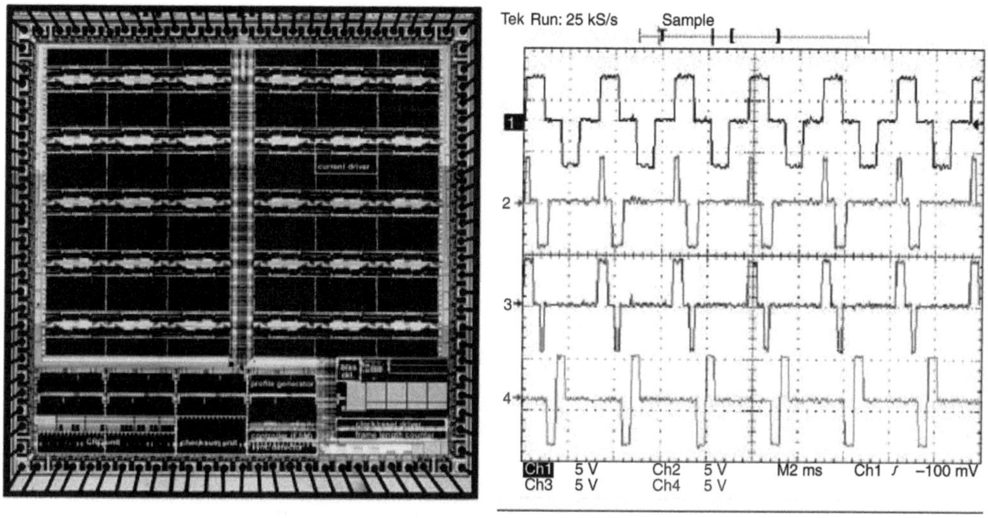

Fig. 6. Stimulator chip and biphasic-stimulation waveforms.

CONCLUSIONS AND FUTURE DIRECTIONS

The success of the cochlear implant is often cited when making the case for a retinal prosthesis to stress the fact that the restoration of a function in human sensory system has been demonstrated even with greatly reduced interface sites with the damaged tissue. The human auditory nerve has 50,000 fibers, yet with only six electrodes in the spiral ganglion in the cochlea, implant patients can understand speech at near normal levels *(34)*. If one wishes to draw an analogy for retinal prosthesis for the required number of electrodes for a given number of nerve fibers, it would require hundreds of interface sites with the retina given the fact that optic nerve has about 1×10^6 individual fibers. Whereas the current implants and research provide great hope for such a system, significant challenges lie in the road ahead. Integrating the power telemetry, data telemetry, and stimulator in a single IC with reasonable size for surgery and with a long lifetime is one of the challenges. This will require use of submicron CMOS technology with hybrid fabrication and packaging of the currently bulky off-chip components like inductors and capacitors. Designing and fabricating electrodes are small enough to accommodate hundreds of electrodes in a given area without exceeding the safe stimulation limits is a fundamental challenge for realizing a high-density retinal prosthesis. Technology for routing multiple wires from the stimulator to the electrodes in a compact and reliable fashion is also required. Apart from the device development challenges, a better understanding of the human visual process with insights into the natural algorithms of the retina will greatly help in improving the quality of vision even with a minimum number of interface sites. The concept of retinal prosthesis has grown from acute experiments with a simple system to chronic human implants with complex systems. This has been possible because of the collaboration between experts from biology, engineering, and medicine through research and development activities. The results achieved so far provide great hope that such similar activities will continue leading toward a high-density retinal prosthesis that can restore functional vision.

REFERENCES

1. Margalit E, Sadda SR. Retinal and optic nerve diseases. Artificial Organs 2003;27(11): 963–974.
2. Humayun MS, Prince M, de Juan E Jr, et al. Morphometric analysis of the extramacular retina from postmortem eyes with retinitis pigmentosa. Invest Ophthalmol Vis Sci 1999;40(1):143–148.
3. Kim S, Sadda S, Pearlman J, et al. Morphometric analysis of the macula in eyes with geographic atrophy due to age-related macular degeneration. Retina 2002;22(4):464–470.
4. Humayun MS, de Juan EJ, Weiland JD, et al. Pattern electrical stimulation of the human retina. Vision Res 1999;39:2569–2576.
5. Humayun MS, Weiland J, Fujii G, et al. Visual perception in a blind subject with a chronic microelectronic retinal prosthesis. Vision Res 2003;43(24):2573–2581.
6. Weiland JD, Yanai D, Mahadevappa M, et al. Visual task performance in Blind Humans with Retinal Prosthetic Implants. Conf Proc IEEE Eng Med Biol Soc 2004;6:4172–4173.
7. Cha K, Horch K, Normann RA. Simulation of a phosphene-based visual field: visual acuity in a pixelized vision system. Ann Biomed Eng 1992;20(4):439–449.
8. Cha K, Horch KW, Normann RA, Boman DK. Reading speed with a pixelized vision system. J Opt Soc Am 1992;9(5):673–677.
9. Cha K, Horch KW, Normann RA. Mobility performance with a pixelized vision system. Vision Res 1992;32(7):1367–1372.
10. Hayes JS, Yin JT, Piyathaisere DV, Weiland J, Humayun MS, Dagnelie G. Visually guided performance of simple tasks using simulated prosthetic vision. Artificial Organs 2003;27(11):1016–1028.
11. Dobelle WH, Mladejovsky MG, Girvin JP. Artifical vision for the blind: electrical stimulation of visual cortex offers hope for a functional prosthesis. Science 1974;183(123):440–444.
12. Veraart C, Wanet-Defalque MC, Gerard B, Vanlierde A, Delbeke J. Pattern recognition with the optic nerve visual prosthesis. Artificial Organs 2003;11:996–1004.
13. Peterman MC, Mehenti NZ, Bilbao KV, et al. The artificial synapse chip: a flexible retinal interface based on directed retinal cell growth and neurotransmitter stimulation. Artificial Organs 2003;27(11):975–985.
14. Zrenner E, Stett A, Weiss S, et al. Can subretinal microphotodiodes successfully replace degenerated photoreceptors? Vision Res 1999;39:2555–2567.
15. Eckmiller R. Learning retina implants with epiretinal contacts. Ophthalmic Res 1997; 29(5):281–289.
16. Fink W, Tarbell M, Weiland JD, Humayun MS. DORA: Digital Object Recognition Audio-Assistant for the visually impaired. ARVO. 2004.
17. Weiland JD, Cogan S, Humayun MS. Micro-Machined, Polyimide Stimulating Electrodes with Electroplated Iridium Oxide. Conf Proc IEEE Eng Med Biol Soc 1999;1:13–16.
18. Weiland JD, Guven D, Magrhibi M, et al. Chronic implantation of an inactive poly (dimethyl siloxane) electrode array in dogs. Invest Ophthalmol Vis Sci 2004;45:4210.
19. Xu X, Tai YC, Huang A, Ho C. IC-integrated Flexible Shear-tress Sensor Skin. Technical Digest, Solid State Sensor and actuator Workshop 2002;12(5):740–747.
20. Rose TL, Robblee LS. Electrical stimulation with Pt electrodes. VIII. Electrochemically safe charge injection limits with 0.2 ms pulses. IEEE Trans Biomed Eng 1990;37(11):1118–1120.
21. Weiland JD, Humayun MS, Anderson DJ. In Vitro Electrical Properties for Iridium Oxide vs. Titanium Nitride Stimulating Electrodes. IEEE Trans Biomed Eng 2002;49(12):1574–1579.
22. Yuan F, Wiler JA, Wise KD, Anderson DJ. Micromachined Multichannel microelectrodes with titanium nitride sites. Proc 21st Int Conf IEEE EMBS 1999.
23. Nichols M. The challenges for hermetic encapsulation of implanted devices—a review. Crit Rev Biomed Eng 1994;22(1):39–67.

24. Wise KD, Anderson DJ, Hetke JF, Kipke DR, Najafi K. Wireless implantable microsystems: high-density electronic interfaces to the nervous system. Proc IEEE 2004;92(1):76–97.
25. Stieglitz T, Haberer W, Lau C, Goertz M. Development of an inductively coupled epiretinal visual prosthesis. Conf Proc IEEE Eng Med Biol 2004;2:4178–4181.
26. Bashirullah R, Liu W, JiY, et al. A smart bi-directional telemetry unit for retinal prosthetic device. Proc Int Symp Circuits Syst 2003;5:5–8.
27. Kendir A, Liu W, Wang G, et al. An optimal design methodology for inductive power link with class-E amplifier. IEEE Trans Circuits Syst I 52:857–866.
28. Sokal NO, Sokal AD. Class-E-A new class of high-efficiency tuned single-ended switching power amplifiers. IEEE J Solid-State Circuits 1975;10:168–176.
29. Wang G, Liu W, Bashirullah R, et al. A closed loop transcutaneous power transfer system for implantable devices with enhanced stability. Proc IEEE Int Symp Circuits Syst 2004; 4:17–20.
30. Wang G, Liu W, Sivaprakasam M, Kendir GA. Design and Analysis of an adaptive transcutaneous power telemetry for biomedical implants. IEEE Trans Circuits and Syst I;52:2109–2117.
31. Proakis J. Digital Communications (4th ed.), McGraw-Hill, 2000.
32. Sivaprakasam M, Liu W, Humayun MS, Weiland JD. A variable range bi-phasic current stimulus driver circuitry for an Implantable Retinal Prosthetic Device. IEEE J Solid-State Circuits 2005;41:763–771.
33. Liu W, Humayun MS. Retinal prosthesis. IEEE Int Solid-State Circuits Conf Dig Tech Pap 2004;218–219.
34. Zeng FG. Trends in cochlear implants. Trends Amplif 2004;8(1):1–34.

6
Development of a Visual Prosthesis
A Review of the Field and an Overview of the Boston Retinal Implant Project

Joseph F. Rizzo III, MD, Laura Snebold, and Monica Kenney

CONTENTS

INTRODUCTION

Blindness is a major form of disability in the world. Loss of vision clearly compromises quality of life, and reduces independence for most patients. In addition, blindness often occurs with a host of other medical problems, including depression, obesity, and a variety of systemic conditions such as diabetes *(1)*. The relatively high incidence of blindness and the impact of blindness on quality-of-life compel many researchers to seek new treatments for blindness.

The causes of blindness throughout the world differ greatly. In the nonindustrialized world, where more than 90% of the world's visually impaired live, infection (i.e., trachoma); poor nutrition (i.e., vitamin A deficiency retinopathy), and cataracts are the

From: *Ophthalmology Research: Visual Prosthesis and Ophthalmic Devices: New Hope in Sight*
Edited by: J. Tombran-Tink, C. Barnstable, and J. F. Rizzo © Humana Press Inc., Totowa, NJ

major causes of blindness *(2)*. Over the last several decades, improvements in the infrastructure of medical care have yielded measurable benefits in some countries, although insufficient financial and human resources have prevented a dramatic reduction in the prevalence of blindness throughout most of the world's population. The situation is quite different in the more industrialized countries, where these forms of blindness are relatively insignificant and where the average life-span is more than 80 yr of age and age-related macular degeneration is the dominant form of blindness. Other forms of blindness secondary to disease of the neural structures of the eye and optic nerve, like retinitis pigmentosa (RP), glaucoma, and diabetic retinopathy, also contribute significantly to the prevalence of blindness within the industrialized world. Apart from the human suffering caused by blindness, an enormous financial burden is shouldered by governments because of the need to provide special rehabilitative and assistive services to the visually-impaired, combined with the loss of taxable income that would otherwise be collected by the government *(2)*.

Despite intense efforts, there are no effective therapies for patients who suffer from neural forms of blindness. The pursuit of strategies to restore function to damaged parts of the central nervous system tissue has become one of the great challenges of medical research. Strategies to transplant retinal nerve cells, retinal pigment epithelial cells, and stem cells are all being investigated *(3)*. The most substantial success to date has come from a molecular approach, which has demonstrated recovery of vision in dogs that were blind from a genetic defect that produces a phenotype of the human form of Leber congenital amaurosis *(4)*. Each of these research initiatives offers hope to the visually impaired. However, this chapter will exclusively address a different approach to treat some forms of retinal blindness—implantation of a retinal prosthesis.

THE INTELLECTUAL FOUNDATION OF A VISUAL PROSTHESIS

A prosthesis is an artificial device that is designed to replace the function of a damaged or lost part of the body. The opportunity to use prosthesis to restore vision to the blind is based on perceived opportunities that would arise by stimulating the afferent visual system at one of the several locations. Potential sites for implementation of a visual prosthesis include the subretinal space; epiretinal surface; optic nerve; lateral geniculate body (LGB); and visual cortex (Fig. 1).

Guidelines for considering the use of a retinal prosthesis should require that a patient had normal vision at some point in their life. The fact that vision had been normal provides assurance that the complex series of interconnections between the photoreceptors and primary visual cortex had at one time been properly established. The opportunity to use prosthesis is predicated on the interruption of the afferent visual pathway at some point, proximal to the primary visual cortex. All prosthetics are designed to recruit the surviving elements of the afferent visual pathway to provide visual images to patients who had become blind, by delivering artificial stimulation to create visual images.

Prosthesis could be of theoretical benefit if used at any location beyond the site of nerve damage. For instance, the rationale for development of a retinal prosthesis relates to the relatively selective degeneration of the outer retina in patients with age-related macular degeneration and RP. In both diseases, the significant degeneration of the rods and cones occurs in a somewhat selective manner, such that hundreds of thousands of

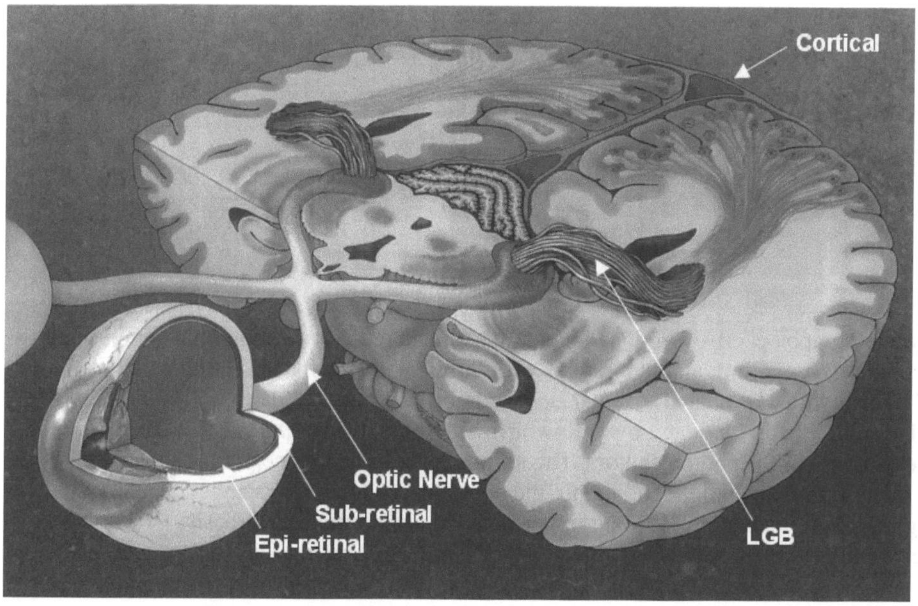

Fig. 1. Illustration of the visual pathway by Steven J Harrison. Visual prostheses are under development for each one of these five sites. LGB, lateral geniculate body.

retinal ganglion cells (RGCs) (which are the only cells from the eye that deliver visual input to the visual cortex) are relatively spared (Fig. 2) *(5–7)*. In these cases, the hope is that stimulation of the remaining nerve cells of the middle and inner retina (using either a sub- or epiretinal approach) might provide neural input to the visual cortex that could produce vision. Stimulation could also be delivered at any point, distal to the photoreceptor level to treat these diseases.

By distinction, many causes of neural blindness damage the inner retina or optic nerve, which would preclude use of a retinal prosthetic, and requires intact fibers within the optic nerve to convey information to the brain. As such, diabetic retinopathy and glaucoma, two relatively common causes of blindness in the industrialized world, would not be amenable to treatment with a retinal prosthesis. Theoretically, these diseases could be treated by stimulation of nerve pathways that are postsynaptic to the termination of the RGCs, either by interfacing with the LGB or the visual cortex.

THE IMPETUS TO DEVELOP VISUAL PROSTHETIC PROJECTS

A great deal of progress has been made in the development of a retinal prosthesis. Perhaps, most notable are the engineering accomplishments that have yielded the development of a number of implantable and electronically sophisticated devices (Fig. 3) *(8–11)*. The following section addresses what might be referred to as Phase I for retinal prosthetic projects, which has been dominated by engineering initiatives and has taken roughly 20 yr.

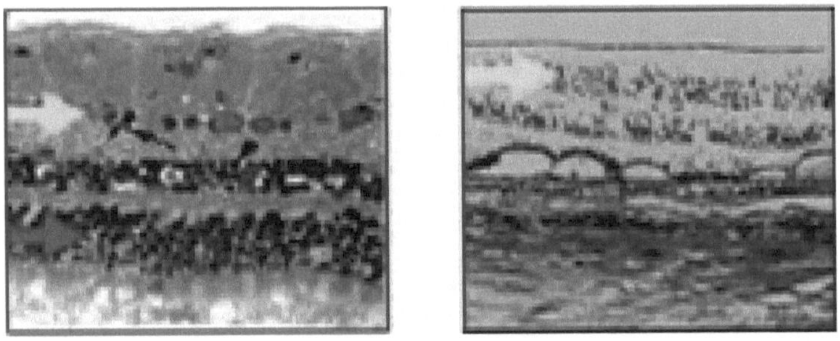

Fig. 2. Photomicrographs of cross-sections of a normal human retina with its normal complement of three cellular layers (Top), and two retinas with different causes for loss of photoreceptors. Retina of human with age-related macular degeneration (Bottom). In diseased retina, the layer of retinal ganglion cells (depicted by yellow arrows), through the optic nerve, are relatively spared despite loss of the photoreceptors. The normal layer of photoreceptors is shown by a red arrow on the left. In the middle section, only the upper two cellular layers are evident. The relative sparing of ganglion cells provides an opportunity to use a retinal prosthesis, which is designed to electrically stimulate the surviving nerve cells, to restore vision for blind patients.

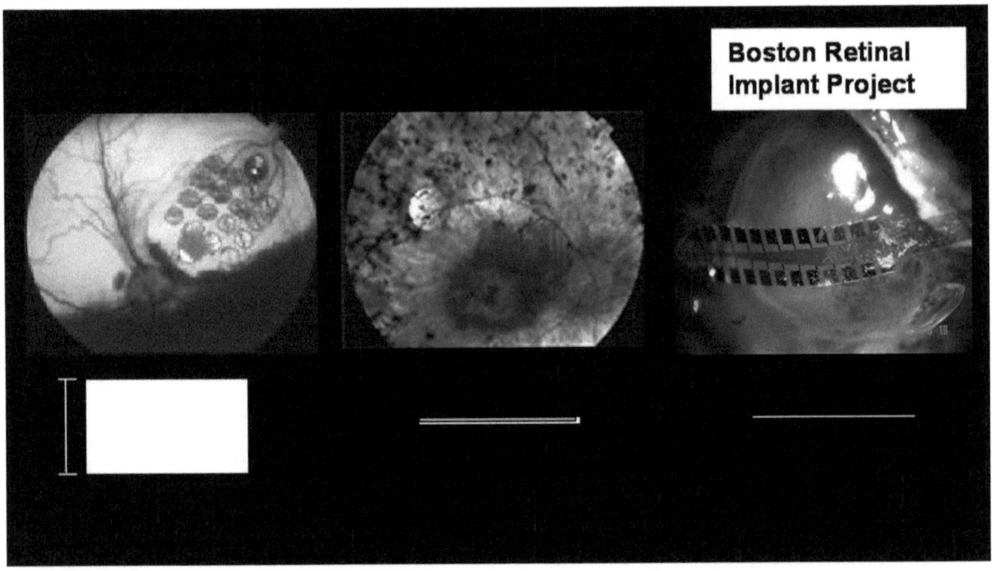

Fig. 3. Representations of intraocular components of the three different retinal prosthetic designs used by the three groups in the United States that have performed human testing. Second sight uses an epiretinal device, whereas both Optobionics and the Boston retinal implant project use a subretinal device. The Boston retinal implant project electrode is only 10 μm thick and extremely flexible.

It is often true in science that a dramatic expansion of research follows the development of new technology (i.e., the polymerase chain reaction) or some great new insight (i.e., recognition of the double helix structure of DNA). However, the field of visual prosthetics was more gradually influenced by three factors. The first significant influence on the field of visual prosthetics may have derived from the scientific accomplishments of two research projects. The work of Hubel and Wiesel *(12)*, which resulted in a 1981 Nobel Prize for Physiology or Medicine, provided a conceptual framework for considering how the brain assembles visual percepts. Their insights provided a straightforward and greatly simplified understanding of the organization of the visual cortex. Additionally, the work of Penfield was instrumental in revealing a wide range of responses that could be elicited by delivering electrical stimulation to the cortex of the human brain, which was performed in a large number of experiments in conscious subjects *(13)*. Collectively, these two research projects may have provided sufficient foundation of knowledge to prompt the mental leap to consider how direct stimulation of the visual cortex might provide a means of restoring vision to the blind. The second significant influence on the field of visual prosthetics came from the remarkable success of cochlear implant projects. Some forms of hearing loss and blindness have homologous pathologies to blindness. Cochlear prosthetic projects began more than 40 yr back and in 1970s began to thrive with advances in digital signal processing. At present, a significant fraction of patients who receive cochlear implants can hear well enough to talk over the telephone, which is a dramatic success given the inability to use lip-reading cues. The third significant influence on the development of visual prosthetics has been the advancement of microelectronic technology, particularly very large scale integration systems. These new integrated circuit design and fabrication methods enabled the development of extremely sophisticated microelectronic devices that could fit within the eye. One interesting footnote to the development of the microelectronics is the fact that Carver Mead, the man who essentially conceived of the notion of very large scale integration systems, was the first person to become interested in the development of a retinal "neuro-morphic" chip (i.e., a computer chip with functional capabilities that are modeled on the basic design architecture of the retina), a field of research that is rapidly expanding in its own right *(14)*.

Harnessing the required engineering capacity to build visual prosthetics has been slow and expensive, but progress has been steadily made by many research groups. There is no longer doubt about whether appropriate devices can be built—many groups have already developed very impressive devices. However, there is a lingering question about whether highly sophisticated devices will be able to restore functional vision to patients who are blind. The research required to learn how to utilize the sophisticated devices to create vision will likely be the primary subject of research as the field progresses.

EARLY HISTORY OF VISUAL PROSTHETICS

The notion of developing visual prosthetics to restore vision to patients with neural forms of blindness dates back to an initial suggestion of Tassiker *(15)*, and then nearly 40 yr back to Brindley and Lewin, who made the first serious efforts to develop an implantable device *(16,17)*. Dobelle later performed similar investigations *(18)*. These researchers made the best use of the relatively crude technology that was available to them to implant stimulating electrodes over the surface of the visual cortex of blind patients. This approach yielded coarse visual percepts, but at the least demonstrated that patients with chronic severe blindness could see something if electrical stimulation were applied to the brain. This basic finding, which now might seem trivial, energized these researchers to make long-term investments to improve their results. But over

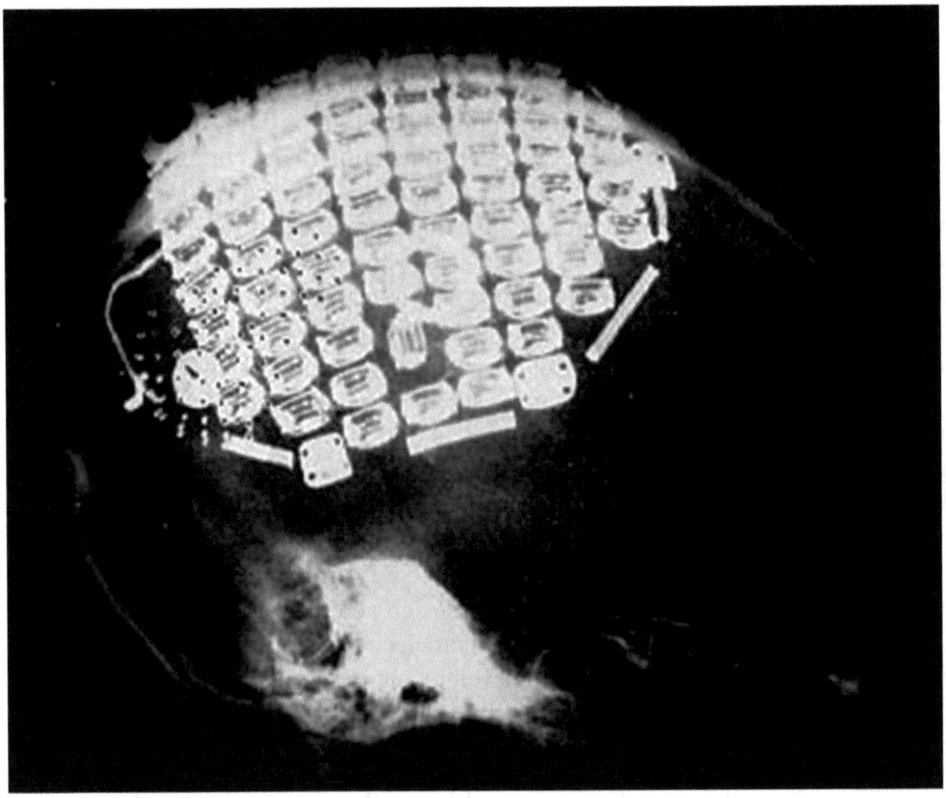

Fig. 4. Visual prosthetic device implanted by Brindley and Lewin *(17)* in a 52 yr old woman who has completely blind. This device had 80 receivers that were positioned directly beneath the skull. Beneath each receiver was an electrode that was placed on the surfaces of the occipital cortex. Delivery of electrical voltage through many electrodes at the same time allowed the patient to perceive many small spots of light.

time, an ever-increasing cynicism of the potential value of a cortical prosthesis developed among the research community. The inability to produce higher quality visual images and the bulkiness of the electronics, which produced a surreal appearance of the patients (Fig. 4), might have contributed to a widespread sense that this approach would never be practical or truly useful to blind patients. The most senior contributor to the field of visual cortical prosthetics is Professor Richard Normann, who for decades has maintained a vigorous research effort at the University of Utah to develop technically sophisticated, cortical electrode arrays. Some of the fruits of this research are now being realized in the implementation of a motor prosthesis *(19)*. The National Institutes of Health established an intramural research program in the late 1970's to develop a visual cortical prosthesis *(20)*. This program now survives as an initiative directed by Philip Troyk, PhD from the Illinois Institute of Technology. Dr. Troyk's team has obtained superb threshold and voluminous psychophysical data from macaque monkeys, which is encouraging *(21)*.

In the late 1980's, two independent research groups began initiatives to develop a retinal prosthesis. The Boston-based group was founded as a collaboration between the Harvard Medical School and the Massachusetts Institute of Technology. Professor John Wyatt of the Massachusetts Institute of Technology has served as co-Director with one of us since that time. Around the same time, doctors DeJuan and Humayun began in a similar project at North Carolina State University. They moved shortly thereafter to Johns Hopkins University and then ultimately to the University of Southern California. Once there, they joined forces with a private biomedical entrepreneur (Alfred Mann) and began the Second Sight Co. to develop their prosthesis. Both of these projects use a retinal approach: This group has chosen a subretinal approach, whereas the Second Sight Co. has chosen an epiretinal approach. Shortly thereafter, the brothers Alan and Vincent Chow entered the field with the idea of using a photodiode-based array under the retina. The Chow brothers founded Optobionics Inc.

Other companies dedicated to developing a retinal prosthesis have also emerged. In Germany, Intelligent Implants (Bonn) is developing an epiretinal prosthesis. A southern German company (i.e., Retina Implant GmbH, based in Tübingen) led by E. Zrenner utilizes a subretinal approach similar to our own. Both German companies have performed chronic human implants (*see* below). In Japan, Nidek Inc. is sponsoring development of a prosthesis that will position a stimulating electrode array in the supra-choroidal space, roughly 1 mm from the retina.

Other research groups include those based at the Kresge Eye Institute in Detroit (for development of a cortical and retinal prosthesis, including exploration of a chemically-based stimulation method). A group headed by Daniel Palanker at Stanford University is attempting to develop a retinal prosthetic system that will beam into the eye a visual image, which will then be captured by implanted photodiode arrays. Over the last 5 yr, additional research groups have developed (in chronological order) in Korea, Australia, and China. A consortium has developed recently in Europe (based in Alicante, Spain), which is working in association with Professor Normann (University of Utah) to develop a cortical prosthesis.

There are now a total of 22 research groups attempting to develop some type of visual prosthetic (Table 1). No group has merged with any other, and no group has ceased to exist. By comparison, there are a total of four companies worldwide that manufacture cochlear prosthetics, although the number of deaf individuals in the world is substantially less than the number of patients with some form of neural blindness.

AN OVERVIEW OF VISUAL PROSTHETIC SYSTEMS AND PROJECTS

The large number of research groups and the varied approaches might convey a lack of significant connection among the groups. In fact, all of the projects share two fundamental needs to be able to create artificial vision. First, the loss of visual input to the brain must be replaced by capturing the elements of the visual scene that a patient cannot see. Second, the information from the visual scene must be utilized to provide some form of artificial stimulation to the nerve tissue. For most groups, the favored approach is the use of electricity to drive the nerve cells. However stimulation is provided, there is a need to learn how to deliver that stimulation in a manner that will produce the desired result, which is to improve the quality of life for visually-impaired patients.

Table 1
Listing of Current Visual Prosthetic Initiatives in the World

Approach	*Research group/ company*	*Principal investigators/group leaders*	*Headquarters*
Subretinal	Electrical stimulation		
	Biomedical Physics and Ophthalmic Technology	Daniel Palanker, PhD	Stanford University, CA
	Boston Retinal Implant Project	Joseph F. Rizzo III, MD John Wyatt, PhD	Massachusetts Eye and Ear Infirmary-Harvard Medical School, Massachusetts Institute of Technology, Boston VA Health Care System, Boston/Cambridge MA
	Optobionics Corporation	Alan Chow, MD Vincent Chow, PhD	Naperville, Illinois
	SUBRET Consortium/ Retina Implant GmbH	Eberhart Zrenner, MD	Tübingen/Reutlingen, Germany
	Biohybrid Retinal Implant	Tohru Yagi, PhD Masami Watanabe, PhD	Tokyo Institute of Technology, Tokyo, Japan
	Japan Retina Implant Group/NIDEK Co.	Yasuo Tano, MD, Yasushi Ikuno, MD, Jun Ohta, PhD	Osaka, Japan
	C-Sight: Chinese Project For Sight	Xiaoxin Li, MD Quishi Ren, PhD	People's Hospital-Peking University Medical School, Beijing, China
	Eye Clinic, University Hospital of Geneva	Marco Pelizzone, MD	Geneva, Switzerland
Epiretinal	Intraocular Retinal Prosthesis Group/ Second Sight Medical Products, Inc.	Mark Humayun, MD James D. Weiland, PhD R. J. Greenberg	Doheny Eye Institute, University of Southern California, LA/Sylmar, CA
	Ligon Research Center Of Vision	Raymond Iezzi, MD Greg Auner, PhD Gary Abrams, MD	Wayne State University, Detroit, Michigan
	Intelligent Medical Implants	Gisbert Richard, MD	University of Bonn, Germany
	Three Dimensions Stacked Retinal Prosthesis	Makoto Tamai, MD, Hiroshi Tomita Mitsumasa Koyanagi	Tohoku University, Tohoku, Japan
	Nano Bioengineering System Research Center/Nano Artificial Vision Center	Sung June Kim, PhD Hum Chung, PhD	Seoul National University Hospital-Seoul National University School of Medicine; Seoul, Korea
	Australian Vision Prosthesis Group	Nigel Lovell PhD Gregg Suaning, PhD	University of New South Wales/University of Newcastle, Sydney, Australia

(Continued)

Table 1 ***(Continued)***

Approach	*Research group/ company*	*Principal investigators/group leaders*	*Headquarters*
Optic nerve	Neural Rehabilitation Engineering Laboratory, Brussels; Université Catholique de Louvain, Belgium	Jean Delbeke, MD, PhD, Claude Veraart, PhD	Brussels, Belgium
LGB	Reid lab	John S. Pezaris, PhD, R. Clay Reid, MD, PhD, Emad N. Eskandar, MD	Massachusetts General Hospital-Harvard Medical School, Boston, MA
Cortical	Utah Visual Neuroprosthesis Program	Richard Normann, PhD.	Salt Lake City, Utah
	Bionic Eye Research Project	Vivek Chowdhury, MD, Minas T. Coroneao, PhD	Prince of Wales Hospital, Randwick, Australia
	Cortivis	Eduardo Fernandez, MD	Universidad Miguel Hernandez; Alicante, Spain
	Intracortical Visual Prosthesis/Illinois Institute Of Technology	Phil Troyk, PhD	Illinois Institute of Technology, Chicago, IL
	PolySTIM Research Group	Mohamad Sawan, PhD	Polystim Neurotechnologies Laboratory, ECOLE Polytechnique, University of Montreal, Montreal, Canada
Neuro-transmitter	Photoelectric Dye-Coupling/ Hahashibara Co.	Toshikhiko Matsuo, MD	Okayama City, Japan

An attribute that is shared by most of the groups is the assembly of remarkably diverse, multidisciplinary research projects. The development of retinal prosthetics has led to very substantive interactions between engineers and biologists. From an engineering standpoint, specialists in materials science, microcircuit design, microfabrication, mechanical engineering, metallurgy, and polymer sciences have contributed to the development of a retinal prosthesis. On the biological side, specialists in anatomy, physiology, ocular surgery, and visual psychophysics have contributed to the development of a retinal prosthesis. Increasingly, the field of visual prosthetics is attracting a wide range of basic scientists who are contributing substantially to the understanding of the consequences of retinal degeneration *(22)*, and preferred methods to stimulate the nerve tissue *(23–26)*. Visual prosthetic research projects are also serving as extraordinary opportunities for young researchers in biomedical science.

COMPARISON OF THE RELATIVE MERITS AND DISADVANTAGES OF VARIOUS PROSTHETIC APPROACHES

Each of the locations being considered as a site for a visual prosthesis (Fig. 1) has certain advantages and disadvantages. There are, quite understandably, very strong opinions about which approach might be best. At present there is not enough data to even begin to exclude or favor any particular approach. We believe that each approach is plausible and has merit. In general, the more central the placement of prosthesis is, the greater the range of diseases that could be treated. For instance, a retinal prosthesis could not be used to treat glaucomatous blindness, but prosthesis at either the LGB or visual cortical would offer a rational option for treatment of glaucoma. However, the convoluted topography of the visual cortex and the need for a substantial neuro-surgical procedure to implant a cortical device impose real but surmountable challenges on the use of visual cortical prosthetic devices.

In general terms, the surgical approach on the eye would be less complicated than the neuro-surgical approach that would be required to implant a visual cortical prosthesis. But the consequences of surgery on the eye and the presence of a foreign object within the eye might be more prone to incite chronic and potentially damaging inflammation compared with the relatively minimal responses that have occurred in the brain *(27)*. The concept of delivering stimulation to the optic nerve is advantageous in that the surgical approach would be relatively easy, but the extremely dense packing of the RGC axons within the optic nerve (i.e., many hundreds of thousands of axons compacted into a roughly 2 mm diameter structure) would seem to limit the degree of spatial detail of the induced images that could be achieved with this approach. More specifically, stimulation delivered to the epiretinal surface should make it possible to selectively stimulate small clusters of surviving RGCs because at this location these surviving cells are distributed over a very wide area. However, it is of interest that human psychophysical results to date (*see* Overview of Human Testing following) are not consistent with the intuitive assumption that induced images would appear cruder if stimulation is delivered to the optic nerve. The pursuit of LGB prosthesis is relatively new. This approach has the advantage that the LGB cells are spatially segregated with respect to their location within the visual field and also the physiological type (i.e., M- vs P-cells *[28]*), but this approach would require a "deep brain" surgical approach and it would be very challenging to implant hundreds of electrodes in this way. This number of electrodes could easily be implanted at the level of the retina or visual cortex.

There are also many more subtle advantages and disadvantages for each approach. A more detailed discussion of the relative merits is beyond the scope of this chapter. Ultimately, intuitions and conjectures about which approaches might be best will have to be substantiated by the results of long-term implants in humans. Until such time as sufficient data becomes available to draw more objective conclusions, it is prudent to remain open-minded about which approach might prove best.

OVERVIEW OF THE BOSTON RETINAL IMPLANT PROJECT

This review of the progress of the Boston retinal implant project is meant to convey a general perspective on the types of research that are relevant to the development of a visual prosthesis. This overview is divided into electronic and biological components.

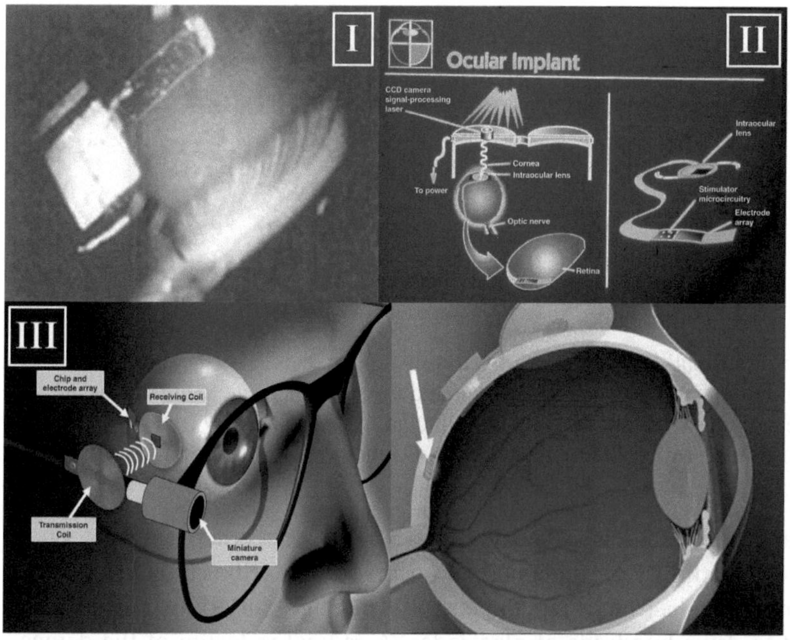

Fig. 5. Graphic images showing the evolution of the retinal prosthetic design concepts over 18 yr. The first two versions (shown across the top row) were both epiretinal devices. The second version was designed to place the bulky electronic components behind the iris, rather than on the retina itself. The third version (lower two images), which are currently testing in animals, is a subretinal prosthesis. The side-view image on the right reveals a distinctive advantage of the current approach—essentially the entire bulk of the device rests outside of the eye. Only the ultra-thin, flexible electrode array (arrow) is placed inside of the eye, under the retina.

Electronic Initiatives

Over the nearly two decades of our Project, three design architectures have been experimented (Fig. 5). The first design was relatively straightforward in that the entire device was placed directly onto the inner (i.e., epiretinal) side of the retina. A cantilever was used to position the stimulating electrode array away from the location of the chip, where it was assumed that some retinal damage would occur because of fixation of the device to the retina. Despite the small and low mass nature of the microelectronics, the device exerted too much force on the retina, especially given the counter-current movements of the vitreous body that occur with eye movements. The second version was an improvement over the first design in that the electronic components were removed from the retina and positioned on a platform just behind the iris (Fig. 5, upper right), which mimicked the approach routinely used to implant intraocular lenses following removal of cataracts. However, this approach required use of an elongated electrode array that needed to be secured over a fairly long area of the retinal surface, which proved to be quite challenging. Additionally, the anatomy behind the iris constrained the size of the radiofrequency secondary coil that was to power the prosthesis to a diameter of 8–10 mm, which by the calculations would have limited the ability to drive the large number of electrodes that might be eventually used.

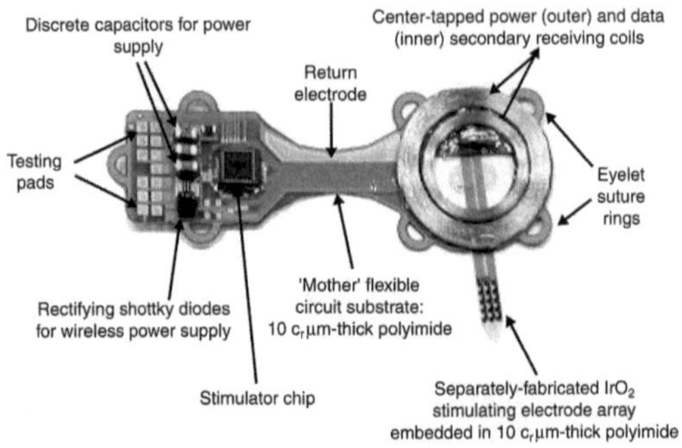

Fig. 6. Photograph of current design of the animal retinal prosthesis. The secondary coils for power and data transmission and the integrated circuit and discrete electronic components are all mounted on a flexible, polyimide substrate. Only the stimulating electrode array (red arrow) enters the eye, where it is positioned within the subretina space. Calibration bar (lower left): 10 mm.

Roughly 5 yr back, the researchers became increasingly concerned with potential problems of the biocompatibility of implanted intraocular devices, the challenge of hermetic encapsulation of the electronics and the potential damage caused by transmission of power sufficient to drive hundreds of electrodes. A radical redesign of the device (i.e., version III), which substantially mitigates the aforementioned concerns, was undertaken. Version III is designed to maintain almost the entire bulk of the device outside of the eye (Fig. 5, lower right). This approach allows us to take advantage of the relatively spacious orbit (i.e., eye socket) that can accommodate a titanium case to provide hermetic encapsulation of the electronics.

A fully assembled first generation wireless prosthetic device (Fig. 6) has been completed. The foundation of the implant is a flexible, 10-µm thick substrate into which wires and electrodes are microfabricated. The stimulator chip, several other discrete electronic components, the data and power receiver coils, and the electrode array are then attached by a variety of means to the thin substrate. After assembly, the device maintains sufficient flexibility to enable it to match the curvature of the posterior sclera. It has been verified that the implant works as designed with testing on the bench by delivering wireless signals to the device (Fig. 7).

More specific details of the device are as follows. The device was designed by the MIT-based engineering team with the perspective that creation of detailed visual images would require a relatively large bandwidth for data transmission. High data rates require a high frequency carrier, but power transmission at high frequencies is inefficient. Therefore, power is transmitted very efficiently at a relatively low frequency (125 kHz), whereas visual scene data is transmitted at a relatively high frequency (13.56 MHz). The system employs a high-efficiency class D oscillator to transmit power; a lower efficiency class A amplifier is sufficient to transmit data.

The core of the electronic system is the IC "stimulator chip." This chip, which contains ≈30,000 transistors was designed and tested entirely by Luke Theogarajan *(29)*. The chip employs aggressive strategies to achieve ultralow power performance—the chip dissipates only about 1.5 mW at low data rates (~100 kilobyte/s), and about 2.5 mW at higher data rates (~500 kilobyte/s) *(30)*. The chip is capable of providing 800 µA for

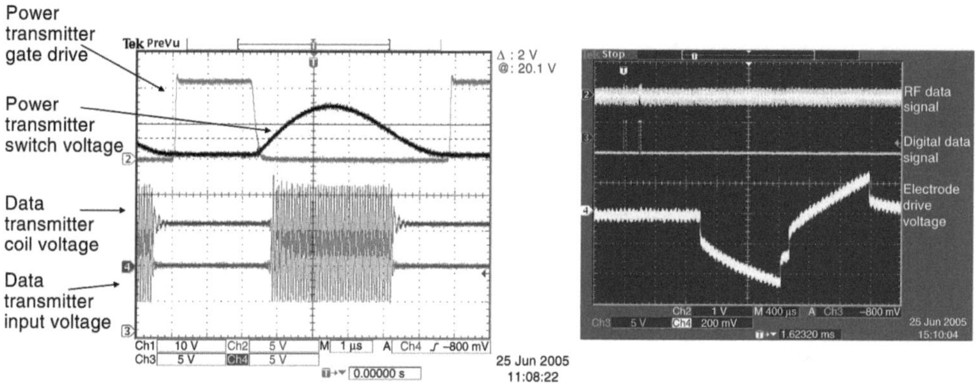

Fig. 7. Operation of the power and data transmission systems. (Left) Transmitter side. Power transmitter gate drive (top trace, light blue) falls periodically at 125 KHz, causing the transmitter switch voltage (dark blue) to rise periodically at the same frequency. Data transmitter responds to a modulated 13.56 MHz input signal (green) by driving the data primary coil, producing the coil voltage shown in red. (Right) Receiver and implant side. Upper trace (blue) shows data input signal to stimulator chip, which was transmitted by the primary coil (shown on left in red, with a different time scale) and received by the data secondary coil. Middle trace shows digital signal extracted by the chip from this input. These signals control current drive to the microelectrodes, which produces the electrode voltage pattern shown in the bottom (green) trace.

each of the 15 electrodes. The pulse width is externally controlled and shared by all electrodes on a given stimulation cycle. The chip produces variable current pulse durations, amplitudes, and interpulse intervals, and it can address individual electrodes. The design is readily expandable to address as many electrodes as will be included in future generations of the prosthesis.

The team performed all of the microfabrication of the device. The design incorporates a flexible circuit, which forms the main substrate onto which all electronic components are assembled. The electrode array and the test integrated circuit are attached to the "mother" flexible circuit by a gold stud bump, flip-chip bonding technique *(31)*. Two generations of flexible circuits have been developed that have been microfabricated with wire-bondable, electroplated, gold traces (50 μm wide; 3 μm thick) within either a polyimide or parylene substrate. Parylene is a biocompatible polymer with low water absorption, which makes it ideally suited for implantation because of its ability to serve as both substrate for microfabrication and hermetic encapsulant for the embedded circuitry. Given the goal of developing methods that will be transferable to commercial manufactures (which is necessary for a intricate, implantable medical device), strategic alliances have been developed with fabrication, assembly, and packaging vendors with experience in implantable devices.

The device uses iridium oxide electrodes, which have substantially greater charge-carrying capacity than do platinum electrodes, which is the current standard for bioimplantable devices (Fig. 8). The use of iridium oxide or similar low impedance coating is considered to be necessary for a retinal prosthesis because this approach will permit use of arrays with much smaller electrodes that will be able to deliver higher levels of

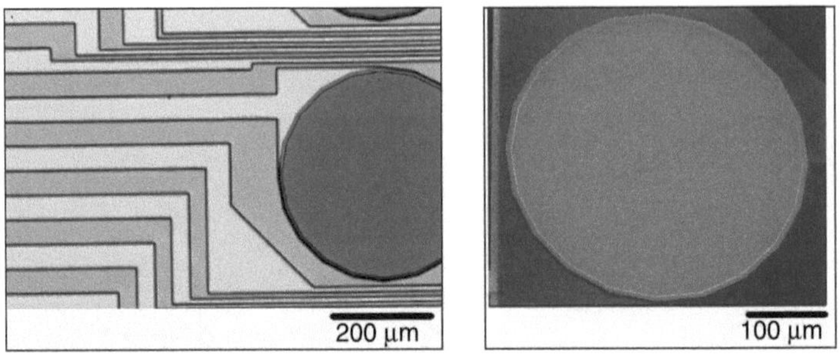

Fig. 8. Scanning electron microscopy of 400 μm diameter, stimulation electrode with sputtered iridium oxide coating after 1 wk of pulsing with no indication of damage.

charge more safely than would be the case with electrodes made with noble metals *(32)*. However, the methods of reliably applying the oxide coating and the methods for sustaining the integrity of these electrodes are not widely known or practiced, and they require considerable technical "know-how" to attain reproducible fabrication results.

One major concern still faced by some research groups is the need to demonstrate that active electronic components that are implanted in and around the eye can survive within the salt-water environment of the body. Technologies have been developed for implantable devices elsewhere in the body, but such technologies would have to be scaled down to fit in or around the eye.

BIOLOGY

Heavy investments have been made in performing retinal physiological experiments to find out about how retinal neurons respond to artificial electrical stimulation. The in vitro single-cell studies have provided the most detailed assessment of how activation thresholds vary with respect to two and three-dimensional displacement of the stimulating electrode from the somas of the RGCs (Fig. 9) *(33)*. This work, and subsequent work from the laboratory on human retina *(34,35)*, revealed the lowest thresholds ever obtained from normal retinas (<0.1 nC), which has subsequently been confirmed by two other researchers *(25,26)*.

After the prosthetic design was converted to a subretinal approach (*see* above), neuronal responses to subretinal electrical stimulation was studied. These experiments have provided the most significant physiological finding—the presence of a substantial difference in the activation thresholds for ON- vs OFF-RGCs that depends on the polarity of the stimulus (Fig. 10) *(23)*. The earlier work with epiretinal stimulation did not reveal such a separation of thresholds between these two cell classes. The fact that the polarity of stimulation delivered to the subretinal space can strongly bias neuronal activation toward certain cell types offers the hope of being able to tailor the appearance of induced visual percepts. It is anticipated that creating spatially-detailed visual percepts will require somewhat selective neuronal activation, but achieving this goal will not be easy given that electrical stimulation has a strong tendency to activate all neurons simultaneously, which is quite unlike the normal behavior of the retina.

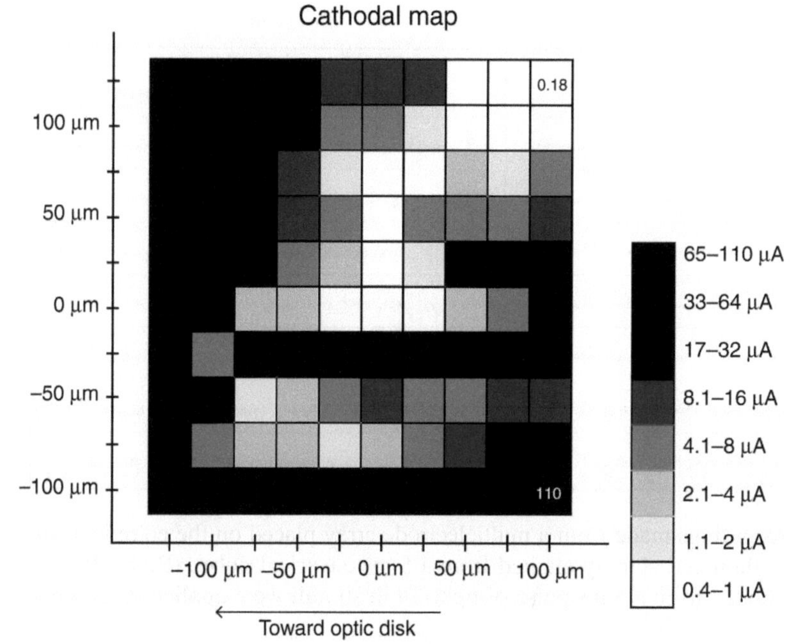

Fig. 9. High spatial density thresholds in the x–y plane around one rabbit retinal ganglion cell body. Thresholds were measured within a grid at 25 μm displacements from the soma, following cathodal and anodal (not shown) stimulation of the epiretinal surface. Thresholds are plotted using identical gray-scale values in both plots. These and other results from this study show that cathodal stimulation gives lower thresholds and more. localized stimulation.

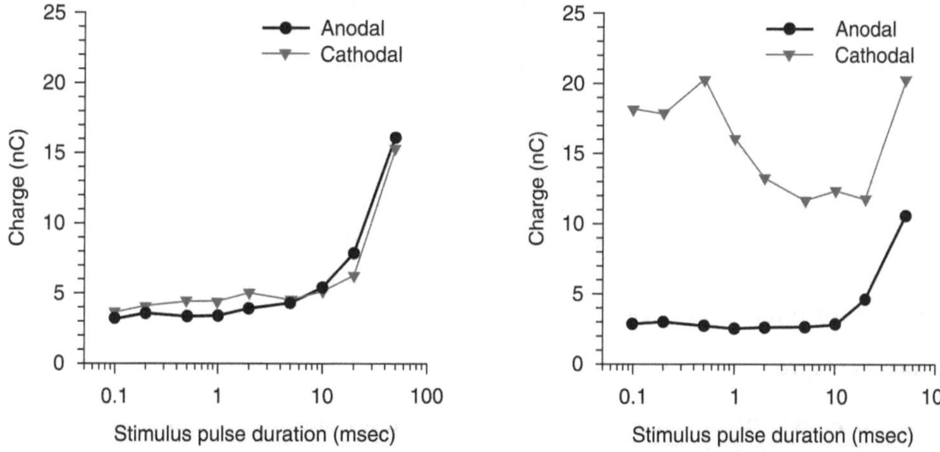

Fig. 10. Threshold charge of ON- (Left) and OFF- (Right) retinal ganglion cells as a function of stimulus pulse duration for both cathodal and anodal current pulses delivered to the subretinal side of normal rabbit retina. Notice the marked separation of thresholds that can be achieved for the OFF-cells by varying stimulus polarity.

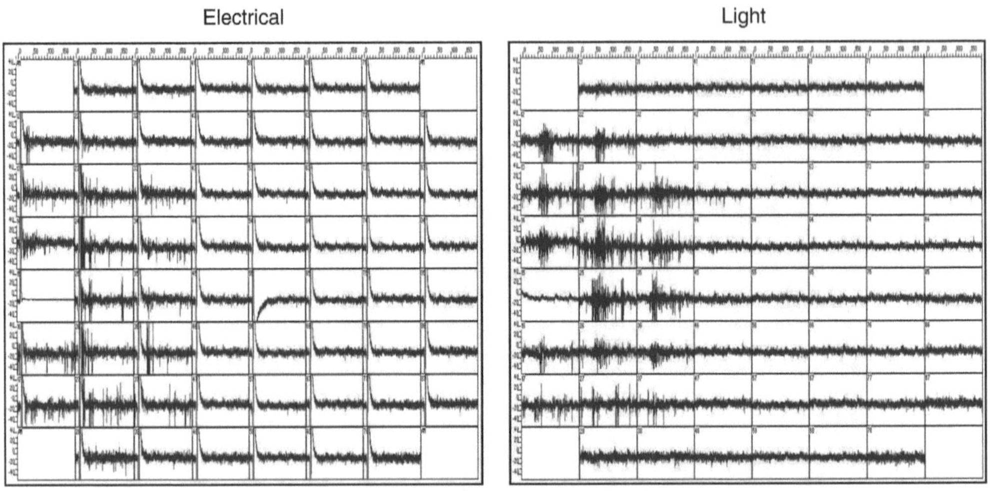

Fig. 11. Electrical recordings made from a multielectrode array placed on the epiretinal surface. (Left) Retinal ganglion cell activity elicited from a 5 ms electrical pulse. (Right) Retinal ganglion cell activity elicited from a 5 ms pulse of light. Both stimuli were applied to the same area of retina.

More recently, we have studied the electrically-induced population responses of up to 60 RGCs in both normal and degenerated retinas by recording with a multichannel electrode array (Fig. 11). Recording "population" responses will help us understand more about how a complex tissue like the retina creates vision. Ultimately, the goal is to emulate the normal light-induced responses of the retina by using artificial electrical stimulation delivered by prosthesis.

The experiments described earlier are valuable in that they provide insights into how one might want to deliver electrical stimulation to activate neurons. Whatever methods are chosen, the stimulation methods must not be injurious. There has been consistent evidence that the electrical stimulation thresholds of degenerated retinas are very elevated compared with normal retinas *(36–38)*. The elevated thresholds are of great concern because of the risk that the use of high stimulating currents will damage the retina or the stimulating electrodes, or both. There are several plausible explanations for the high thresholds, and presently there is no sufficient understanding of the cause of the elevated thresholds. The need to develop stimulation methods that could be used for many years without damaging the neurons that are the intended targets of stimulation is an important subject for future biological experimentation.

SURGICAL METHODS

The design of the *ab externo* device is advantageous in that it minimizes the amount of foreign material that needs to be placed into the eye. However, the design approach of placing the device behind the back of the eye and placing the electrode array under the retina creates a more demanding challenge for the surgeons. A surgical approach that achieves the goals has been developed (Fig. 12), but doing so has required the development of methods to minimize bleeding from the choroid as the subretinal space

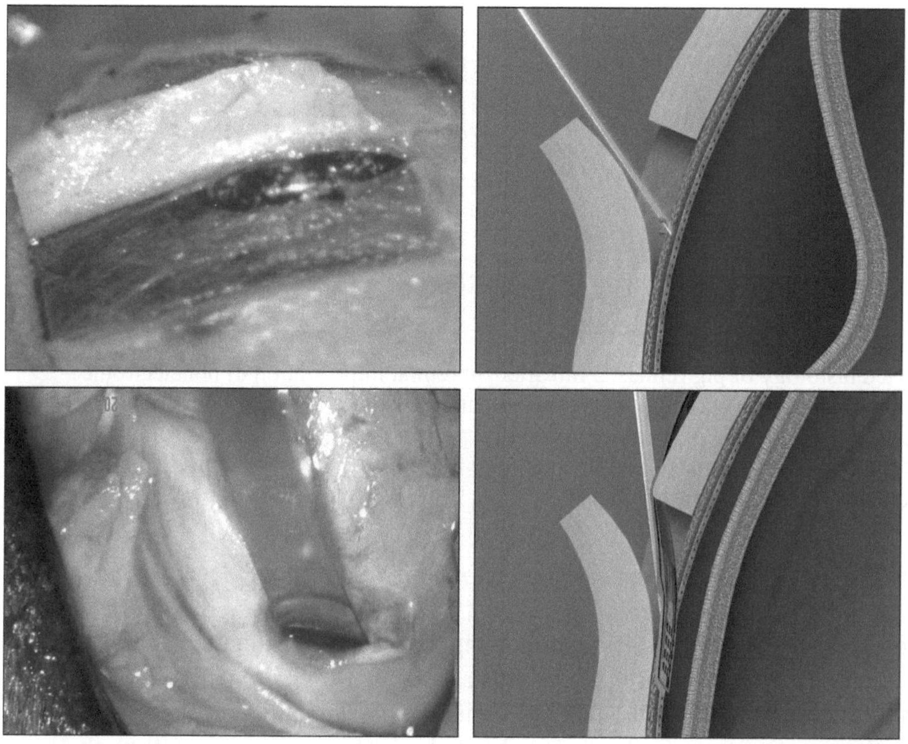

Fig. 12. Pictorial overview of our standard *ab externo* surgical approach. (Upper Left) Creation of a scleral flap through the back wall of the eye (i.e., the sclera). (Upper right) Artist's image of a retinal bleb, which is raised to reduce the potential of retinal damage. (Lower left) Insertion of a mechanical guide through the choroid and under the retina. (Lower right) Artist's image of the insertion of the electrode array through the scleral flap and choroid, under the retina, which has returned to its original position.

was approached, and to prevent injuring the retina as the electrode was inserted into the subretinal space. The methods have still not been perfected, although the creation of a bleb under the retina has consistently protected the retina from direct injury during the surgery.

The surgical method requires much less extensive intraocular surgery to implant the device that is needed to place a device on the inner retinal surface. Reducing the amount of intraocular surgery will probably enhance biocompatibility because postoperative inflammation is likely to occur to whatever foreign material is implanted. It is important to document the degree to which foreign materials implanted into the eye incite inflammation (*see* next section).

BIOCOMPATIBILITY OF IMPLANTED MATERIALS

A moderately in–depth study of biological reactions induced by implantation of one of six materials into the subretinal space has been performed. The most extensive study included the assessment of six different materials that are candidates either to serve as a substrate for microfabrication or as a coating to (hopefully) provide barrier protection

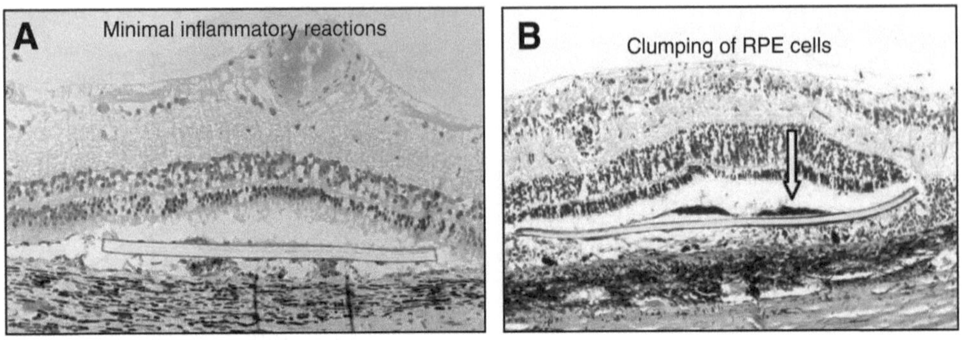

Fig. 13. Selected histology of pig retinas, 3 mo after subretinal implantation of nonelectronic implants through a retinotomy. (Left) Relatively little anatomical alteration following implantation of parylene. (Right) Example of clumping of retinal pigment epithelial cells over implant, following implantation with polyimide. All implants were made as 0.5×5 mm^2 strips that were 10 μm thick. Hematoxylin and eosin staining.

against penetration by either water vapor or ions into the region of the embedded electronics (Fig. 13).

Fifteen histological criteria were assessed following 3 mo long implantations in pigs, with specific attention to the responses of the retina or retinal pigment epithelium or the presence of any inflammatory cells. In brief summary, all implants produced some pathology, but the reactions were always < 10 μm thick (Fig. 13) *(39)*. Although there was a fear of more aggressive tissue reactions producing a dense covering over the devices, it was pleasing to learn that the responses were not more significant. However, it remains to be determined if these responses produce any untoward consequences on the electrical behavior of the device. For instance, it is possible that even a relatively thin membrane might increase the impedance of electrodes. Higher levels of stimulating current will have to be delivered, which would impose a greater demand on the available power for the prosthesis and potentially increase the risk of inducing damage to the electrodes.

Human Studies

Electrical stimulation studies have been performed on five blind patients with RP and one normally-sighted subject. These studies were "acute" studies in which an electrode array was placed onto the epiretinal surface for a period of not longer than several hours. The two primary goals were to obtain a detailed assessment of perceptual thresholds, and to test the hypothesis that there would be a correlation between the geometry of the electrical stimulus and the geometry of the induced percepts *(37,38)*.

In roughly 50% of stimulation trials with the last four patients, a match between the geometry of the stimulus and the induced percept was achieved, although only for very basic stimuli, such as a spot or a line. In the very best cases more spatially detailed percepts were produced, when multiple electrodes were driven (Fig. 14). Of special interest was the fact that it was unable to achieve higher quality vision in one normally sighted patient (This patient had orbital cancer and her eye had to be removed to treat the cancer).

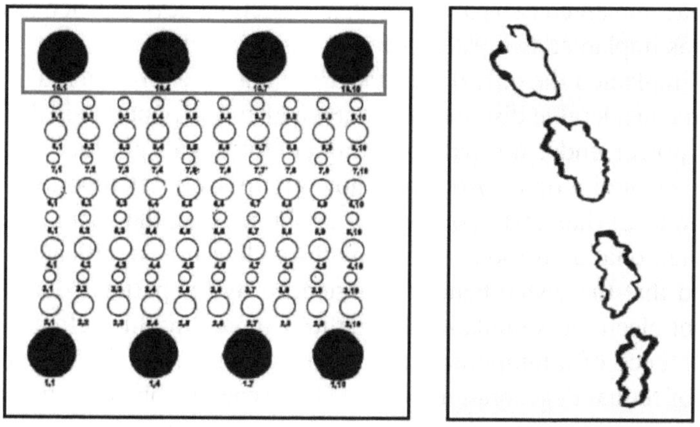

Fig. 14. Psychophysical result from one trial of electrical stimulation of the retina in one of our legally-blind patients. (Left) Schematic of the electrode array, showing the pattern of stimulation delivered to the retina through four (400 μm diameter) electrodes, shown within the red box. (Right) Drawing by the patient of the percept that was elicited by the stimulus.

The inability to achieve better vision in the normally sighted patient is very instructive, as it suggests that the ignorance of effective stimulation methods compromised the results. This interpretation continues to motivate the study of retinal physiology and prompts to begin a comprehensive program to develop a better understanding of the normal patterns of neural transmission within the afferent visual system. The goal is to try to learn how to predictably influence visual perception with artificial electrical stimulation.

OVERVIEW OF HUMAN TESTING

This section will provide a very brief review of some of the human testing performed to date. The intent of this section is to provide a brief snapshot of some of the more significant and qualitative results. The all-important quantitative results of stimulation thresholds have shown very wide variation among the groups, which to some extent can be explained by the difference in testing methods. A presentation of the details of the threshold results is beyond the scope of this chapter. The reader is advised to seek the original articles to learn many more details about the full range of findings that these experiments have provided.

Four groups have performed chronic human testing. The first to perform testing was the company headed by Alan Chow team. This group (i.e., Optobionics, Inc.) has utilized subretinal approach to implant a photodiode array, which they have implanted into the eyes of 12 patients who were blind from RP. The second group was led by Humayun, currently at the University of Southern CA and the Second Sight Medical Products. This company has performed chronic implants in six humans. Their device is a modified cochlear prosthesis (obtained from Advanced Bionics Co., an affiliate of their company through the Al Mann Foundation) that utilizes a 1 mm thick, hand-made (i.e., nonmicrofabricated) cable. This cable runs from behind the ear, into the orbit and through the wall of the eye (just behind the cornea) and internally is fixed to the epiretinal surface with a metal tack. More recently, both German companies have performed chronic implants.

The Bonn-based group has implanted two patients with an epiretinal device; whereas the Tübingen-based group has implanted two patients with a subretinal device.

The Chow group has implanted the only device that does not have any external connections. There has been considerable discussion about whether this photodiode device can generate sufficient power under normal operating conditions to produce power needed to stimulate degenerated retinas *(40)*. The Chow group itself has reported that infrared light at 87 mW/cm *(2)* (i.e., just very slightly less bright than the summer sun) was required to record neuronal activation *(41)*. Notwithstanding this significant concern, Chow has reported that the visual benefit experienced by his patients does not result from point-to-point electrical stimulation but rather from a "trophic" effect. His group has presented evidence of a temporary (i.e., months-long) trophic effect in at least one animal model of retinal degeneration *(42,43)*. This type of tropism, originally discovered in experiments on retinal transplantation *(44)* has now been linked to a purified soluble factor *(45)*. Any tropism induced by subretinal surgery or implantation might prove advantageous in conjunction with prosthesis.

Most recently, the Zrenner group announced results from testing their two implanted patients, both of whom were able to deliver electrical stimulation to an array of 16 electrodes through hard-wired connections. In one patient, activation of these 16 electrodes produced predictable images in both patients, including the perception of pea-sized percepts when single electrode stimulation was used *(46)*, which is qualitatively very similar to the findings in the epiretinal experiments *(37,38)*. This patient was also able to discern the orientation of a line of stimulating electrodes *(47)*. Retina Implant GmbH's human testing results contribute to the rapid momentum already incurred by this field, making it more likely that visual prostheses will return vision to the blind. The epiretinal German group has implanted four patients, three of whom were able to identify the location of spots and lines. One of their patients could detect movement in response to sequential stimulation of electrodes across the array *(47)*.

There can be little doubt that the results of human testing have revealed that: (1) blind patients, even if legally-blind for decades, can see phosphenes in response to electrical stimulation; and (2) modulation of electrical stimulation can alter the appearance of phosphenes. These findings support the hope that artificial stimulation could produce "useful" vision for the blind. However, none of these experiments produced reliable and detailed percepts that reliably matched the stimulation patterns *(48,49)*.

However, despite great technical advances and greatly increased knowledge, there has been only modest improvement in the quality of vision induced by electrical stimulation since the seminal experiments by Brindley *(48)*. The lack of good insights into how best to stimulate the retina is one likely impediment to achieving better quality results. The significant pathologies that develop in degenerating retinas *(22)* must also be contributing to the challenge of producing better quality vision.

SUMMARY

There have been very significant accomplishments in the building of retinal prosthetics. The inability to create better percepts at this stage of research is not unlike the history of cochlear implants, which required many years of work to discover methods that ultimately produced great success in helping many deaf patients. It is suspected that intensive study of biological methods will provide more detailed insights into methods to lower stimulation thresholds and to create better vision, even in retinas that have undergone degeneration.

At this point in the evolution of the field of visual prostheses research, it is reasonable to experience a cognitive dissonance about the potential value of these devices. There are compelling reasons to have substantial optimism, and there are substantial unresolved questions that can raise uncertainty about the long-term prospects of these devices. The substantial critical mass of investigators that are contributing to the field makes it likely that, at the least, severely blind patients will experience functionally useful benefits from visual prosthetics.

REFERENCES

1. Margolis M, Coyne K, Kennedy-Martin T, Baker T, Schein O, Revicki D. Vision-Specific Instruments for the Assessment of Health-Related Quality of Life and Visual Functioning. Pharmacoeconomics 2002;20:791–811.
2. Dowling J. Artificial Human Vision. Expert Rev Med Devices 2005;2:73–85.
3. Huang Q, Xu P, Xia X, Hu H, Wang F, Li H. Subretinal transplantation of human fetal lung fibroblasts expressed cliliary neurotrophic factor gene prevent photoreceptor degeneration in RCS rats. Zhonghua Yan Ke Za Zhi 2006;42:127–130.
4. Bennett J. Gene therapy for Leber congential amaurosis. Novartis Found Symp 2004; 255:195–202.
5. Humayun MS, Prince M, de Juan E, et al. Morphometric analysis of the extramacular retina from postmortem eyes with retinitis pigmentosa. Invest Ophthalmol Visual Sci 1999;40:143–148.
6. Santos A, Humayun MS, de Juan E Jr, et al. Preservation of the inner retina in retinitis pigmentosa. A morphometric analysis. Arch Ophthalmol 1997;115.
7. Stone JL, Barlow WE, Humayun WS, et al. Morphometric analysis of macular photoreceptors and ganglion cells in retinas with retinitis pigmentosa. Arch Ophthalmol 1992;110:1634–1639.
8. Humayun MS, Weiland JD, Fujii GY, et al. Visual perception in a blind subject with a chronic microelectronic retinal prosthesis. Vision Res 2003;43:2573–2581.
9. Chow A, Chow VY, Packo K, Pollack J, Peyman GA, Schuchard R. The artifical silicon retina microchip for the treatment of vision loss from retinitis pigmentosa. Arch Ophthalmol 2005;122:460–469.
10. Gekeler F, Szurman P, Grisanti S, et al. Compound subretinal prostheses with extra-ocular parts designed for human trials: successful long-term implantation in pigs. Graefes Arch Clin Exp Ophthalmol 2006 [Epub ahead of print].
11. Hornig R, Laube T, Walter P, et al. A method and technical equipment for an acute human trial to evaluate retinal implant technology. 2005;2:S129–S134.
12. Hubel D, Wiesel TN. Receptive fields, binocular interactions, and functional architeture in the cat's visual cortex. J Physiol 1962;160:106–154.
13. Penfiled W, Perot P. The brain's record of auditory and visual experience. A final summary and discussion. Brain 1963;86:595–696.
14. Douglas R, Mahowald M, Mead C. Neuromorphic Analogue VLSI. Ann Rev Neurosci 1995; 18:255–281.
15. Tassiker G. US Patent 2,760,483, 1956.
16. Brindley GS. Effects of electrical stimulation of the visual cortex. Hum Neurobiol 1982;1:281–283.

17. Brindley GS, Lewin WS. The sensations produced by electrical stimulation of the visual cortex. J Physiol 1968;196:479–493.
18. Dobelle WH, Mladejovsky MG, Evans JR, Roberts TS, Girvin JP. "Braille" reading by a blind volunteer by visual cortex stimulation. Nature 1976;259:111–112.
19. Donoghue J, Nurmikko A, Friehs G, Black M. Development of neuromotor prostheses for humans. Suppl Clin Neurophysiol 2004;57:592–606.
20. Hambrecht FT. Neural prostheses. Ann Rev Biophys Bioeng 1979;8:239–267.
21. Bradley D, Troyk P, Berg J, et al. Visuotopic mapping through a multichannel stimulating implant in primate V1. J Neurophysiol 2005;93:1659–1670.
22. Marc RE, Jones BW, Watt CB, Strettoi E. Neural remodeling in retina degeneration. Prog Retinal Eye Res 2003;22:607–655.
23. Jensen RJ, Rizzo JF. Thresholds for activation of rabbit retinal ganglion cells with a subretinal electrode. Exp Eye Res 2005;submitted.
24. Jensen RJ, Ziv OR, Rizzo JF. Responses of rabbit retinal ganglion cells to electrical stimulation with an epiretinal electrode. J Neural Eng 2005;2:S16–S21.
25. Fried S, Hsueh H, Werblin F. A method for generating precise temporal patterns of retinal spiking using prosthetic stimulation. J Neurophysiol 2006;95:970–978.
26. Sekirjinak C, Hottowy P, Sher A, Dabrowski W, Chichilnisky E. Electrical stimulation of mammalian retinal ganglion cells with multi-electrode arrays. J Neurophysiol 2006; 95:3311.
27. Normann RA, Maynard EM, Rousche PJ, Warren DJ. A neural interface for a cortical vision prosthesis. Vision Res 1999;39:2577–2587.
28. Rizzo J. Embryology, anatomy, and physiology of the afferent visual pathway. In: Miller N, Newman N, eds. Walsh and Hoyt's Clinical Neuro-Ophthalmology 6th ed., Philadelphia: Lippincott, Williams, and Wilkins, 2005;3–82.
29. Theogarajan LS. A low power wireless 15-channel implantable retinal stimulatorchip. Cambridge, MA: Massachusetts Institute of Technology Department of Electrical Engineering and Computer Science, 2005.
30. Kelly S, Wyatt J. A power-efficient voltage-based neural tissue stimulator with energy recovery. IEEE Int Solid-State Circuits Conf: (ISSCC), 2004.
31. Gingerich M, Shire D, Karcich K, Schulz C, Wyatt J, Rizzo J. Assembly and packaging developments for an *ab externo* retinal prosthesis. ARVO. Fort Lauderdale, FL, 2004.
32. Robblee LS, Lefko JL, Brummer SB. Activated Ir: An electrode suitable for reversible charge injection in saline solution. J Electrochem Soc 1983;130:731–732.
33. Jensen R, Rizzo JF, Ziv O, Grumet A, Wyatt J. Thresholds for activation of rabbit retinal ganglion cells with an ultra-fine, extracellular microelectrode. Invest Ophthalmol Visual Sci 2003;44:3533–3543.
34. Grumet AE, Wyatt JL, Rizzo JF. Multi-electrode stimulation and recording in the isolated retina. J Neurosci Methods 2000;101:31–42.
35. Grumet A. Electric stimulation parameters for an epi-retinal prosthesis. 1999.
36. Humayun MS, de Juan E Jr, Dagnelie G, Greenberg RJ, Propst RH, Phillips DH. Visual perception elicited by electrical stimulation of retina in blind humans. Arch Ophthalmol 1996;114:40–46.
37. Rizzo J, WJ, Loewenstein J, Kelly S, Shire D. Methods and perceptual thresholds for short-term electrical stimulation of human retina with microelectrode arrays. IOVS 2003.
38. Rizzo JF, Wyatt J, Loewenstein J, Kelly S, Shire D. Perceptual efficacy of electrical stimulation of human retina with a microelectrode array during short-term surgical trials. Invest Ophthalmol Visual Sci 2003;44:5362–5369.
39. Montezuma SR, Loewenstein J, Scholz C, Rizzo JF. Biocompatibility of Subretinal Materials in Yucatan Pigs. Invest Ophthalmol Visual Sci 2004; submitted.
40. Zrenner E. Will Retinal Implants Restore Vision. Science 2002;295:1022.

41. McCall MA, DeMarco PJ, Crosby AL, et al. Visually Evoked Activity from a Subretinally placed artificial silicon retina. Assoc Res Vision Ophthalmol 2005.
42. Ball SL. Temporary trophic effect in two animal models (Chow group). In: Rizzo JF, ed., 2005.
43. Cheng Y, Yin H, Fernandes A, et al. Long-term neuroprotective effect of the subretinal implant in the RCS rat. Assoc Res Vision Ophthalmol 2004.
44. LaVail MM, Yasumura D, Matthes MT, et al. Protection of mouse photoreceptors by survival factors in retinal degenerations. Invest Ophthalmol Visual Sci 1998;39:592–602.
45. Husson-Danan A, Leveillard S, Mohand-Said S, Chalmel F, Poch O, Sahel JA. Rod-derived cone viability factor/Txnl-6 expression in the transgenic P23H rat, an autosomal dominant model of retinitis pigmentosa. Invest Ophthalmol Visual Sci 2005;46:E-Abstract 5185.
46. Wickelgren I. A Vision for the Blind. Science 2006;312:1124–1126.
47. Zrenner E. Subretinal stimulation for retinal prosthesis. ARVO. Fort Lauderdale, FL, 2006.
48. Loewenstein J, Montezuma S, Rizzo JF. Outer retinal degeneration: an electronic retinal prosthesis as a treatment strategy. Arch Ophthalmol 2003.
49. Loewenstein J, Montezuma SR, Rizzo JF. Outer retinal degeneration: an electronic retinal prosthesis as a treatment strategy. Arch Ophthalmol 2004;122:587–596.

7
The Minimally Invasive Retina Implant Project

Heinrich Gerding, MD, FEBO

CONTENTS

INTRODUCTION

The idea of using electrical stimulation to treat defects of the visual system is by far not a new one *(1)*. As early as 1755 LeRoy *(2)* elicited electrically evoked phoshenes in blind people, 36 yr before the first description of the principles of bioelectric stimulation by Galvani *(3)*. Krause and Schum *(4)* and Foerster *(5)* proved the possibility to stimulate the human visual system reproducibly and potentially useful with reliable topographical coordinates. Button and Puttnam *(6)* were the first to test chronically implanted epicortical electrodes in blind patients followed by the famous and impressing trial of the Brindley group applying an epicortical multielectrode system *(7)* to the visual cortex of two patients.

The idea of an electrically active implant for direct stimulation of the retina in blind patients was first proposed by Tassiker in his 1956 patent description and publication *(8,9)*. This idea was evolved in a period of intensive investigation on electrical stimulation of the eye and retina mainly intended to perform physiological eye research and to develop new diagnostic techniques. Potts and Inoue *(10)* proved that eyes of patients with late stage retinitis pigmentosa are still susceptible to electrical stimulation. On the other hand Brindley *(11)* had already demonstrated that externally applied *trans*-scleral stimulation of the globe seemed not to be sufficient to provide a spatial resolution useful for the rehabilitation of blind people. The introduction of modern vitrectomy by Machemer *(12)* created an essential prerequisite for the development of intraocular retinal prostheses. Applying this technique Dawson and Radtke *(13)* reported in 1977 successful results of a long-term experiment with an epiretinal multielectrode array implantation in a cat eye. They already demonstrated biocompatibility of their arrays,

From: *Ophthalmology Research: Visual Prosthesis and Ophthalmic Devices: New Hope in Sight*
Edited by: J. Tombran-Tink, C. Barnstable, and J. F. Rizzo © Humana Press Inc., Totowa, NJ

stable implant position, and a reasonable threshold for cortical responses at a current level of 30 μA and a charge transfer of 2.4×10^{-8} C, respectively. At that time furthermore development was discontinued for technological reasons.

With recent technological breakthroughs in the production of silicone chip devices and the creation of new applicable microelectrodes and biocompatible encapsulation materials, the past two decades have witnessed a renaissance in the development of visual system implants. These new premises led to the creation of new concepts for sub- and epiretinal implants in the first half of the 1990s *(14–18)*. Initial activities were followed by the establishment of a series of worldwide collaborative groups projecting epiretinal *(19–24)*, subretinal *(25–31)*, intrapapillary *(32)*, suprachoroidal *(33)*, extraocular *(34–36)*, or complex *(37)* retina implant systems. Besides using electrical charge to stimulate the retina, other concepts were designed focusing on the microinjection of retinal neurotransmitters to mediate visual information *(28,38–40)*.

RETINAL PROSTHESIS: CONCEPTS AND UNSOLVED PROBLEMS

The common principle of all retina implant concepts is that of a functional bypass of distal retinal cell layer degeneration. There are at least four basic requirements to be met in order to provide a clinically useful device:

1. A sufficient and long-lasting survival of contactable proximal retinal neurons.
2. The possibility to elicit visual qualities of perception by electrical stimulation or neurotransmitter release that will provide more than unstructured visual sensations.
3. Technical feasibility of implantation, mechanical fixation, and explantibility of implants with acceptable risk profile.
4. Biocompatibility of the devices with survival of the contacted retina, and a permanently reproducible visual perception.

The survival of contacted retinal neurons and RPE (retinal pigment epithelium) has broadly been discussed elsewhere *(41–50)*. Although, data about this issue are by far not completely conclusive, it can be summarized that probably there will be a sufficient proportion of retinal neurons surviving distal retinal degenerative diseases for a long time. Nearly the same level of evidence is provided concerning the question of visual sensations. Stimulation was tested in acute quasi-realistic human and animal stimulation experiments *(51–56)* and first data with permanently implanted electrode arrays are available *(57,58)*. Recent results of human in vivo and simulation tests show that devices with very few electrodes can mediate a surprising quality of visual perception *(57,59,60)*.

Implantation, mechanical stabilization, potential secondary reactions toward a foreign structure and the survival of contacted retina are crucial biomedical issues for any kind of retinal implant. The new challenge to intraocular surgery is to place solid implants epi- or subretinally in a precise, long-term highly stable, and atraumatic way. It seems that many of the essential surgical techniques and concepts for the implantation of epi- or subretinal implants were successfully developed and tested so far and that there is a series of modified and combined surgical procedures at hand *(24,61–68)*. In contrast to the successful search for primary surgical solutions it seems that problems of intraocular implant biocompatibility and tissue reactivity are still unsolved, especially, when large-scale intraocular implants are applied *(69,70)*. Successful long-term results are only available for relatively small, or electrically inactive, or incomplete sub- and epiretinal implants or for implants, which leave main components outside the globe *(24,62,64,66,70)*. Experimental series applying complete epiretinal implants with anterior segment IOL-receivers and posterior segment microelectrode arrays (MA) were resulting in a relatively high percentage of severe intraocular complications like cyclitic, retroiridal, or retrociliary membrane formation, severe PVR (proliferative vitreoretinopathy) and/or total retinal detachment *(69–71)*. Severe tissue reactions against large diameter anterior segment implants are well known from the experimental and clinical use of intraocular lenses or iris diaphragms *(72)*. It seems that implants extending the diameter of conventional intraocular lenses with major contact to the iris or ciliary body tend to cause adversive reactions in the anterior segment or in the anterior part of the posterior segment. Problems obviously are potentiated by implants extending to the posterior segment as can be concluded from a series of experimental implantations showing especially intensive proliferations along microcables extending to the posterior segments *(69)*.

Facing the problems that may be caused by these large scale retina implant system are probably the reason why several groups have changed their implant design in a way so that the need of intraocular deposition of implant material is reduced considerably *(57,67,68,73,74)*. In these attempts the intraocular components of the system were limited to a stimulating device (microelectrode or multiphotodiode arrays) placed either sub- or epiretinally. The disadvantage of this technology is the necessity to connect intraocular structures by *trans*-scleral/*trans*-choroidal cables with other electronic components outside the eye. The penetration of cables again may cause severe local proliferative reactions and necessitate perhaps a limitation of ocular motility.

A maximal reduction of intraocular implant material is the design principle of a suprachoroidal or extraocular device designed by two groups *(35,75)*. Here, stimulating electrodes are positioned in the suprachoroidal space, a scleral pocket, or in episcleral position leaving any intraocular structures untouched. This approach reminds of the early trial of external stimulation by Brindley *(11)*. Theoretical disadvantages are the increasing distance between electrodes and target neurons probably resulting in a lower resolution and higher necessary charge delivery to reach stimulation thresholds *(76)*.

MINIMAL INVASIVE RETINAL IMPLANT

The intention of the minimal invasive retinal implant (miRI) is to combine the need for a close spatial contact of stimulating electrodes with retinal neurons and the intention to minimize the intraocular deposition of implant components. The only way of combining these two basic criteria at a maximum level is the design of an implant with all components outside the eye and stimulating electrodes penetration the sclera and choroid. This principle is obviously in conflict to the general paradigm that the choroid has to be regarded as a critical structure, which might cause severe complications after single and even more multiple penetration. It was the aim of the miRI project to examine the feasibility of the retina implant system with these characteristic features and, if possible, to develop medical devices for the rehabilitation of blind people with distal retinal degenerations. Early results of feasibility studies and experimental data will be mentioned later. Before that the basic technological concept will be exemplified.

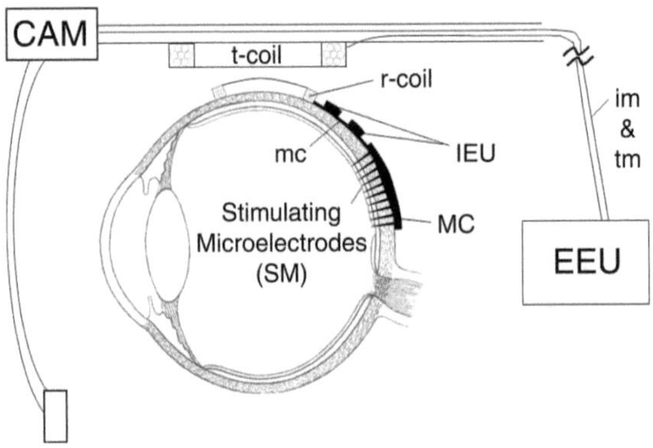

Fig. 1. Principle construction of the miRI.

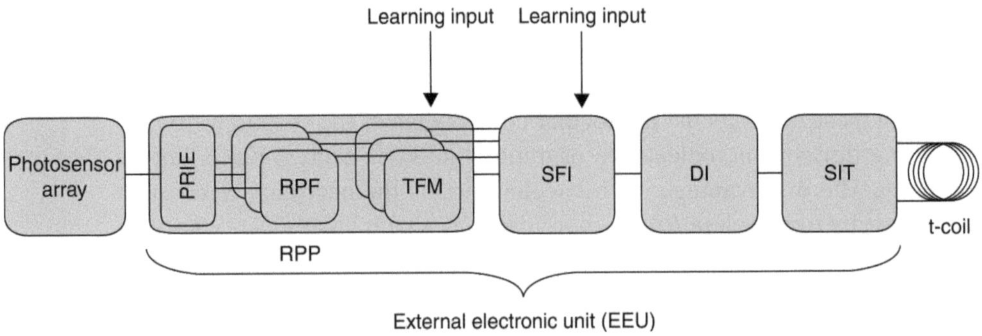

Fig. 2. Functional construction of the EU.

The principle construction of the miRI is demonstrated in Fig. 1. It consists of two main components: the external unit (EU) and the implant unit (IU). The EU (Figs. 1 and 2) includes a camera (CAM), the external electronic unit (EEU), a transmitter coil (t-coil), and energy supply. The CAM is completely integrated into a normal-style glass frame. Image microcables within the temples are connecting the CAM with the EEU. The EEU can be fixated underneath and covered by normal cloths. Transmitter microcables return parallel to image microcables through the temples connecting the t-coil. The t-coil is integrated into a distension of the temple so that it is centered to the horizontal midline axis of the globe. Minor frame modifications and a cable leaving the end of the temple behind the auricle will be the only visible signs of the implant carrier status. The dimension of the EEU can be minimized so that the patient will not have to carry any bulky or noticeable equipment.

The IU (Figs. 1 and 3) consists of the receiver coil (r-coil), microcable connections (mc), the internal electronic units, microelectrode carrier (MC), and penetrating stimulating microelectrodes. Stimulating electrodes are the only component of the implant that penetrates the sclera. The rest of the IU is placed on the outer surface of the sclera and in this position mechanically stabilized by conventional intrascleral sutures comparable with those used in buckling procedures.

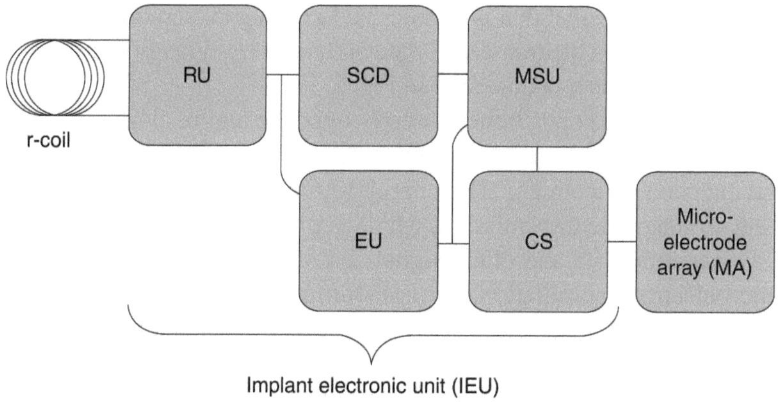

Fig. 3. Functional construction of the IU.

Figures 2 and 3 depict the functional construction of the miRI schematically. The external CAM uses a photosensor array consisting of 10,000–50,000 pixel units at the input side and transmits the data stream of primary image information through a cable connection to the EEU. The EEU transforms the primary image information into a stream of data that can be transmitted wirelessly through the t-coil to the IU. In principle the EEU represents a strong reduction of the natural information processing in the bypassed retinal layers. It has to include sophisticated information technology reducing possible image ambiguity and simulating small eye movements *(17,77)* The technical information process is primarily based on the assumption of virtual bypassed retinal layers with normal function, but has to be corrected for the deviant connectivity and processing of diseased retina, especially, in the contacted afferent retinal structures that naturally have changed with the progress of disease *(48,49,78)*. The necessary modality is implemented in the functional layers of the retinal parallel processor (RPP). First the parallel retinal encoder recalculates the two dimensional matrix of image information. Then the primary matrix information undergoes a process of parallel retinal image encoding in the RPP. Each layer of parallel processing retinal parallel filter (RPF) provides a set ($n > F \times \Sigma$ stimulation electrodes, $F = 1$–8) of filter functions that represent approximated implementations of the normal primate retina function at different cellular layers. One of the main differences in layer modality is the specific use of either bipolar or ganglion cell receptive field function. A very important feature of the RPF is the tuning and learning modality. This allows a selection of different characteristic filter functions in the first layer of the RPF and a modulation of filter parameters in the second layer, the transfer function modulator. Filter selection and parameter tuning can be performed during training sessions with patients after implantation and at any time of implant use. It is the aim of these training sessions to match a reported subjective perception elicited by the implant so that it will come close to an expected or optimized visual sensation.

Signal output of the RPP is combined in the signal filter integrator in order to synthesize a temporal arrangement of transmitted information for each stimulating microelectrode. Then data are transferred to the data integrator, which combines all incoming streams of information and a clock signal that is transferred to the signal and information transmitter. The transmitter combines the transfer of data and necessary energy to run the IU and is connected with the t-coil as demonstrated in Fig. 3.

The coil in the IU (r-coil), which is positioned directly opposite and in close approximation to the t-coil (r-coil), and the receiver unit are receiving the electromagnetically transferred energy and encoded data stream. The energy to run the implant is supplied by the basic wave of electromagnetic transmission, which is processed and stored in the internal energy unit. Transmitted data and clock signals are decoded in the signal and clock decoder. The microelectrode stimulation unit transforms the incoming data into a definite stimulation signal for each microelectrode, which will be realized by the current source and applied to the MA. The definite current curve applied by single microelectrodes is representing a combination of slow basal amplitudes directed to outer retinal (mainly bipolar cells) and fast spike trains for inner retinal neurons (mainly ganglion cells).

By the injection of complex patterns of stimulation one electrode can be used to elicit different visual effects. With the introduction of more complex electrodes consisting of independent conductive layers, which can individually be connected, the multiuse function of each penetrating electrode can be expanded so that the principle of "one electrode–one pixel" can be overcome and the number of resulting stimulation effects will be the number of electrodes multiplied by a factor X of multiuse capacity.

The construction of the miRI-system allows the use of more than one MC. By this the penetration of larger numbers of electrodes can be facilitated. Microelectrode choice and application can be individualized to meet the needs of implant carriers. The number and density of microelectrodes supported by each MC can be modified.

miRI DEVELOPMENT AND FIRST RESULTS

The idea of a miRI with penetrating electrodes was evolved from data generated in several experimental series performed in our laboratories. One of these was focussing on the long-term outcome of completely intraocular implanted epiretinal devices with an anterior IOL-receiver and a posterior MA. Results of this series were rather disappointing, as mentioned earlier, with severe complications, like proliferative vitreoretinopathy, retinal detachment, or cyclitic membrane formation in the majority of cases *(69)*. A meta-analysis of several approaches using different epiretinal devices was supporting these results indicated that only isolated posterior segment epiretinal implants or small diameter implants could be applied without major complications *(70)*. Still a high percentage of local adversive effects (fibrous interface formation, secondary retinal damage) were observed in the series with small epiretinal devices.

The second input for the creation of the miRI concept was generated from experiments focussing on biocompatibility test of devices for the intraocular mechanical fixation of epiretinal retina implant systems *(79–81)*. In these experiments large diameter conventional retinal tacks and newly designed small diameter microtacks were inserted *ab interno* in a way that the retina, choroid, and sclera were completely penetrated and the peak of the device was positioned outside the eye. Long-term observations showed that theses structures, although permanently penetrating all layers were relatively well tolerated. Surprisingly, the insertion of these devices could be performed without significant choroidal hemorrhage, choroidal detachment, or severe proliferative vitreoretinal reactions postoperatively. Furthermore testing of smaller elements in the micrometer scale clearly provided evidence that intraocular and especially, retinal reactions decreased in first approximation exponentially with the reduction in implant diameter. The retina remained intact according to histological criteria in the vicinity of implants when applying these devices *ab interno (81)* although still some undesired epiretinal proliferations were observed. The next step of progression was the penetration of the eye *ab externo* and this was leading to further reduction of secondary proliferations and retinal disintegration of neuronal structures *(82)*. As the preliminary results of *ab externo* implantation of microelectrodes in rabbits were favorable, a study program was worked out for the development of a definitive miRI.

Shape and dimension of microelectrodes with a relatively high aspect ratio (ratio of length divided by diameter) were optimized in several series of biomechanical analysis using scleral tissue of different species (rabbit, pig, monkey, human donor eyes). Selection criteria were the force necessary for the penetration of electrodes and on the other hand the mechanical stability of electrodes. Deformation of electrodes or the apex of electrodes during scleral penetration was valuated as absolute exclusion criterion. It was demonstrated in these experiment that the examined scleral tissue can be simultaneously penetrated by multielectrode arrays *(83)*. As expected, the number and density of electrodes proved to be a critical parameter severely limiting the range of applicable array designs. Anyhow, forced *ad hoc* penetration of electrodes revealed to be relatively traumatic, leading to immediate retinal damage around the site of penetration. A solution to this was found in the principle of very slow electrode penetration. It is applied by suturing electrode arrays to the scleral surface in way that a continuous force into the direction of desired penetration is applied so that a slow invasion of electrodes is resulting. The process of electrode penetration is finalized once the inner surface of the electrode carrier, which is a precise negative of the eye curvature, gets into touch with the outer scleral surface.

One of the most critical questions of the project so far was the problem of principle biocompatibility of the implantation procedure and long-term tissue reactions toward the implant. In order to examine this, a surgical standard procedure was successfully developed and tested in a series of in vitro procedures *(83)*. In order to come as close as possible to the biology of human eyes in these experiments, it was decided to test electrically inactive multielectrode implants in nonhuman monkeys. The focus of these experiments was to study if the principle of slow penetration of electrodes is feasible, if chorioidal hemorrhages or detachment may develop during penetration, if undesired retinal effects might occur or vitreoretinal proliferations would set a critical limit. Electrode length in these experiments was chosen so that a proportion of elements were totally penetrating all layers extending into the posterior vitreal cortex. Results of this series of experiments so far indicate that the surgical procedure and penetration of electrodes are very well tolerated *(83)*. Figure 4 shows an example of a 10 electrode implant 2 mo after implantation. The principle of slow electrode penetration, covering a postoperative period of several weeks in individual cases proved to be a well-working principle and revealed to be very atraumatic *(84,85)*. The retina around slowly penetrating electrodes was relatively well preserved and the vitreous did not develop adverse proliferative

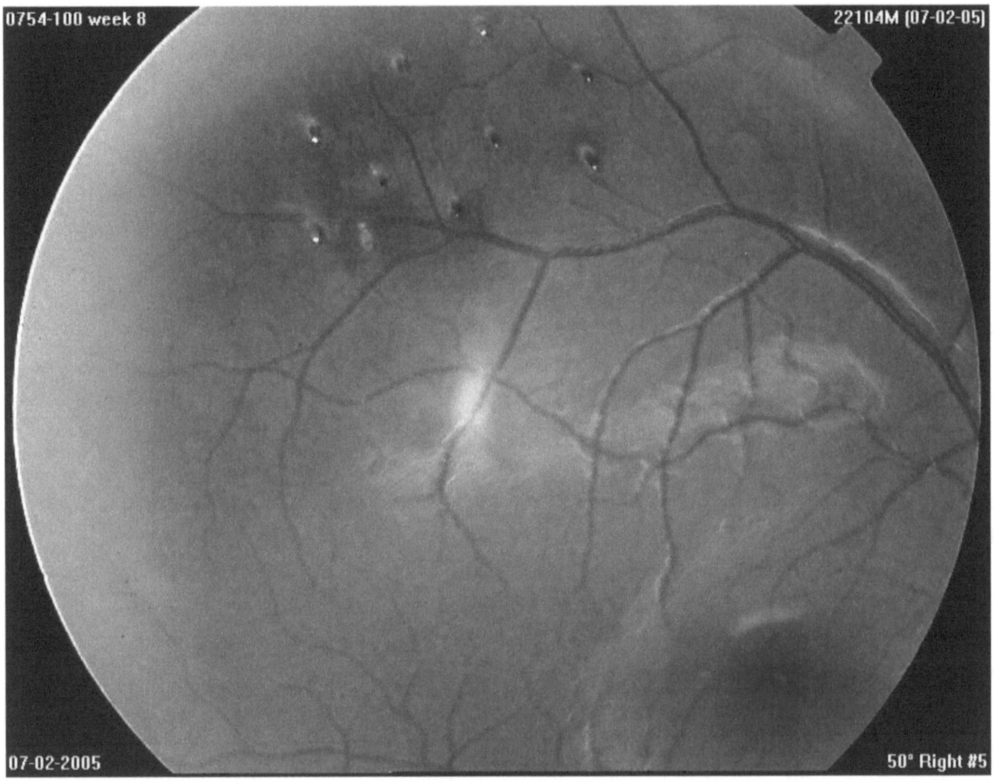

Fig. 4. Fundus image of a monkey eye (*Macaca fascicularis*) 8 wk after implantation of a 10 electrode miRI-prototype.

reactions, even in cases of total retinal penetration *(84–86)*. Over all, the principle of the miRI proved to be a feasible option of a nearly atraumatic retina implant system.

REFERENCES

1. Uhlig CE, Taneri S, Benner FP, Gerding H Electrical stimulation of the visual system. From empirical approaches to the development of visual implants. Ophthalmologe 2001;98: 1089–1096.
2. LeRoy C. O 'u' L'on rend compte de quelques tentatives que L'on a faites pour guérir plusieurs maladies par l'électricité. Hist. Acad. Roy Sciencies (Paris), Mémoires Math Phys 1755;87–89.
3. Galvani L. De viribus electricitatis in motu musculary, commentaries. De Bononiensi Scientiarum et Artium Instituto atque Academia 1791;7:363–418.
4. Krause F, Schum H. Die epileptischen Erkrankungen, In: Kuttner H, ed., Neue Deutsche Chirurgie, Enke, Stuttgart, 1931:482–486.
5. Foerster O. Beiträge zur Pathophysiologie der Sehbahn und der Sehsphäre. J Psychol Neurol 1939;39:463–485.
6. Button J, Puttnam T. Visual responses to cortical stimulation in the blind. J Iowa St. Med Soc 1962;52:17–21.
7. Brindley GS, Rushton D. Implanted stimulators of the visual cortex as visual prosthetic devices. Trans Am Acad Ophthalmol Otolaryngol 1974;78:741–745.
8. Tassiker GE. Retinal stimulator, US patent, 351/1G0R, #2760483.1996.8.

9. Tassiker GE. Preliminary report on a retinal stimulator. Br J Physiol Opt 1656;13:102–105.
10. Potts AM, Inoue J. The electrical evoked response of the visual system (EER). Effect of adaptation and retinitis pigmentosa. Invest Ophthalmol Vis Sci 1969;8:605–612.
11. Brindley GS. The site of electrical excitation of the human eye. J Physiol 1962;127: 189–200.
12. Machemer R, Buettner H, Norton EWD, Parel JM. Vitrectomy, a pars plana approach. Trans Am Acad Ophthalmol Otolaryngol 1971;75:813–820.
13. Dawson WW, Radtke ND. The electrical stimulation of the retina by indwelling electrodes. Invest Ophthalmol Vis Sci 1977;16:249–252.
14. Chow AY. Artificial retina device. United States Patent no. 5,016,633, issued May 21, 1991.
15. Humayun MS, Probst RH, Hickingbotham D, de Juan E, Dagnelie G. Visual sensation produced by electrical stimulation of the retinal surface in patients with end-stage retinitis pigmentosa. Invest Ophthalmol Vis Sci 1993;34:S659.
16. Rizzo J, Socha A, Edell D, Antkowiak B, Brock D. Development of a silicon retinal implant: surgical methods and mechanical design. Invest Ophthalmol Vis Sci 1994;35:S1535.
17. Eckmiller R. Learning retina implants with epiretinal contacts. Ophthalmic Res 1997;29:281–289.
18. Zrenner E, Miliczek KD, Gabel VP, et al. The development of subretinal microphotodiodes for replacement of degenerated photoreceptors. Ophthalmic Res 1997;29:269–280.
19. Wyatt J, Rizzo J. Ocular implants for the blind. IEEE Spectrum 5/1996;47–53.
20. Rizzo JF, Wyatt J. Prospect for a visual prosthesis. The Neuroscientist 1997;3:251–262.
21. Kerdraon YA, Downie JA, Suaning GJ, Capon MR, Coroneo MT, Lovell NH. Development and surgical implantation of a vision prosthesis model into the ovine eye. Clin Experiment Ophthalmol 2002;30:36–40.
22. Chung H, Yu H, Yu Y, et al. Development of polyimide photodiode electrode array system for laser signaling retinal prosthesis. ARVO abstract no. 4470 at www.arvo.org 2002; accessed December, 2006.
23. Hornig R, Laube T, Walter P, et al. A method and technical equipment for an acute human trial to evaluate retina implant technology. J Neural Eng 2005;2:126–134.
24. Walter P, Mokwa W. Epiretinal visual prosthesis. Ophthalmologe 2005;102:933–940.
25. Chow AY Chow VY. Subretinal electrical stimulation of the rabbit retina. Neurosci Lett 1997;225:13–16.
26. Peyman G, Chow AY, Liang C, Chow VY, Perlman JI, Peachey NS. Subretinal semiconductor microphotodiode array. Opthalmic Surg Lasers 1997;29:234–241.
27. Zrenner E. Will retina implants restore vision? Science 2002;295:2213.
28. Peterman MC, Mehenti NZ, Bilbao KV, et al. The artificial synapse chip: a flexible retinal interface based on directed retinal cell growth and neurotransmitter stimulation. Artif Organs 2003;27:975–985.
29. Leng T, Wu P, Mehenti NZ, et al. Directed retinal nerve cell growth for use in a retinal prosthesis interface. Invest Ophthalmol Vis Sci 2004;45:4132–4137.
30. Palanker D, Vankov A, Huie P, Baccus S. Design of a high-resolution optoelectronic retinal prosthesis. J Neural Eng 2005;2:105–120.
31. Palanker D, Huie P, Vankov A, et al. Migration of retinal cells through a perforated membrane: implications for a high-resolution prosthesis. Invest Ophthalmol Vis Sci 2004;45:3266–3270.
32. Fang X, Sakaguchi H, Fujikado T, et al. Electrophysiological and histological studies of chronically implanted intrapapillary microelectrodes in rabbit eyes. Graefes Arch Clin Exp Ophthalmol 2005;244:364–375.
33. Sakaguchi H, Fujikado T, Fang X, et al. Transretinal electrical stimulation with a suprachoroidal multichannel electrode in rabbit eyes. Jpn J Ophthalmol 2004;48:256–261.
34. Chowdhury V, Morley JW, Coroneo MT. Feasibility of extraocular stimulation for a retinal prosthesis. Can J Ophthalmol 2004;40:563–572.

35. Chowdhury V, Morley JW, Coroneo MT. Stimulation of the retina with a multielectrode extraocular visual prosthesis. ANZ J Surg 2005;75:697–704.
36. Chowdhury V, Morley JW, Coroneo MT. Evaluation of exraocular electrodes for a retinal prosthesis using evoked potentials in cat visual cortex. J Clin Neurosci 2005;12:574–579.
37. Yagi T, Watanabe M. A computational study on an electrode array in a hybrid retinal implant. Proc 1998 IEEE Int Joint Conf Neural Networks 1998;780–783.
38. Iezzi R, Safadi M, Miller J, McAllister JP, Auner G, Abrams GW. Feasibility of retinal and cortical prosthesis based upon spatiotemporally controlled release of L-glutamate. Invest Ophthalmol Vis Sci 2001;42:S941.
39. Peterman MC, Bloom DM, Lee C, et al. Localized neurotransmitter release for use in a prototype retinal interface. Invest Ophthalmol Vis Sci 2003;44:3144–3149.
40. Peterman MC, Noolandi J, Blumenkranz MS, Fishman HA. Localized chemical release from an artificial synapse chip. Proc Natl Acad Sci USA 2004;101:9951–9954.
41. Stone JL, Barlow WE, Humayun MS, de Juan E, Milam AH. Morphometric analysis of macular photoreceptors and ganglion cells in retinas with retinitis pigmentosa. Arch Ophthalmol 1992;110:1634–1639.
42. Santos A, Humayun MS, deJuan E, Greenberg RJ, Marsh MJ, et al. Preservation of the inner retina in retinitis pigmentosa. Arch Ophthalmol 1997;115:511–515.
43. Humayun MS, Prince M, deJuan E, et al. Morphometric analysis of the extracellular retina from post-mortem eyes with retinits pigmentosa. Invest Ophthalmol Vis Sci 1999;40:143–148.
44. Cursiefen C, Holbach LM, Schlotzer-Schrehardt U, Naumann GOH. Persisting retinal ganglion cell axons in blind atrophic human eyes. Graefes Arch Clin Exp Ophthalmol 2001;239:158–164.
45. Kim SY, Sadda S, Humayun MS, deJuan E, Melia BM, Green WR. Morphometric analysis of the macula in eyes with geographic atrophy due to age-related macular degeneration. Retina 2002;22:464–470.
46. Kim SY, Sadda S, Pearlman J, et al. Morphometric analysis of the macula in eyes with disciform age-related macular degeneration. Retina 2002;22:471–477.
47. Gerding H. Artificial human vision. MEJO 2003;11:22–33.
48. Marc RE, Jones BW, Watt CB, Strettoi E. Neural remodelling in retinal degeneration. Prog Retin Res 2003;22:607–655.
49. Jones BW, Watt CB, Marc RE. Retinal remodelling. Clin Exp Optom 2005;88:282–291.
50. Wu HJ, Li XX, Dong JQ, Pei WH, Chen HD. Effects of subretinal implant materials on the viability, apoptosis and barrier function of cultured PRE cells. Graefes Arch Clin Exp Ophthalmol 2007;245:35–42.
51. Gerding H, Eckmiller RE, Hornig R, Ortmann V, Kolck A, Taneri S. Safety assessment and acute clinical tests of epiretinal retina implants. 2002 Annual Meeting Abstract (on CD-ROM and www.arvo.org). Association for Reserach in Vision and Ophthalmology. ARVO abstract no. 4488 at www.arvo.org 2002; accessed December, 2006.
52. Rizzo JF, Wyatt J, Loewenstein J, Kelly S, Shire D. Perceptual efficacy of electrical stimulation of human retina with a microelectrode array during short-term surgical trials. Invest Ophthalmol Vis Sci 2003;44:5362–5369.
53. Rizzo JF, Wyatt J, Loewenstein J, Kelly S, Shire D. Methods and perceptual threshold for short-term electrical stimulation of human retina with microelectrode arrays. Invest Ophthalmol Vis Sci 2003;44:5355–5361.
54. Humayun MS, de Juan E, Weiland JD, et al. Pattern stimulation of the human retina. Vision Res 1999;39:2569–2576.
55. Hornig R, Laube P, Velikay-Parel M, et al. A method and technical equipment for an acute human trial to evaluate retinal implant technology. J Neural Eng 2005;2:29–34.

56. Walter P, Kisvarday ZF, Gortz M, et al. Cortical activation via an implanted wireless retinal prosthesis. Invest Ophthalmol Vis Sci 2005;46:1780–1785.
57. Humayun MS, Weiland JD, Fujii GY, et al. Visual perception in a blind subject with a chronic microelectronic retinal prothesis. Vision Res 2003;43:2573–2581.
58. Guven D, Weiland JD, Fujii G, et al. Long-term stimulation by active epiretinal implants in normal and RCD1 dogs. J Neural Eng 2005;2:65–73.
59. Hayes JS, Yin VT, Piyathaisere D, Weiland JD, Humayun MS, Dagnelie G. Visually guided performance of simple tasks using simulated prosthetic vision. Artif Organs 2003;27: 1016–1028.
60. Dagnelie G, Barnett D, Humayun MS, Thompson RW. Paragraph text reading using a pixelized prosthetic vision stimulator: parameter dependence and task learning in free-viewing conditions. Invest Ophthalmol Vis Sci 2006;47:1241–1250.
61. Majii A, Humayun MS, Weiland JD, Suzuki SD, Anna SA, de Juan E. Long-term histological and electrsphysiological results of an inactive epiretinal electrode array implantation in dogs. Invest Ophthalmol Vis Sci 1999;40:2073–2081.
62. Walter P, Szurman P, Vobig M, et al. Successful long-term implantation of electrically inactive epiretinal microelectrode arrays in rabbits. Retina 1999;19:546–552.
63. Husain D, Loewenstein JI. Surgical approaches to retinal prosthesis implantation. Int Ophthalmol Clin 2004;44:105–111.
64. Chow AY, Chow VY, Pacho KH, Pollack JS, Peyman GA, Schuchard R. The artificial silicon retina microchip for the treatment of vision loss from retinitis pigmentosa. Arch Ophthalmol 2004;122:460–469.
65. Schanze T, Sachs HG, Wiesenack C, Brunner U, Sailer H. Implantation and testing of subretinal film electrodes in domestic pigs. Exp Eye Res 2006;82:1156–1157.
66. Mahadevappa M, Weiland JD, Yanai D, Fine I, Greenberg RJ, Humayun MS. Perceptual threshold and electrode impedance in three retinal prosthesis subjects. IEEE Trans Neural Syst Rehabil Eng 2005;13:201–206.
67. Sachs HG, Gekeler F, Schwahn H, et al. Implantation of stimulation electrodes in the subretinal space to demonstrate cortical responses in Yucatan minipig in the course of visual prosthesis development. Eur J Ophthalmol 2005;15:493–499.
68. Sachs HG, Schanze T, Brunner U, Sailer H, Wiesenack C. Transscleral implantation and neurophysiological testing of subretinal polyimide film electrodes in the domestic pig in visual prosthesis development. J Neural Eng 2005;2:57–64.
69. Kolck A, Mueller-Kaempf S, Sellhaus B, Taneri S, Gerding H. Experimental implantation of combined anterior/posterior segment retinal prosthesis in rabbits: results of long-term observation. ARVO-abstract no. 4220 at www.arvo.org 2004; accessed December, 2006.
70. Ezelius H, Gerding H. The minimal invasive Retinal Implant (miRI) project: risk analysis of different retinal prosthetic devices and design of a new concept. ARVO abstract no. 3176 at www.arvo.org 2006; accessed December, 2006.
71. Büchele Rodrigues E. Retina implant project: chronic implantation of active epiretinal implants. Doctoral thesis, Medical Faculty, University of Marburg, 2003; available at www.d-nb.de, code: urn:nbn:de:hebis:04-Z.2004-1489.
72. Taneri S, Gerding H. Retinal detachment and phthisis bulbi after implantation of an iris prosthetic system. J Cataract Refract Surg 2003;29:1034–1038.
73. Rizzo JF, Wyatt JL, Loewenstein J, et al. Development of a wireless, ab externo retinal prosthesis. ARVO abstract no. 3399 at www.arvo.org 2004; accessed December, 2006.
74. Hornig R, Velikay-Parel M, Feucht M, Zehnder T, Richard G. Early clinical experience with a chronic retinal implant system for artificial vision. ARVO abstract no. 3216, 2006 (www.arvo.org; last accessed December, 2006).

75. Kamei M, Fujikado T, Kanda H, et al. Suprachoroidal-transretinal stimulation (STS) artificial vision system for patients with retinitis pigmentosa. ARVO abstract no. 1537, 2006 (www.arvo.org; last accessed December, 2006).
76. Yamauchi Y, Franco LM, Jackson DJ, et al. Comparison of electrically evoked cortical potential thresholds generated with subretinal or suprachoroidal placement of a microelectrode array in the rabbit. J Neural Eng 2005;2:48–56.
77. Eckmiller R, Neumann D, Baruth O. Tunable retina encoders for retina implants: why and how. J Neural Eng 2005;2:91–104.
78. Cottaris NP, Elfar SD. How the retinal network reacts to epiretinal stimulation to form the prosthetic visual input to the cortex. J Neural Eng 2005;2:74–90.
79. Taneri S, Bollmann FP, Uhlig C, Thelen U, Gerding H. The Retina Implant—Project: in vitro and in vivo testing of different tack types for intraocular fixation of retina implants. Invest Ophthalmol Vis Sci 1999;40(S1):733.
80. Gerding H, Taneri S, Benner FP, Reichelt R, Thelen U, Uhlig CE. Successful long-term evaluation of intraocular titanium tacks for the mechanical stabilization of posterior segment ocular implants. Mat-wiss u Werkstofftechn 2001;32:903–912.
81. Thelen U, Gerding H. The minimal invasive Retinal Implant (miRI) project: experimental testing of electrodes completely penetrating the sclera, choroid, and retina in rabbits. ARVO abstract no. 3214 at www.arvo.org 2006; last accessed December, 2006.
82. Gerding H. et al., unpublished data.
83. Stupp N, Niggemann B, Gerding H. The minimal invasive Retinal Implant (miRI) project: development of surgical techniques and experimental testing in a series of primate implantations. ARVO abstract no. 3191 at www.arvo.org 2006; last accessed December, 2006.
84. Gerding H, Ezelius H, Niggemann B. The minimal invasive Retinal Implant (miRI) project: a novel approach toward the restoration of vision in patients with degenerative retinal diseases. ARVO abstract no. 3214 at www.arvo.org 2006; last accessed December, 2006.
85. Niggemann B, Weinbauer GF, Gerding H. The minimal invasive Retinal Implant (miRI) project: first series of implantation with long-term follow-up in nonhuman primates. ARVO abstract no. 1031 at www.arvo.org 2006; last accessed December, 2006.
86. Friederichs-Gromoll S, Niggemann B, Gerding H. The minimal invasive Retinal Implant (miRI) project: histological results after long-term follow-up of implants in the nonhuman primate model. *ARVO abstract* no. 3163 at www.arvo.org 2006; last accessed December, 2006.

8

A Retinal Implant System Based on Flexible Polymer Microelectrode Array for Electrical Stimulation

Jongmo Seo, Jingai Zhou, Euitae Kim, Kyo-in Koo, Jang Hee Ye, Sung June Kim, Hum Chung, Dong-Il Dan Cho, Yong Sook Goo, and Young Suk Yu

CONTENTS

INTRODUCTION

Photoreceptor loss as a result of retinal degenerative diseases, such as age-related macular degeneration and retinitis pigmentosa is a leading cause of blindness in adult *(1)*. Despite a near-total loss of the photoreceptors, the inner nuclear and ganglion cell layers survive at fairly high rates in the patients with retinitis pigmentosa *(2)* and age-related macular degeneration *(3)*. Retinal prostheses have great potential in alleviating the problems and disabilities produced by these diseases. The feasibility of the electrical stimulation of the remaining retinal neurons is supported by clinical studies which showed that controlled electrical signals applied to a small area of the retina of a blind volunteer through a microelectrode resulted in the perception of a small spot of light *(4)*.

The first effort to form a retina prosthesis team in South Korea was done by Prof. Sung June Kim, PhD, in the department of ophthalmology, Seoul National University School of Electrical Engineering and Computer Science, and by Prof. Hum Chung, MD, in SNU School of Medicine. This was recognized by the Korean Science and Engineering Foundation in 2000 and they, together with 20 faculty members with diverse backgrounds, were given a 9-yr grant, 1 million USD/yr, to establish an Engineering Research Center (ERC) on NanoBioSystems at SNU, with emphasis on

From: *Ophthalmology Research: Visual Prosthesis and Ophthalmic Devices: New Hope in Sight*
Edited by: J. Tombran-Tink, C. Barnstable, and J. F. Rizzo © Humana Press Inc., Totowa, NJ

the visual and other neural prostheses using micro- and nanotechnologies. Because successful development of visual prosthesis requires multidisciplinary research, including electrical engineering, biomedical engineering, ophthalmology, and physiology, NanoBioSystem ERC provided a good starting point. More recently, another 6-yr grant, 0.6 million USD/yr, for the development of an artificial retina was given by the ministry of Wealth and Welfare of Korea in 2005.

Korean research group for the development of artificial retina made polyimide electrode array for retinal stimulation, silicon retinal tack for epiretinal fixation of the electrode array, and the stimulation chip for animal experiment. Both sub- and epiretinal approaches are investigated. In this chapter, electrical stimulation system made up of microelectrode array (MEA) with stimulation chip, micromachined silicon retinal tack, surgical techniques for electrode implantation, and the electrophysiological work will be introduced.

FLEXIBLE POLYIMIDE ELECTRODE ARRAY

After the historical experiment of Dawson and Radtke *(5)*, electrical retinal stimulation has been considered as the most feasible method to elicit vision-mimicking phosphene. To stimulate the retina electrically, epiretinal, subretinal, and *trans*scleral suprachoroidal approaches are proposed. Epiretinal approach is targeting retinal ganglion cells for electrical stimulation, bypassing intraretinal neural network for the ease of surgery *(6)*. Subretinal approach has an idea of replacing the photoreceptor function, thus stimulating bipolar cells, aiming to utilize the intraretinal signal processing, even though the surgical approach to the subretinal space is always challenging work *(7,8)*. *Trans*scleral suprachoroidal approach has the same idea as subretinal approach, and developed to reduce the risk of choroidal or retinal damage during subretinal approach *(9)*.

In author's team, both sub- and epiretinal approaches are investigated *(10)*. For epiretinal implantation, the electrode should be tightly faced onto the retina, thus the electrode array should be flexible and be securely fixed onto the soft, fragile retina. For subretinal implantation, electrode must be thin and have blunt edge not to hurt retinal layers. Polyimide was chosen as the base material for the retinal stimulator to meet these requirements. It is cheap, flexible, and well known by its biocompatibility and its resistance to acid or solvent *(11)*. It also recovers its original shape even after rolling or folding. Most of all, it can be used in silicon microfabrication process, which enables bulk production of the polyimide electrode in various design with the high density of microelectrodes *(12)*. Gold metal is widely used as an electrode material for neural stimulation, but shows weak adhesion to polyimide base. To overcome this problem, thin titanium layer was presputtered between polyimide base and gold layer. Figure 1 shows the microfabrication process for polyimide MEA.

Various shapes of polyimide MEAs are designed to reduce the tissue damages and to make close contact to the retina. To prevent tearing of edge, MEA is designed to have rounded corners and circular holes for retinal tack. The rounded corners may also reduce the retinal tissue damage. Polyimide (PI2525, HD Micro Systems, Parlin, NJ) was prepared following the manufacturer's specification and MEA was fabricated based on semiconductor manufacturing technique. Strip-shaped polyimde MEA is 16 μm in thickness and 4×4 mm^2 in head size for covering macula. Manufactured polyimide

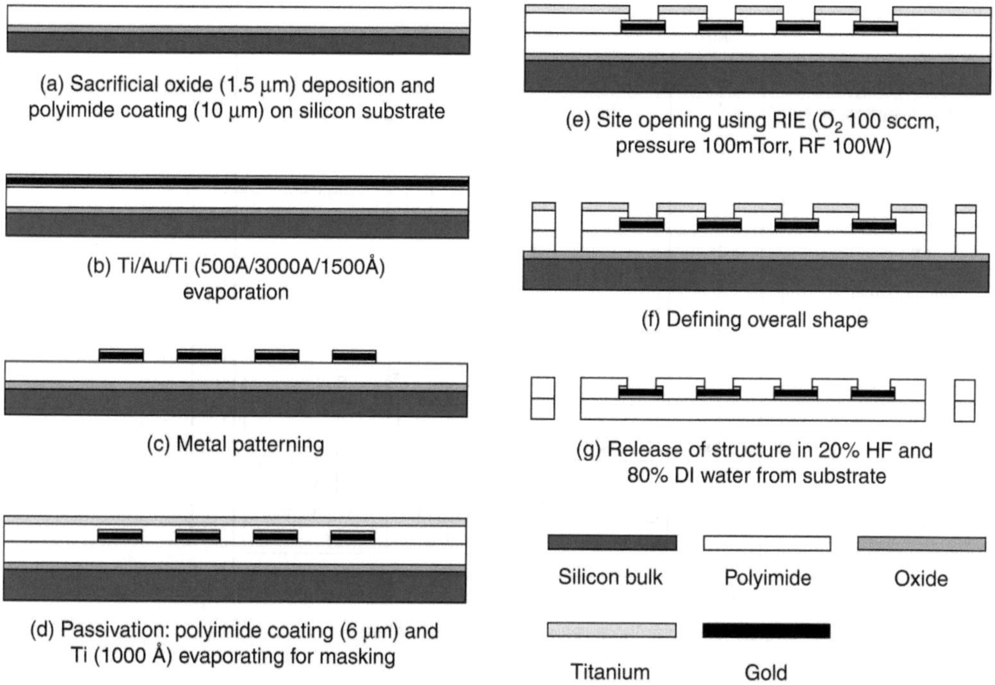

Fig. 1. Fabrication process of polyimide electrode array. Note that the final release requires careful time control to protect upper titanium layer preserved on gold layer.

MEA's were checked for electrode impedances. Impedance for a strip-shaped electrode site ($960 \times 450\ \mu m^2$) was typically 800 Ω in phosphate-buffered solution (pH 7.4) measured at 1 kHz with a potentiostat (Zahner Elektrik IM6e, Germany) *(13)*.

In vitro and in vivo biocompatibility of a polyimide MEA was tested by coculture with human retinal pigment epithelial (RPE) cells and in the rabbit eyes. RPE cells showed good affinity to MEA and there seemed no abnormal morphological changes or no piled-up growth. Also, there was no histological difference between control and operated rabbit eyes except damaged photoreceptors in subretinally implanted group. Electroretinography (ERG) revealed no difference between the transplanted and the control eye *(14)*.

IMPLANTABLE CURRENT STIMULATION SYSTEM

A prototype of implantable retinal stimulation system was proposed for animal experiments (Fig. 2A). This prototype consists of an external and an internal unit. Receiver/stimulator in the internal unit is built around four IC chips (Fig. 2B). Chip I (data/power receiver chip including data decoding and voltage regulation function) and Chip II (current stimulation chip) are custom IC's designed by us, and Chip III (parameter memory chip), and Chip IV (battery charge chip) are off-the-shelf commercial chips. Whereas chip II and III are powered by a small rechargeable Li-ion battery for stimulation, Chip I is powered by the instantaneous radiofrequency (RF) transmitted power *(15)*.

In the design, the external unit is needed for two purposes only: battery charging and parameter passing. Once the parameters for pulse shape and electrode selection are

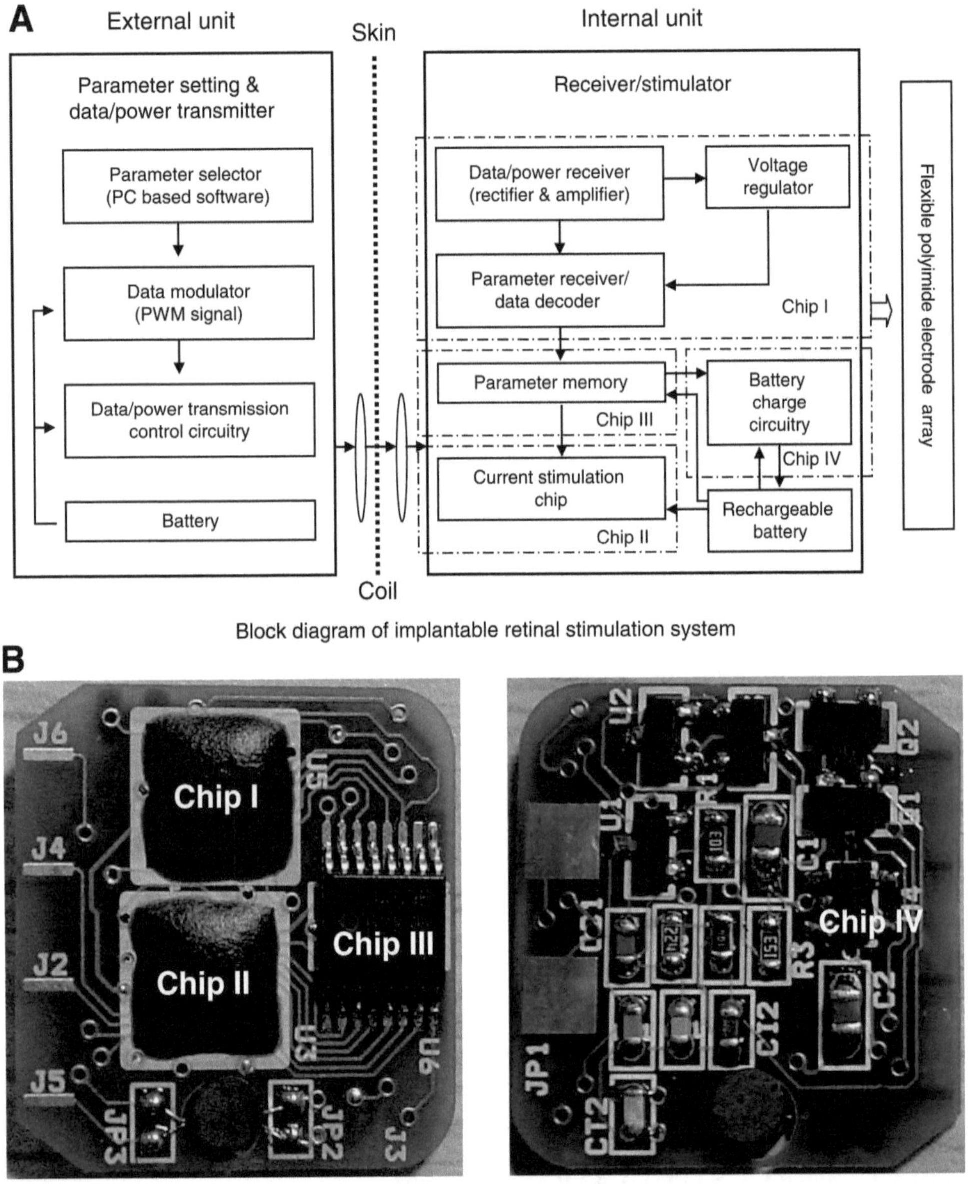

Fig. 2. Implantable retinal stimulation system. **(A)** Block diagram of implantable retinal stimulation system. **(B)** Receiver/stimulator circuitry (left: upside; right: downside). Rechargeable battery is overlaid on the downside surface and is not shown here.

passed to the parameter memory (Chip III in the internal unit) using the RF transmission through coil, the external unit is no longer needed. Therefore, there is no need for experimental animal to be harnessed during stimulation. This reduces the risk of the damage by the experimental animal itself. It also has a small effect on power savings because chip I is not powered during stimulation.

Current stimulation pattern was adopted as the electrical signal for neural stimulation because the current stimulation can provide more stable stimulation level to the electrode–retina interface than the voltage stimulation regardless of the impedance change at the interface *(16)*. Charge-balanced biphasic electrical stimulation was used as an electrical-signal pattern. Pulse amplitude, duration, frequency, and selective activation of the specific electrode can be modified by wireless communication between external unit and implanted circuit.

External Unit

External unit consists of parameter selector, data modulator, data-power transmission control circuitry, rechargeable battery, and external coil. Parameter selector is a PC-based software, and offers stimulation parameters, such as amplitude, duration, stimulation rate, and channel selection. These data can be delivered to a data-power transmitter by RS-232 cable in serial data format. The data-power transmitter consists of an oscillator (2.5 MHz), a Class-E amplifier, and an external coil working as a primary coil. Data for electrical signal were changed to pulse width modulation signals before being carried by 2.5 MHz sinusoidal wave using amplitude shift keying modulation.

Internal Unit

The power of data/power receiver chip (Chip I) is provided by external unit through the inductive coupling coils. Chip I and Chip II are fabricated by a foundry (Austria Microsystems, Austria) using a 0.8 μm complementary metal oxide semiconductor (CMOS) technology. Chip III (SN54AHC595, Texas Instruments) and Chip IV (LTC4054L, Linear Technology, Milpitas, CA) are commercial chips. Current stimulation chip (Chip II) has current bias circuitry, current generator, timing logic circuitry for pulse duration and pulse rate, biphasic waveform generator, and charge cancellation circuit. This chip provides charge-balanced biphasic waveform to stimulation electrode. Battery charge circuitry (Chip IV) is for monitoring the received parameter and for charging up the small rechargeable battery through inductive coil pair when the battery recharging parameter is received.

System Packaging and Test

Internal unit was hermetically housed in biocompatible titanium alloy (Ti-6Al-4V ELI) and weighs less than 20 gm including rechargeable battery. The connection between receiver-stimulator and stimulation electrode array was established with platinum–iridium alloy wire, 0.025 mm in diameter and coated with teflon. All parts of the internal unit were coated with biocompatible silicone elastomer except the stimulation channel of electrode array. The whole system was tested in white rabbit, and the detailed surgical technique will be introduced in next subchapter. The receiver and stimulator unit were operated as designed and the stimulation chip can deliver stable currents from 8 μA to 2 mA. Both pulse width and interpulse delay can be modulated up to 3 ms. The stimulation rate can be increased up to 1000 pulses/s per electrode.

SILICON RETINAL TACK FOR EPIRETINAL PROSTHESIS

For the epiretinal fixation of the stimulation electrode array, biocompatible glues and conventional retinal tacks were tried *(17)*. However, glues are hard to manipulate in the vitreous cavity and easily induce glial cell proliferation and neuronal degeneration, even

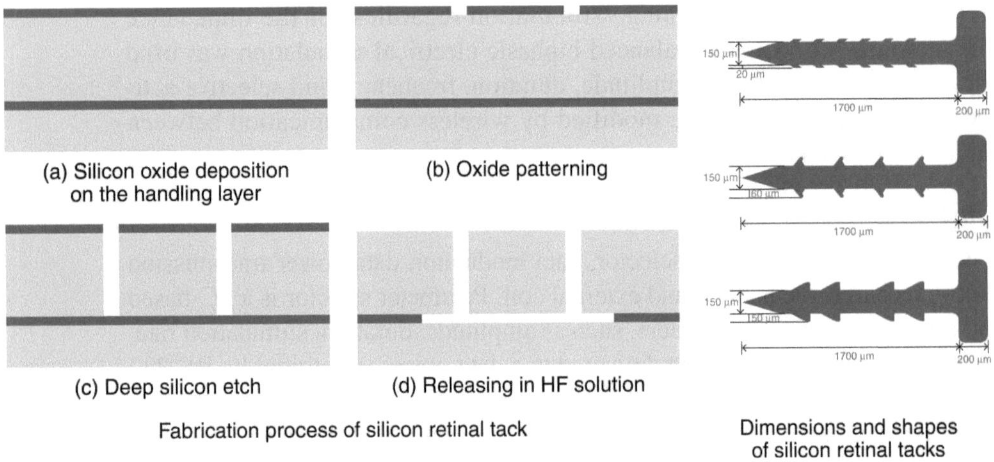

Fig. 3. Fabrication process flow and designs and dimensions of silicon retinal tack.

though they showed good properties and biocompatibility when applied onto the skin or onto the cornea. Conventional titanium retinal tacks are too hard and large for use with the soft and delicate retinal stimulator, which may result in distortion or damage of the MEA. Using microelectromechanical systems technology, microscaled silicon retinal tacks can be easily produced in mass and their design can be easily modified or customized *(18)*.

To evaluate the usability of the silicon retinal tack, various types of tacks are designed as shown in Fig. 3. The first design has barb structures in its shaft to improve positional stability, and the other design has a staple-shaped feature for enhancing the holding ability. Originally, it was designed to have 3 mm length and 300-μm width shanks, but it was found that half-scaled tacks with 1.5 mm length and 150-μm width shanks showed better performance and durability in vivo.

The fabrication process is very simple and highly reproducible as shown in Fig. 3. The main substrate is a silicon-on-insulator wafer with 100 μm of device layer, 2 μm buried oxide layer, and 300 μm of handle layer. Two micrometer of silicon dioxide was deposited on the backside by plasma-enhanced chemical vapor deposition, and the structure was defined by photolithography. After silicon dioxide was patterned, deep silicon etching was performed to the bottom of the handle layer, and the buried oxide layer of the silicon-on-insulator wafer was etched by the concentrated (49%) hydrogen fluoride (HF) solution to release the structure as shown in Fig. 4 *(19)*.

Silicon is brittle, which can cause the gripping site to break. This drawback can be overcome by surface modification of silicon retinal tacks or surgical instruments. It includes coating of tacks with silicon dioxide or parylene layer and coating of the surface of surgical instruments with silicone rubber or other soft materials. Some retinal tacks were oxidized to make the silicon surface hydrophilic. The thickness of the silicon dioxide is 100 nm. Three μm-thick parylene was deposited on the surfaces of other retinal tacks to enhance the durability and the chronic biocompatibility of the retinal tack. Barbing made the tack more easily inserted and stable after implantation. However, the barbs could be broken into pieces in the case of removal of the tack by force. The half-scaled silicon retinal tack showed better penetration of the sclera and caused negligible

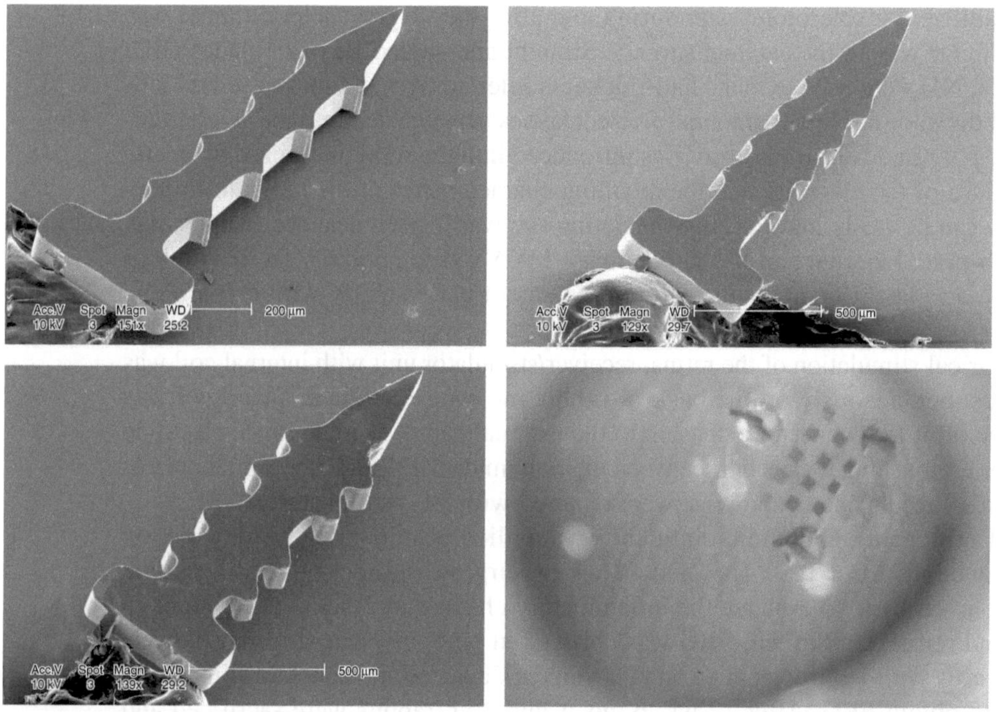

Fig. 4. Fabrication results and epiretinally implanted electrode with silicon retinal tack in the rabbit eye.

fragmentation during manipulation. However, as the size of the tack gets smaller, it would be more difficult to handle the tacks because of the size and the static electric force. The most effective and appropriate size of the silicon retinal tack should be investigated in further study.

SURGICAL TECHNIQUE FOR EPIRETINAL AND SUBRETINAL APPROACH

New Zealand white rabbits were used for in vivo evaluation of the polyimide electrode array, retinal tacks, and retinal current stimulation system. All procedures conformed to the Association for Research in Vision and Ophthalmology Statement on Use of Animals in Ophthalmic and Vision Research. For epiretinal implantation of the polyimide MEA with silicon retinal tacks, lens-sparing three port pars planar vitrectomy was done in white rabbits under general anesthesia achieved by repetitive intramuscular injection of 25 mg ketamine and 6 mg xylazine per kg of body weight. The right eye of each rabbit was used for the test and the left eye was kept intact for the control. After vitrectomy, polyimide MEA was curled and inserted into the eyeball through sclerotomy site. The silicon retinal tack was handled with end-gripping vitreous forceps and the polyimide MEA was fixed epiretinally near the visual streak using the silicon tacks. Sclerotomy site and conjunctiva were repaired with 8-0 vicryl suture and the external part of the polyimide electrode array was fixed onto the sclera with 6-0 prolene suture.

For subretinal implantation, polyimide MEA was introduced into subretinal space through *trans*scleral approach from the limbus toward the posterior pole of the rabbit eye. To stabilize the sclerotomy site during and after the surgery, scleral tunnel was made in similar way in the cataract surgery. Straight and angled Beaver® blades (BD, Franklin La, NJ) were used to make half-thickness sclerotomy and 2 mm of scleral tunnel toward the choroid. Small amount of viscoelastics was injected into the scleral tunnel and the polyimide electrode array was introduced into the subretinal space by gentle snap. Because of the elasticity and the recoiling characteristics of the polyimide, electrode array can be easily introduced. Stimulation site was located near the visual streak of posterior pole. External part of the polyimide MEA was temporarily fixed with biocompatible acrylate glue and sutured onto the sclera permanently with 6-0 prolene suture *(10)*.

For electrical stimulation of the retina, receiver/stimulator unit with internal coil was implanted subcutaneously on the back of rabbit and long subcutaneous tunnel was made to the orbit for connector extension to the external part of the polyimide electrode array. After establishing connection between the polyimide MEA and receiver/stimulator unit, Tenon's capsule and conjunctiva were repaired with 8-0 vicryl suture.

Topical ofloxacin ophthalmic ointment was applied postoperatively for 1 wk on operation wound. After 1, 2, 4, 8, and 12 wk of surgery, indirect ophthalmoscopic examination was done to evaluate the inflammatory changes or other complications in vitreous and retina. On 8th wk, ERG was checked in both operated and control eyes, after mydriasis with 1% tropicamide and 30 min dark adaptation. Ganzfeld stimulation was used to obtain ERG. After 12 wk, 6 mo, 1 and 2 yr, rabbits were sacrificed and their eyes were enucleated to evaluate cataract and other morphological changes of eyeball. The histological change of retina was evaluated under light microscope with hematoxylin and eosin stain.

Indirect ophthalmoscopic examination revealed that polyimide MEA had not induced haziness or inflammatory change of vitreous for 2 yr after the operations. ERG showed no functional change in the operated eye. Dissection of eyes also certified that there was no retinal detachment, vitreous haziness, cataract changes in both implantation techniques. There was no displacement of epiretinally fixed polyimide MEA or subretinally implanted array during follow-up period. No melting or fragmentation of the silicon retinal tacks was found on the scanning electron microscope after 1 yr of implementation. No inflammatory change or significant alteration was found on the histological examination of the retina except photoreceptor damage in the subretinally implanted group, and there was a fibrous encapsulation grown around the retinal tack on the posterior surface of the sclera *(14)*.

Electrically evoked potential (EEP) was recorded and was compared with the visual evoked potential (VEP) by flash light in rabbits with subretinal polyimide MEA in their eyes, 1 mo after surgery. VEP was recorded under full-field flash light stimulation by 0.6 J at 2 Hz, whereas EEP was recorded under electrical current stimulation of the retina by MEA. EEP more than 2.5 mA was similar to VEP, indicating that MEA is suitable for the development of the artificial retina prototype. However, N1, P100, and N2 of EEP showed shorter latencies than those of VEP, by 15.1 ms (32.9%), 76.8 ms (38.3%), and 99.7 ms (60.1%) respectively, and these findings should be considered in the parameters of the electrical stimulation of the retina in the future study *(20)*.

UNDERSTANDING OF RETINAL NEURAL CIRCUIT WITH MULTIELECTRODE ARRAY TECHNOLOGY

Electrophysiological work, such as evoked potential and ERG can provide objective responses to the electrical stimuli in the experimental animals. In understanding and analyzing the neuronal circuit in the retina, physiology also shows its powerful ability. To design and modify the electrical stimuli in artificial retina system, fundamental knowledge of the way how the retinal circuit encodes and processes visual information is important. Simultaneous recording of the multiple signals carried by a number of retinal neurons can be great help to address these issues. Multielectrode array technology has been established as a powerful tool for this particular purpose *(21)*.

One of the interesting features in the retinal circuit is that there are about 100,000,000 photoreceptors, hundred times more than ganglion cells, and this implies that the information generated from the photoreceptors are being compressed to a great degree into the small-numbered ganglion cell activities *(22,23)*. Nowhere else the visual system is represented with as few neurons as in the ganglion cells. This means that the ganglion cells can hardly function as independent channels for visual information processing.

To understand neuronal circuitry in the retina, extracellular recording in the isolated rabbit retina was performed using 8 × 8 multielectrode array (Multi Channel Systems GmBH, Germany), with each electrode diameter of 30 μm and the interelectrode distance of 200 μm *(24,25)*. Followed by spike sorting, the patterns of interaction between pairs of neurons in the retinal ganglion cells were studied by examining the cross-correlation function, which represented the firing rate of one neuron as a function of time before or after a spike from the other neuron. Flat cross-correlation function with time appears when two neurons fire independently, whereas a peak near zero time represents synchronous firing of two neurons.

Correlated Firing

As in the salamander retina *(26)*, three different types of correlation were observed in the rabbit retina with different time-scales; broad (firing synchrony within 100 ms), medium (~10 ms), and narrow (~1 ms) correlation (Fig. 5). The broad correlation could be blocked by $CdCl_2$ (Fig. 6A), which blocked synaptic vesicle release, while the medium and narrow correlations persisted. The medium and narrow correlations still remained after the glutamatergic transmission was postsynaptically blocked with specific glutamate receptor blocker 6-cyano-7-nitroquinoxaline-2,3-dione and AP-7 (Fig. 6 B,C), demonstrating that these correlations must have been mediated by electrical synapse through gap junction channels. To evaluate this hypothesis, the gap junction channel blocker, heptanol was applied, and this led to partially blocked narrow correlation (Fig. 6D). These findings correspond well with others *(26,27)*, but a major difference was found in the fraction of narrow correlations. In the salamander retina, the spikes paired with millisecond synchrony constituted only a small fraction of each cell's spike train; on average, 8.1% of 61 cells were involved in narrow correlations. And in the cat retina Y cell action potential caused a spike in a coupled Y cell with a probability of only 1–4%. But in this experiment with the rabbit retina, more than 50% of 31 cells were involved in narrow correlations. Possible explanation may reside in the gap junction channel density varying with different species between the

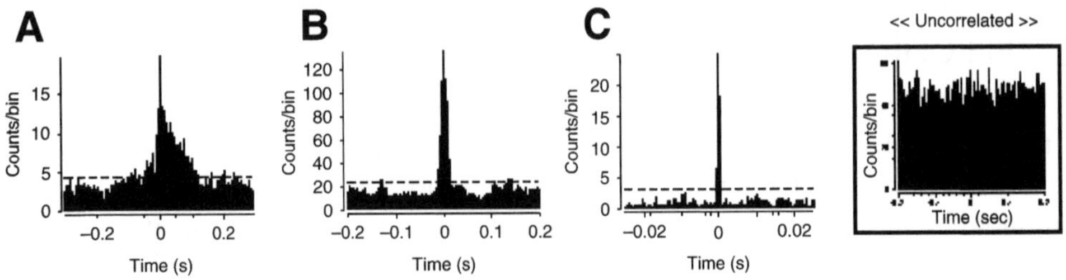

Fig. 5. Three types of correlations in retinal ganglion cell firing. Different time-scale is introduced in each cross-correlogram. **(A)** Broad correlation (bin: 1 ms). **(B)** Medium correlation (bin: 1 ms). **(C)** Narrow correlation (bin: 0.8 ms). Inset shows a flat uncorrelated example. Dashed line indicates 99% confidence limits.

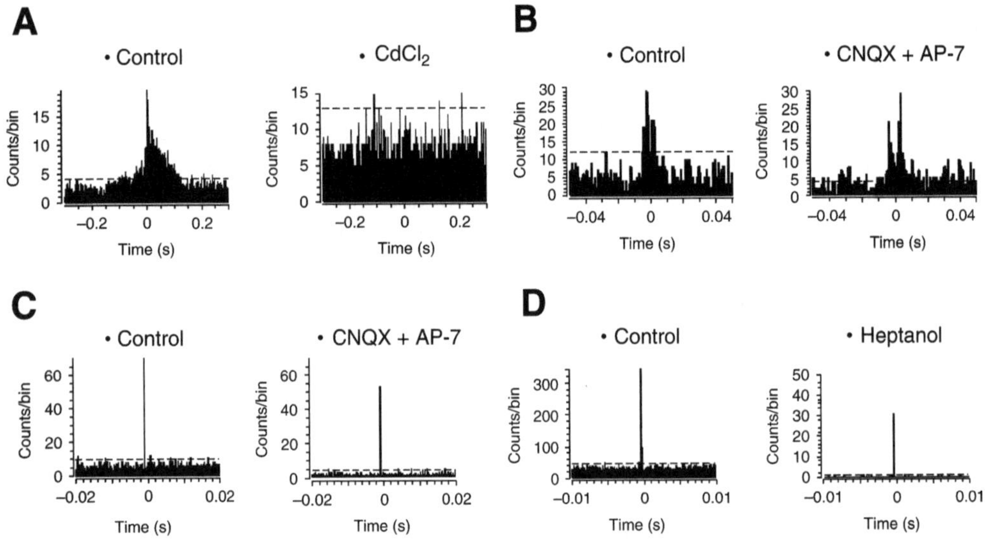

Fig. 6. Correlation was blocked with antagonists. **(A)** Broad correlation disappeared when presynaptic vesicle release was blocked with $CdCl_2$. (Lt) Control, (Rt) after treatment of 200 μ*M* $CdCl_2$. **(B,C)** Medium and narrow correlation still persisted when glutamate receptor was blocked with antagonist. (Lt) Control, (Rt) after cotreatment of AMPA/Kainate receptor antagonist, 50 μ*M* 6-cyano-7-nitroquinoxaline-2,3-dione and NMDA receptor antagonist, 100 μ*M* 2-amino-phosphonoheptanonic acid (AP-7). **(D)** Narrow correlation was partly blocked with gap junction channel blocker, heptanol. The ordinate scale is different in left and right. (Lt) Control. (Rt) After treatment of 1 m*M* heptanol. Bin width was 1 ms in A, B and 0.8 ms in C, D. Dashed line indicates 99% confidence limits.

rabbit and the salamander. The difference in functional cell types might also have been attributed. Our 31 cells belonged to off-center ganglion cells, whereas the functional cell types of 61 cells recorded in the Brivanlou's study were not classified *(26)*.

Retinal ganglion cell cluster map; Changing the reference neuron in cross-correlation analysis enabled separation of reacting target neurons, showing narrow correlations with each reference neuron. Connecting each reacting neurons formed a retinal ganglion cell cluster map, which showed synchronous firing pattern. The cluster map consisted of

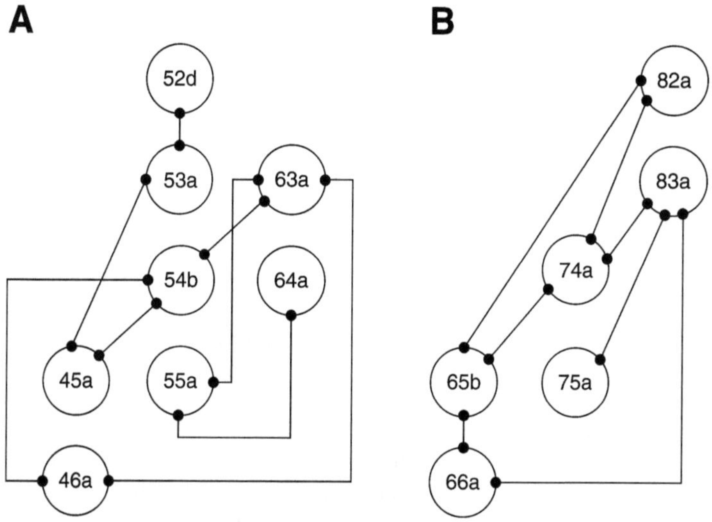

Fig. 7. Cluster map of synchronized firing cells drawn from the correlation data. **(A)** Cluster A. **(B)** Cluster B. The neuron 45a represents the first neuron recorded and sorted on fourth column and fifth row of 8 × 8 array electrode.

four to eight nearby cells (Fig. 7). Retinal ganglion cell's firing was nearby identified in significant synchrony from multineuron recording with multielectrode array. This population coding of visual information in the ganglion cell layer is perhaps regarded as a more efficient way than independent cell coding for information transmission *(22,28)*.

CONCLUSION

The Korean retinal implant project is based on a polyimide electrode and an electrical stimulation system. The electrode was manufactured using silicon fabrication process and was proven biocompatible in cell culture studies with RPE cells. Implantable current stimulation system consists of an external and an internal part, where the external part is only needed during battery charging and for parameter passing. Thus, for normal stimulation operation, the animal can be freely moving. Surgical techniques were developed for both epi- and subretinal implantation of electrodes. Micromachined tacks were developed for fixing of the epiretinal electrode. The electrodes were implanted for durations up to 2 yr and the conditions of the devices as well as tissues were checked at the end of experiments. EEPs were recorded and compared with visually evoked potentials recorded from the same animals. All evidences proved suitability of the developed devices and surgical techniques. Retinal neural circuits are studied with multielectrode arrays to understand the cross-connectivity of the cells and to obtain parameters for effective stimulations. The devices and stimulation system are being improved for chronic implantations for animals and human patients.

ACKNOWLEDGMENTS

This work was supported by the Nano Bioelectronics and Systems Research Center of Seoul National University, whereas an ERC supported by the Korean Science and Engineering Foundation and Nano Artificial Vision Research Center supported by Korea Health 21 R&D Project, MOHW Grant No. A050251.

REFERENCES

1. Hyman L. Epidemiology of eye disease in the elderly. Eye 1987;1(Pt 2):330–341.
2. Humayun MS, Prince M, de Juan E Jr., et al. Morphometric analysis of the extramacular retina from postmortem eyes with retinitis pigmentosa. Invest Ophthalmol Vis Sci 1999;40(1):143–148.
3. Medeiros NE, Curcio CA. Preservation of ganglion cell layer neurons in age-related macular degeneration. Invest Ophthalmol Vis Sci 2001;42(3):795–803.
4. Humayun MS, de Juan E Jr., Dagnelie G, Greenberg RJ, Propst RH, Phillips DH. Visual perception elicited by electrical stimulation of retina in blind humans. Arch Ophthalmol 1996;114(1):40–46.
5. Dawson WW, Radtke ND. The electrical stimulation of the retina by indwelling electrodes. Invest Ophthalmol Vis Sci 1977;16(3):249–252.
6. Humayun MS, Weiland JD, Fujii GY, et al. Visual perception in a blind subject with a chronic microelectronic retinal prosthesis. Vision Res 2003;43(24):2573–2581.
7. Zrenner E. Will retinal implants restore vision? Science 2002;295(5557):1022–1025.
8. Chow AY, Chow VY, Packo KH, Pollack JS, Peyman GA, Schuchard R. The artificial silicon retina microchip for the treatment of vision loss from retinitis pigmentosa. Arch Ophthalmol 2004;122(4):460–469.
9. Nakauchi K, Fujikado T, Kanda H, et al. Transretinal electrical stimulation by an intrascleral multichannel electrode array in rabbit eyes. Graefes Arch Clin Exp Ophthalmol 2005;243(2):169–174.
10. Chung H, Seo JM, Kim KA, et al. Comparison between the epiretinal and subretinal implantation of the polyimide electrode array for the electrical stimulation of the retina. Invest Ophthalmol Vis Sci 2006;47:E-Abstract 3179.
11. Richardson RR Jr., Miller JA, Reichert WM. Polyimides as biomaterials: preliminary biocompatibility testing. Biomaterials 1993;14(8):627–635.
12. Kim HK, An SK, Jun SB, Kim SJ, Chung H, Yu YS. Polyimide-based microelectrode arrays for epi-retinal implant. Proc Int Sensor Conf, Seoul, Korea, ISSN 1225–3278, 2001;107–108.
13. Zhou JA, Kim ET, Seo JM, Chung H, Kim SJ. A Seven Segment Electrode Stimulation System for Retinal Prosthesis. Invest Ophthalmol Vis Sci 2006;47:E-Abstract 3178.
14. Seo JM, Kim SJ, Chung H, Kim ET, Yu HG, Yu YS. Biocompatibility of polyimide microelectrode array for retinal stimulation. Mat Sci Eng C-Bio S 2004;24:185–189.
15. Park SI. A study on Next Generation Deep Brain Stimulation System for Parkinson's Desease. PhD Dissertation, School of Electrical Engineering and Computer Science, Seoul National University, Korea. 2006.
16. Sivaprakasam M, Liu W, Wang G, Weiland JD, Humayun MS. Architecture tradeoffs in high-density microstimulators for retinal prosthesis. IEEE Trans. Circuits Syst. I-Regul. Pap. 2005;52(12):2629–2641.
17. Loewenstein J, Rizzo JF, Shahin M, Coury A. Novel retinal adhesive used to attach electrode array to retina. Invest Ophthalmol Vis Sci 1999;40(4):Abstr. 3874.
18. Paik SJ, Cho DI, Seo JM, et al. Development of silicon-micromachined retinal tacks for artificial retina implantation. 1st Int IEEE EMBS Conf Meural Eng, Capri Island, Italy, 2003:185–188.
19. Chung H, Paik SJ, Lee AR, et al. Revised Silicon Retinal Tack for Epiretinal Fixation of the Electrode. Invest Ophthalmol Vis Sci 2005;46:E-Abstract 1514.
20. Kim SH, Seo JM, Kim SJ, et al. Analysis of electrically evoked potential in rabbits with polyimide retinal stimulator. J Korean Ophthalmol Soc 2004;55(8):1363–1369.
21. Meister M, Pine J, Baylor DA. Multi-neuronal signals from the retina: acquisition and analysis. J Neurosci Methods 1994;51(1):95–106.

22. Meister M. Multineuronal codes in retinal signaling. Proc Natl Acad Sci USA 1996; 93(2):609–614.
23. Meister M, Berry MJ 2nd. The neural code of the retina. Neuron 1999;22(3):435–450.
24. Nisch W, Bock J, Egert U, Hammerle H, Mohr A. A thin film microelectrode array for monitoring extracellular neuronal activity in vitro. Biosens Bioelectron 1994;9(9–10):737–741.
25. Egert U, Schlosshauer B, Fennrich S, et al. A novel organotypic long-term culture of the rat hippocampus on substrate-integrated multielectrode arrays. Brain Res Brain Res Protoc 1998;2(4):229–242.
26. Brivanlou IH, Warland DK, Meister M. Mechanisms of concerted firing among retinal ganglion cells. Neuron 1998;20(3):527–539.
27. Mastronarde DN. Interactions between ganglion cells in cat retina. J Neurophysiol 1983;49(2):350–365.
28. Thorpe S, Delorme A, Van Rullen R. Spike-based strategies for rapid processing. Neural Netw 2001;14(6–7):715–725.

9
Electrodé Architecture

Meeting the Challenge of the Retina–Electrode Interface

Lee J. Johnson and Dean A. Scribner

CONTENTS

INTRODUCTION

Retinal prosthesis presents a unique design challenge: how to form an electrical stimulation interface to a curved surface? The challenge is primarily the lack of fabrication methods to pattern conductive material onto a spherical or curved surface. The proximity of the electrodes to the retina is important because increased distance requires increased stimulation charge and reduces acuity. Even flexible substrates, such as polyamide, which can flex naturally into a cylindrical shape do not readily transform into a spherical shape. With the future of retinal prosthesis design moving toward implanted rigid silicon chips, the problem becomes clear; how to bridge the gap between a flat silicon chip and a curved retina. In this chapter, the various geometries, structures, and materials under consideration in the field of retinal prosthesis are reviewed. This chapter is concerned primarily with electrical stimulation, so the interesting neuro-chemical or transplant based systems will not be addressed. In addition, some of the devices presented in this chapter may no longer be in active development, but are included here for the positive lessons that can be learned from their designs.

REQUIREMENTS FOR A CURVE SURFACE ARRAY

An adult human eye has a nominal radius of curvature of 12.7 mm. An obvious geometrical problem arises when placing a flat surface against a curved retina. For example,

From: *Ophthalmology Research: Visual Prosthesis and Ophthalmic Devices: New Hope in Sight*
Edited by: J. Tombran-Tink, C. Barnstable, and J. F. Rizzo © Humana Press Inc., Totowa, NJ

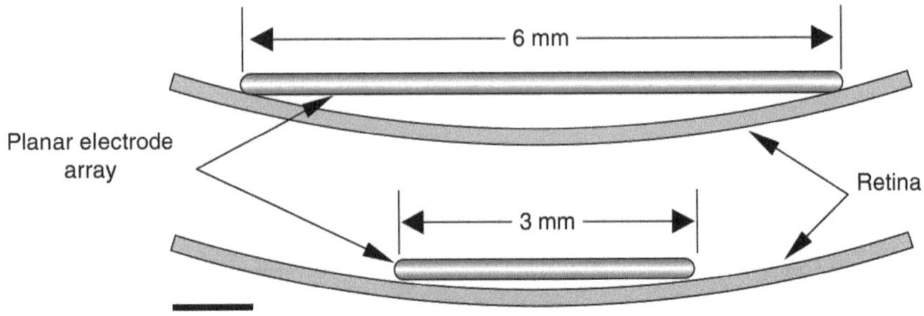

Fig. 1. The gap between a flat electrode array and the curved retina results in reduced efficacy of the retinal prosthesis. In this schematic the radius of curvature is 12.7 mm. The two devices shown have lengths of 3 mm and 6 mm and the maximum distances between the array and the retina are 100 and 360 μm, respectively. (Bar = 800 μm.)

for a flat array 3 mm across resting on the epiretinal surface, the electrodes in the center of the array will be 100 μm from the retina as shown in Fig. 1. Note that a device with a 3 mm width will provide a field of view of approx 10° and thus is a reasonable width for demonstration purposes. Although, the 100 μm distance may intuitively seem small, the span will have a significant effect on device performance, especially, high-resolution devices. There are two primary ways in which the distance acts on the device function, increased threshold current and loss of resolution. The neuron stimulation threshold in terms of current density is dependant on distance because the charges or current, moving from a source on a two-dimensional insulating surface can move in 2π steradians. As the area through which the charges flow is proportional to the radius squared, the charge density is reduced as function of distance in proportion to the radius squared. This means that to provide the same charge density at a larger distance the total current and the current density at the electrode surface must increase. The resolution is lost because electrical current will disperse, essentially causing a blurring effect.

A recent report by Palanker et al. *(9)* estimates the electrode current required to stimulate a neuron. Specifically their result, shown here as Eq. (1), gives the current density, j_e, for a hemispherical electrode as a function of electrode radius, r_0, cell diameter, L, estimated cell threshold voltage, ΔV, saline impedance, γ and distance between the electrode and the cell, Y (*see* Fig. 2A for diagram).

$$je = \frac{E_0}{\gamma}\frac{\Delta V}{\gamma \cdot L} = \left(1 + \frac{L}{r_0} + \frac{Y^2 + Y(2r_0 + L)}{r_0^2}\right) \quad (1)$$

$$\Delta V = \frac{j_e \gamma \cdot L}{\left(1 + \frac{L}{r_0} + \frac{Y(2r_0 + L)}{r_0^2}\right)} \quad (2)$$

Palanker et al. used the equation to create a graph of the effect of distance, Y, on the threshold current as shown in Fig 3. They define threshold current as that required to generate a 30 mV drop across a 10 μm long cell. Palanker graphed the current for electrode radii of 5, 15, 50, and 150 μm, respectively.

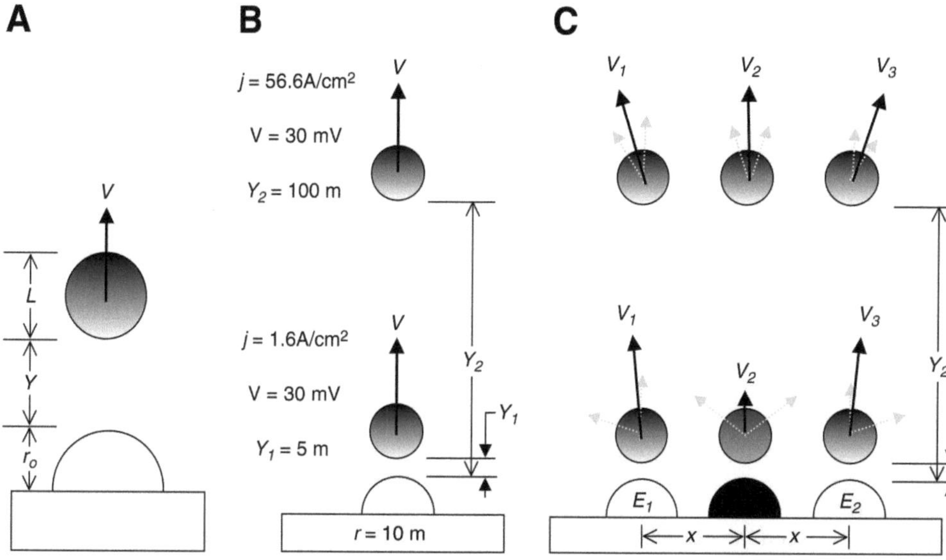

Fig. 2. (A) Diagram of the electrode-cell geometry for Eq. (1). **(B)** Increasing distance from $Y_1 = 5$ to 100 µm increases the required threshold current density at the electrode. Threshold current is the current required to create a ΔV equal to 30 mV. **(C)** The effect of distance on resolution. Each cell (circles) experiences a ΔV from each active electrode (dotted arrows). These ΔV are the result of the electric field from each electrode and thus, add vectorally to result in a total ΔV_k (black arrows) for each cell. The center cell in the case of Y_1 experiences a $\Delta V_{1,}$ which is lesser than $\Delta V_{2,3,}$ thus the retina resolves the two discrete electrodes. This is not so for the more distant case of Y_2, here the ΔV_1 is about the same as $\Delta V_{2,3}$. (*See* text for details.)

Figures 2B,C present the two effects of distance, increased current density at the electrode surface and loss of resolution. Figure 2B presents an example of the increased current density required to maintain the desired $\Delta V = 30$ mV. Using Eq. (1), for a 20 µm diameter electrode a 95 µm increase in distance, from $Y = 5$ to 100 µm, will cause the required current density to increase by a factor of 35, from 1.6 to 56.6 A/cm^2. The resolution effect is based on the fact that the voltage difference, ΔV, across a cell is dependant on all the current sources in the vicinity of a given cell. In Fig. 2C, for the two active electrodes, $E_{1,2}$, each cell (circles) experiences a ΔV from each active electrode (dotted arrows). These ΔV are the result of the electric field from each electrode and thus, add vectorally to result in a total ΔV_k (black arrows) for each cell. The cells in each row and the electrodes are separated by a distance *x,* which is used in determining the vectors ΔV_k. For the three cells at a distance Y_1 from electrode array, the difference between the total ΔV vector (sum of two dotted lines on the cell) of the center cell (ΔV_2) from those of the side cells ($\Delta V_{1,3}$) is sufficient such that for the correct current the center cell will not be stimulated whereas the side cells are. The three cell "retina" can resolve that two separate electrodes are on.

This is not the case for the cells at Y_2, the larger distance. Here the total ΔV experienced by each cell is virtually the same because the distance of side cells and the center cell from the electrodes is almost the same. The three cell "retina" perceives the two active

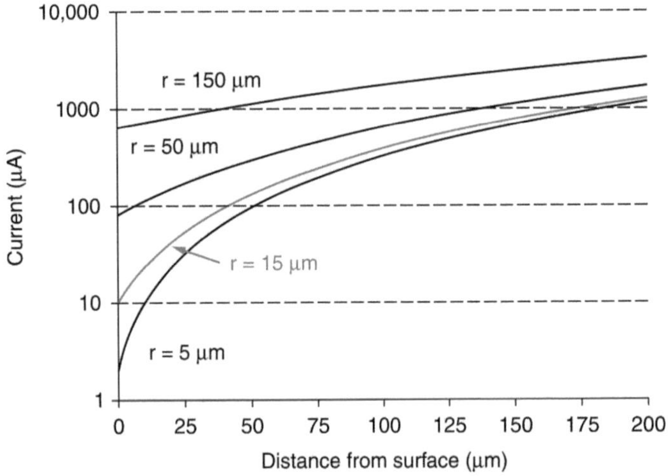

Fig. 3. Threshold current required to generate 30 mV drop across a 10 μm long cell, plotted as a function of distance between the cell and the electrode surface. The four curves correspond to current calculated for electrode radii of 5, 15, 50, and 150 μm.

electrodes as one. As an example, for $x = 30$ μm, $Y = 5$ μm, $j_e = 1.6$ A/cm^2, $\gamma = 70$ W, and $L = 10$ μm eq. (2) results in $\Delta V_{1,3}$=0.030 V whereas $\Delta V_2 = 0.002$ V. Changing Y to 100 μm and j_e to 56.6A/cm^2 results in $\Delta V_{1,3}$=0.051 V whereas $\Delta V_2 = 0.053$ V. Of course the use of hemispherical electrodes and 10 μm cells makes a number of simplifications concerning cell shape, axons, and dendrites; the general principle is the same for real retinas and electrodes.

A comparison of two early studies of human retinal stimulation shows that reduced distance between the electrode and the retina may reduce thresholds. In one study a gold weight fixation method was used with arrays of electrodes of 100 or 400 μm diameters *(11)*. They found that the majority of the reported thresholds were lower than in a study using a hand held, more distant from the retina electrode *(12)*, 0.28–2.8 mC/cm^2 vs 0.16–80 mC/cm^2 respectively. Even within the study by Rizzo et al. they found lower thresholds in general for a smaller, weighted 100 μm diameter electrode vs a 250 μm hand held electrode, 50 μA vs 500 μA. A seemingly contrary position is taken in a recent paper by Mahadevappa *(13)*, suggesting that the distance from the retina is not a factor in thresholds in practice until the distance is more than 500 μm. The graph by Palanker et al. indicates that this would likely be the case for a larger 400 μm electrode studied by Mahadevappa. However, 400 μm electrodes will not result in high-resolution retinal prostheses. The smaller electrodes required for a high-resolution retinal prosthesis will require close contact.

EPIRETINAL OR SUBRETINAL PLACEMENT

A primary division in retinal prostheses is between epiretinal and subretinal *(14,15)*. The descriptive difference is the placement of the device against the inner limiting membrane, epiretinal, or between the retina and the sclera, subretinal (*see* Fig. 4.) The physiological differences are many. A key physiological difference is that epiretinal

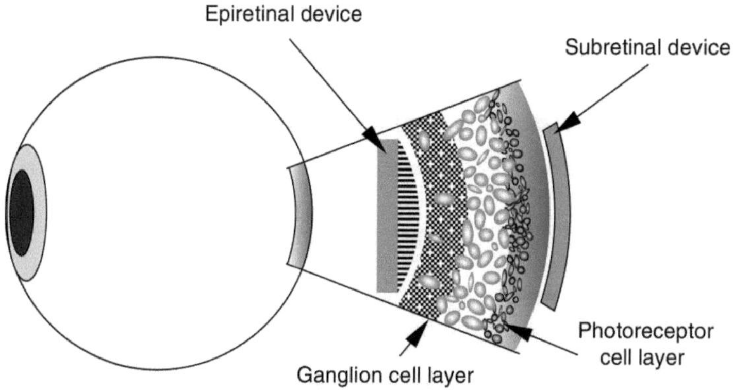

Fig. 4. Diagram indicating the general subretinal and epiretinal placements of retinal prostheses. Not drawn to scale.

devices are closer to the ganglion cells. Thus, they may require preprocessing of the image before stimulation to make up for the bypassed function of bipolar, horizontal, and amacrine cells. Subretinal devices are more likely to stimulate bipolar cells because they are closer to the bipolar cells than to the ganglion cells and thus, in theory, preprocessing may not be required. Another difference is in the potential effect of the diseased retina on the imaging of the implant. In the severest form of age-related macular degeneration (AMD) that likely to lead to blindness, the retina becomes more opaque. This opaqueness would make it difficult for subretinal devices with electronic imaging arrays located behind the retina to collect sufficient light to form an image. The surgical approach is different for epiretinal vs subretinal implantation. The epiretinal approach requires a vitrectomy and entering the eye. The vitrectomy is a standard procedure for a retinal surgeon.

FLEXIBLE DESIGNS

Many of the research groups developing retinal prostheses use flat metal arrays on polyamide. Polyamide has the benefit of flexibility, if only in one direction. This allows the device to conform in one direction to the curvature of the retina. The designs of most devices using polyamide connect the polyamide stimulation array directly to the retina. The limitation of this design is that a lead is required for each electrode. This may reduce the number of electrodes possible because of space limitations and routing complications. Most systems using polyamide currently are using less than 100 electrodes. In fact, the four groups described in this chapter have fewer than 30 electrodes. Using the latest technology DRC, Inc specifies 5 μm as the smallest reliable line size and spacing. Each line therefore requires 10 μm. The limit for a 3 mm wide polyamide cable would be 300 lines. Far fewer than the thousands required for reading and face recognition. Layering signal lines is possible, but could create capacitive coupling problems and reduce polyamide array flexibility.

Polyamide electrodes are formed using standardized methods (Fig. 5) *(16)*. The first step is to spin a layer of polyamide onto a solid substrate, such as a silicon wafer and cure it in an oven. In some facilities, large prefabricated sheets of polyamide are used

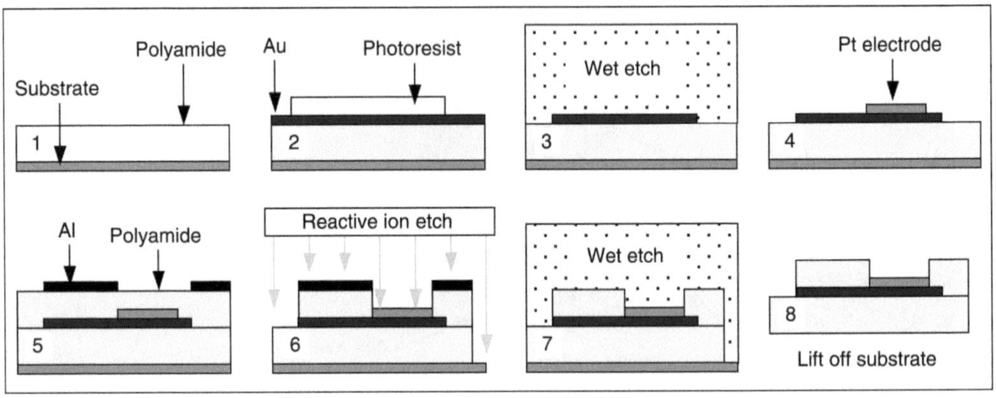

Fig. 5. Steps to fabricate polyamide electrode arrays and cables. Polyamide is spun onto a substrate and gold and/or other metals are patterned on (1–4). Aluminum is patterned as a mask to block the reactive ion etch of polyamide in selected areas (5,6). A thin wet etch removes the Al (7). Finally the device is lifted of the substrate (8).

instead. The thickness of the base layer can vary, but 5–15 μm is typical. The next step is to deposit a layer of metal, typically sputtered gold, to act as the interconnections and other device features. Lift-off technology is used to pattern the interconnections and features in a photoresist and remove excess gold to reveal the desired pattern. An additional metallization step to add patterned biocompatible electrode materials, such as platinum or iridium may be performed if those metals have not been used as the base metal. A second layer of polyamide is spun to provide insulation of the metal features. In some cases, before the second layer of polyamide added, a layer of silicon nitride is used to improve impedance. Sputtered and wet etched aluminum can then be used as a mask to allow the selective reactive ion etching of the polyamide to reveal windows to the electrodes and other polyamide device features.

The Tubingen, Intelligent Implants (IIP), and Boston VA groups all use variations of polyamide with metal traces leading to openings in the top layer polyamide to form the electrode sites (Fig. 6). The Tubingen group published a recent article for a device with 16 electrodes on polyamide *(3,17)*. Their electrode is for subretinal implantation and the active area covers approx 1×1 mm^2. The Boston VA group has had a number of designs in the past for both epi- and subretinal implantation. Their current design is a subretinal array similar to the Tubingen group. They have the ability to make their electrode array from made of either polyamide or parylene-C. They report results with a test version of three device that has 25 electrodes in an 1×1 mm^2 array *(18)*. The IIP group also uses a flat polyamide array with exposed electrodes to connect their electronics to the retina *(6)*. The device, as with the Tubingen and Boston VA groups has a portion of the polyamide, which acts as a cable to transmit current laterally from the electronics before directing current into the retina. The IIP device is different in that they plan to create an epiretinal device.

Two unique designs use micro electro mechanical system (MEMS) type fabrication methods to bring their arrays in close approximation with the retina (Fig. 7). A design of the Boston VA group, in collaboration with the Cornell Nanofabrication Institute and MIT, involves a flexible epiretinal array, which can unfold when inserted into the eye

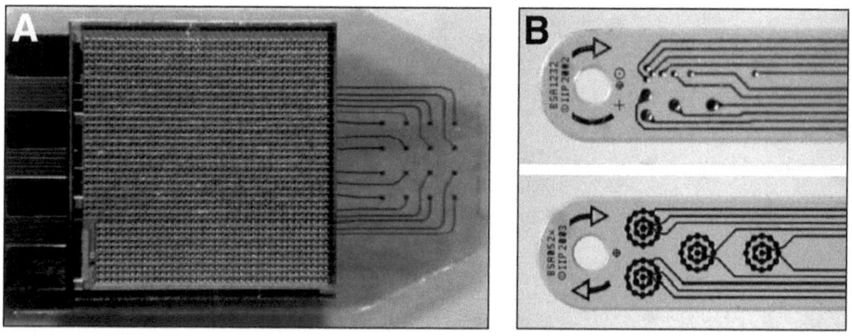

Fig. 6. Three examples of polyamide arrays/cables used for prototype retinal prostheses. **(A)** The Tubingen group device with array attached. **(B)** Two test versions of the IIP device *(3,6)*.

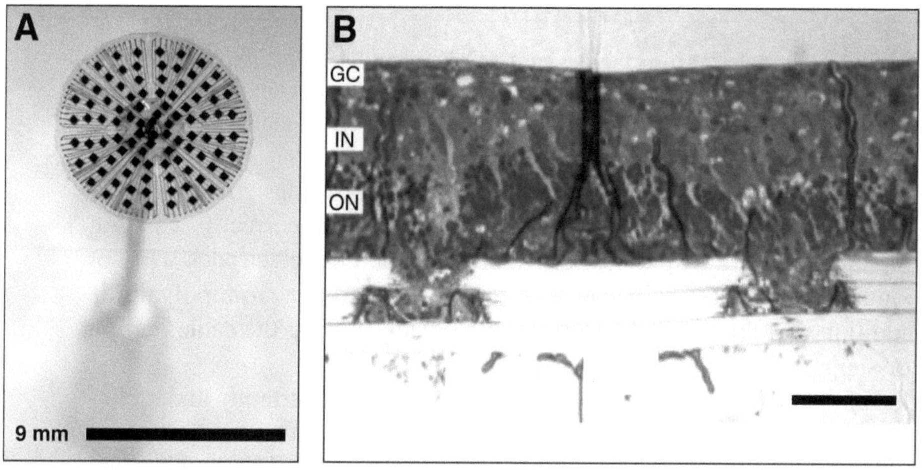

Fig. 7. MEMS type flexible devices with three-dimensional structures. **(A)** Boston VA group inflatable array with 100 electrodes supported by a cannula *(1)* (image courtesy of D. Shire). **(B)** Stanford/Palankar recessed holes design showing migrated retina cells (scale bar = 50 μm) *(2)*.

through a narrow incision. This interesting design allows for a large area of 9 mm diameter to be covered by a device that could be easier to implant because of the small 3 mm incision required to implant. In addition, inflatable channels within the device would gently press the arms of the device and the electrodes against the retina to improve contact. The Stanford/Palanker group has a different approach. They hope to bring the retina closer to the electrodes by creating chambers in a multilayer polyamide membrane. Their results from in vitro culture and in vivo 9 d implantations show that the outer retinal cells will migrate into the recesses of the subretinal device *(2)*. Their concept for a subretinal device would require lower current levels because of the close approximation of the electrodes and the cells. Some issues remain, such as how well the cells will survive in the recesses.

Two other flexible designs use a silicon rubber, or polydimethylsiloxane (PDMS), to create retinal electrode arrays (Fig. 8). Lawrence Livermore National Laboratory has a PDMS device, which is very similar to the flat polyamide devices with gold traces

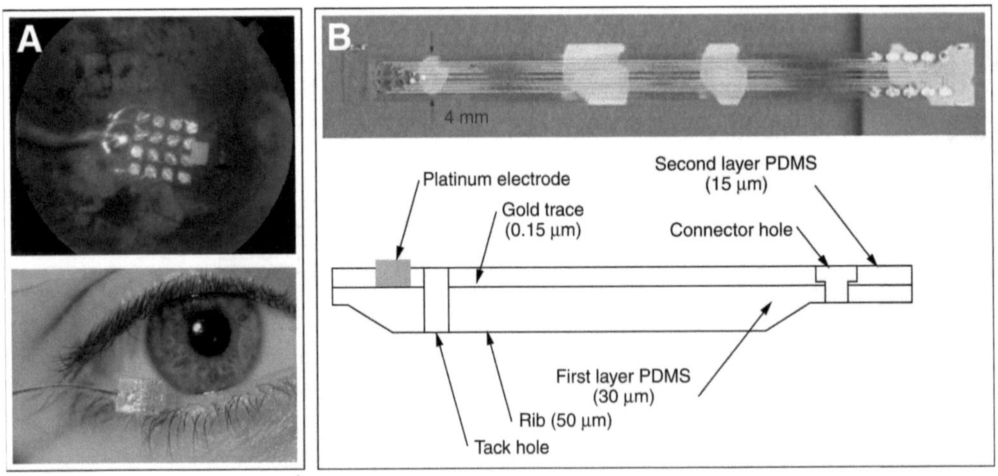

Fig. 8. Two flexible Polydimethylsiloxane devices. (**A**) Doheny/Second Sight 16 electrode device is a thick Polydimethylsiloxane array molded to have a curvature to fit the retina *(7,8)*. (**B**) A polyamide like electrode array developed at the Lawrence Livermore National Laboratory *(10)*.

described earlier *(10)*. As with most polyamide designs, this type of design allows for a cylindrical curvature, but not a spherical curvature to fit the retina correctly. The thin PDMS can be distorted toward a spherical curvature, but then the electrodes will not have regular contact with the retina because of folds. A difference from polyamide devices is the addition of a thicker PDMS "rib," which extends along the cable portion of the device for support.

Second, Sight, Inc has a device made of silicon rubber, which is currently under clinical testing by USC/Doheny Eye Institute *(13)*. This device has 16 electrodes, 500 μm in diameter embedded in the silicon rubber. There is a cable made of silicon rubber and continuous with the electrode pad, which contains the wire leading to the stimulation current source. Unlike flat polyamide or PDMS devices the silicon is molded into a spherical geometry to fit the surface of the retina and make good contact to reduce thresholds. However, while flexible the PDMS array is thick enough that it may not readily conform to the retina. The difficulty in this design is how to scale up to hundreds or thousands of electrodes with a method that does not use silicon fabrication methods or MEMS techniques.

MEMS/HYBRIDE DESIGN

Sandia National Laboratory has a unique design that attempts to meet the challenge of adapting to the spherical curvature of the retina *(4,5)*. Their design uses MEMS methods to create springs for electrode posts "float" on, Fig. 9. The springs are made from micromachined silicon. In the device reported in their 2005 conference paper *(19)*, they detail implantation of 5×5 mm^2 array with 81 electrodes. They reported a few minor problems with inserting the device into the scleral incision, getting all the electrodes in contact with the retina and damage to the electrode posts. They solved some problems with an insertion sleeve and continue to improve their device. The obvious advantage is the ability in theory to match the curvature of any retina if the difference

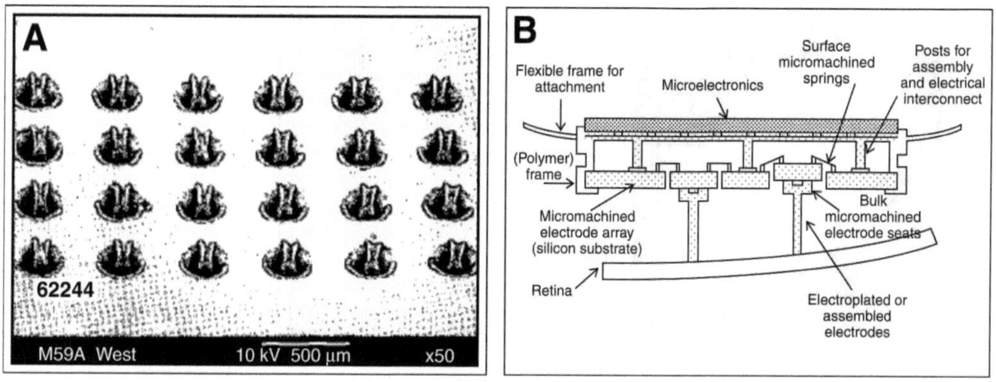

Fig. 9. SEM image of the Sandia device **(A)** and a cross-sectional diagram of the same **(B)**. In the SEM picture, the white bar is 500 µm. The center-to-center spacing of the electrodes is approx 500 µm *(4,5)*.

in surface elevation is less than 100 µm, the maximum deflection distance of the springs. A high-resolution device would require electrodes with a separation of 50 µm or less. The separation between their electrodes is 500 µm. As designed, this could be problematic, as higher resolutions will require closer spring spacing. The geometry of their device indicates that the deflection distance is proportional to the spring width and closer spacing would likely lead to reduced deflection distance. However, as the device uses standard MEMS fabrication methods manufacturing higher resolution devices is possible in theory.

RIGID DESIGNS

One of the earliest rigid designs is that of Optobionics, Inc. (Fig. 10). The prosthesis is a subretinal 25 µm thick silicon chip with 5000 photodiode electrodes. The device is flat and nonconforming to the shape of the retina. However, as the device is implanted subretinally the retina can be pushed toward a flat profile. The device is 2 mm in diameter, which results in a 30 µm distance from the edges of the device to the retina. This distance may be small enough that the increase in current and loss of resolution may be minimal. The Optobionics (Naperville, IL) evice suggests that smaller devices may be way to avoid conformal fit to the shape of the retina. Of course, the field of view for such a small device is limited.

Two groups working at Stanford have produced rigid devices with protruding electrode columns. An interesting rigid design, developed by the Stanford/Harris group *(20)*, involves an array of carbon nanotube (CNT) columns. The columns are grown on a rigid silicon substrate. The CNT electrode array is designed to penetrate the retina to allow for close contacted between the retinal neurons and the electrodes. The CNT columns have some degree of flexibility, which allows them to cause less damage as they penetrate the retina. The design makes some accommodation to the curvature of the retina in that electrodes at the edges of the device could penetrate slightly into the retina to allow center electrodes to get closer to the retina. The columns have been grown to heights of 100 µm, at a height:width aspect ratio of 4:1.

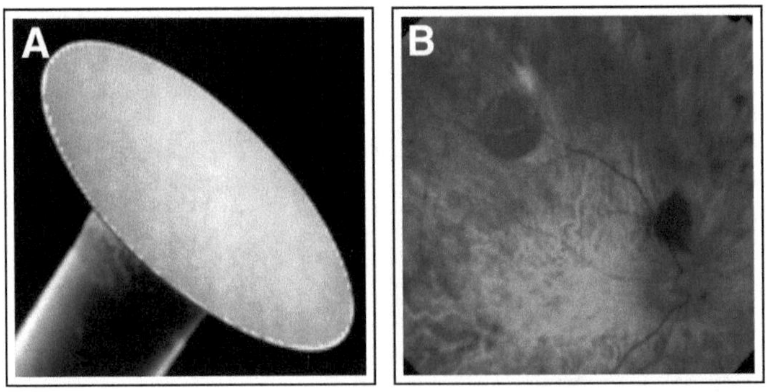

Fig. 10. The Optobionics retinal prosthesis. The 2 mm silicon disk **(A)** is placed behind the retina **(B)**.

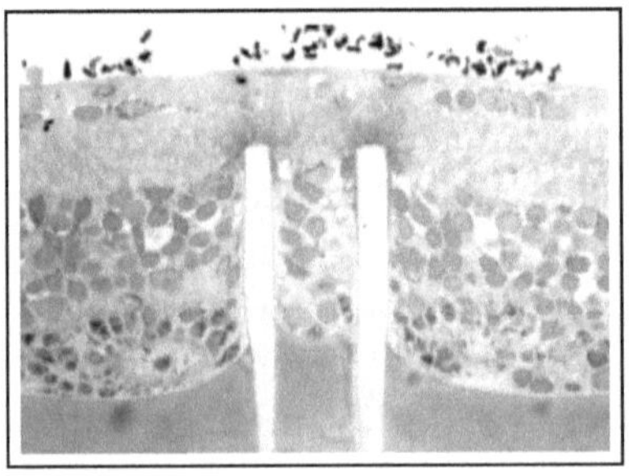

Fig. 11. Rigid columnar retinal prosthesis electrodes. The Stanford/Palanker group lithographically fabricated silicon columns 10 μm wide and 70 μm long penetrate from the photoreceptor cells to the inner nuclear layer *(2)*.

The Stanford/Palanker group has a concept for a pillared electrode array made on a lithographically fabricated silicon wafer (Fig. 11). The device would be 3 mm in diameter. They implanted a prototype device with 10 μm columns in a RCS rat for 15 d. Their experiment demonstrated the ability of the columns to penetrate (Fig. 11C).

As with the Optobionics device, the Naval Research Laboratory (NRL) device is a rigid device based on a silicon chip *(21–24)*. The 3200 electrode, NRL device is 6 mm long and 3 mm wide. The resulting gap between the silicon chip and the retina would be as great as 360 μm at the center. Unlike the Optobionics device, the NRL device is hybridized through indium bump bonds to an array of microwires imbedded in a glass matrix. In addition, the NRL device is an epiretinal device, which requires a conformal fit or a significant space between the center of the device and the retina will result. The microwire/glass array acts as a bridge material to span the gap between the flat silicon chip and the retina. It conducts the stimulus current from electrode to the retina through the microwires, but it insulates one electrode from the others with the glass matrix material.

Microchannel glass is fabricated using glass-drawing procedures that involve bundled stacks of composite glass fibers (Fig. 12 B,C). An acetic acid-etchable glass rod is inserted into a nonacetic acid-etchable glass tube. This pairing of dissimilar glasses is drawn at an elevated temperature into a fiber of smaller diameter. Several thousand of these fibers are

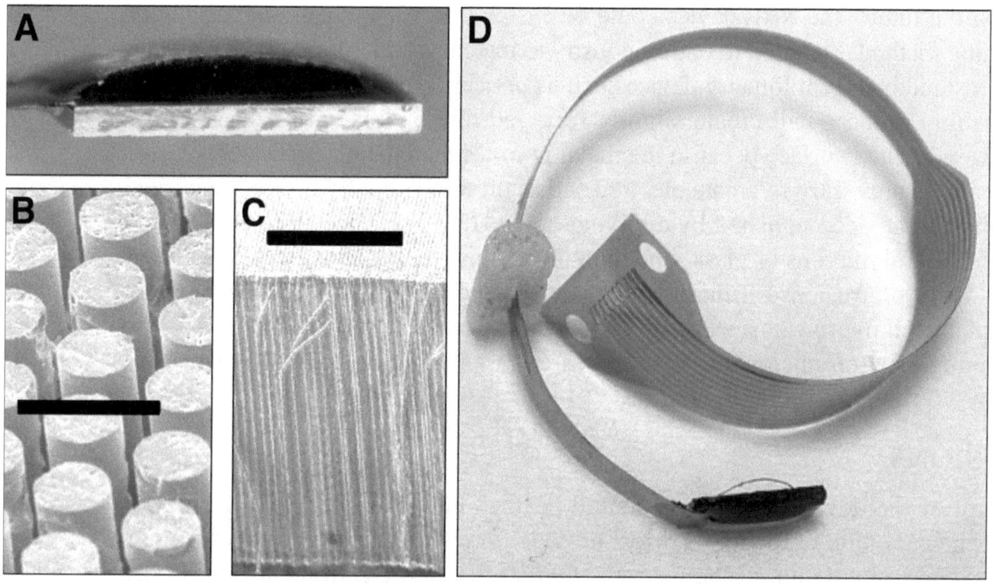

Fig. 12. The Naval Research Laboratory stimulation array. **(A)** The side view of a partially encapsulated device showing the curvature of the device surface. **(B)** A high magnification SEM of the 5.5 μm microwires (bar = 10 μm). **(C)** A low magnification SEM of the side of the microwire glass. A few microwire can be seen to be bent from the glass cleaving process (bar = 100 μm). **(D)** the completed retinal prosthesis. The surgical sponge on the polyamide cable is used to reduce leakage of saline during the acute implant procedure.

then cut and stacked in a hexagonal-close-packed arrangement, yielding a hexagonal-shaped bundle. This bundle is subsequently drawn at an elevated temperature, fusing the individual composite fibers together whereas reducing the overall bundle size. At this stage, the fibers are hexagonal-shaped and contain a fine structure of several 1000 μm sized (5–10 μm diameter) acid-etchable glass fibers in a hexagonal-close-packed pattern. These fibers are then bundled together in a 12-sided bundle and fused together at an elevated temperature. Standard microchannel plate glass is obtained from the combined boule at this point by slicing thin 200–1000 μm thick wafers, which are polished flat. Wafers are then placed in acetic acid to remove the acid-etchable glass. In this way, a glass with extremely uniform, parallel channels is obtained.

Next, electroplating is performed to fill the channels with a metal to create microwires. The high, approx 200:1, aspect ratio of the channel length to channel diameter makes it difficult to electroplate the metal. Thus, the channel filling metal must readily electroplate. One such metal is nickel, which is easier to electroplate than gold or platinum. As nickel is highly electro-chemically active, it is not biocompatible as an electrode. However, this has been found that it is possible to "cap" the nickel microwires with a layer of gold, platinum, or other suitable electrode material.

Microwire glass electrodes used for eventual in vivo testing with the stimulation array will have one side of the electrode curved to create a spherical surface to allow positioning of the high-density electrode array in extremely close approximation to the retinal tissue. The radius of curvature is nominally 12.7 mm to provide a conformal fit against the retina. In the future, the NRL device could be custom made with diamond turning or other polishing methods standard to optics industry to match any curvature of the patient's retina as measured by a profilometry device such as optical computed tomography.

The polishing process will create slightly recessed microwires with respect to the curved microwire glass surface because the metal is softer than the glass. Therefore, further processing is necessary to create electrodes that protrude slightly above the curved surface. This can be accomplished by applying a 5% HF/5% HN3 etch to the surface that removes several microns of glass. The duration of this etching step can be increased to provide longer protrusions of microwires, which reduce the electrode impedance by increasing the metal microwire surface area in contact with the saline environment of the eye. A scanning electron micrograph of microwire glass having channel and microwire diameters of 5.6 μm is shown in Fig. 12.

CONCLUSIONS

Almost all of the device designs presented in this chapter make some attempt to account for the curvature of the retina. Most provide for cylindrical curvature and a few provide for spherical curvature. The true advantage of curved electrode arrays is only apparent for high-resolution devices. A 300 μm distance from the retina is not important for 300 μm diameter electrodes. However, with the exception of the NRL and Optobionics arrays, there appear to be no fully functioning implantable devices with resolution high enough to take advantage of the curvature accommodations of the devices described in this chapter. As the Optobionics device offers little in the way of curvature accommodation and the NRL device has not been tested in humans at the time of this publication, no clear experimental evidence exists to say, which designs described in this chapter are the best. What is clear is that there exists a great diversity of approaches to meeting the challenge of the curved retina. With such diversity it is likely that an effective solution will be found and the challenge met.

REFERENCES

1. Shire DB, Rizzo JF. Microfabrication initiatives at the VA Center for Innovative Visual Rehabilitation (CIVR), in Second Joint Embs-Bmes Conf 2002, Vol. 1–3, Conf Proc—Bioeng—Integrative Methodologies, New Technologies, 2002:2399–2400.
2. Palanker DV, Vankov A, Huie P, et al. Design of a high-resolution optoelectronic retinal prosthesis. Invest Ophthalmol Visual Sci 2005;46:5278.
3. Gekeler F, Kobuch K, Schwahn HN, Stett A, Shinoda K, Zrenner E. Subretinal electrical stimulation of the rabbit retina with acutely implanted electrode arrays. Graefes Arch Clin Exp Ophthalmol 2004;242(7):587–596.
4. Okandan M, Wessendorf K, Christenson, T, et al. MEMS conformal electrode array for retinal implant, in Boston Transducers'03: Digest of Tech Pap, vols. 1 and 2 2003:1643–1646.
5. Okandan M, Wessendorf K, Christensen T, et al. Micromachined conformal electrode array for retinal prosthesis application, In: Becker H, Woias P, eds. Microfluidics, Biomems, and Medical Microsystems, 2003:45–51.
6. Hornig R, Laube T, Walter P, et al. A method and technical equipment for an acute human trial to evaluate retinal implant technology. J Neural Eng 2005;2(1):S129–S134.
7. Humayun MS, de Juan E Jr., Weiland JD, et al. Pattern electrical stimulation of the human retina. Vision Res 1999;39(15):2569–2576.

8. Humayun MS, Weiland JD, Fujii GY, et al. Visual perception in a blind subject with a chronic microelectronic retinal prosthesis. Vision Res 2003;43(24):2573–2581.
9. Palanker D, Vankov A, Huie P, Baccus S. Design of a high-resolution optoelectronic retinal prosthesis. J Neural Eng 2005;2(1):S105–S120.
10. Guven D, Weiland JD, Maghribi M, et al. Implantation of an inactive epiretinal poly (dimethyl siloxane) electrode array in dogs. Exp Eye Res 2006;82(1):81–90.
11. Rizzo JF 3rd, Wyatt J, Loewenstein J, Kelly S, Shire D. Perceptual efficacy of electrical stimulation of human retina with a microelectrode array during short-term surgical trials. Invest Ophthalmol Vis Sci 2003; 44(12):5362–5369.
12. Humayun MS, de Juan E Jr, Dagnelie G, Greenberg RJ, Propst RH, Phillips DH. Visual perception elicited by electrical stimulation of retina in blind humans. Arch Ophthalmol 1996;114(1):40–46.
13. Mahadevappa M, Weiland JD, Yanai D, et al. Perceptual Thresholds and Electrode Impedance in Three Retinal Prosthesis Subjects. IEEE Trans Neural Syst rehabil Eng: a publication of the IEEE Eng Med Biol Soc 2005;13(2):201–206.
14. Yamauchi Y, Enzman V, Franco LM, et al. Subretinal placement of the microelectrode array is associated with a low threshold for electrical stimualtion. Invest Ophthalmol Visual Sci 2005;46:1511.
15. Margalit E, Maia M, Weiland JD, et al. Retinal prosthesis for the blind. Surv Ophthalmol, 2002;47(4):335–356.
16. Stieglitz T, Schuettler M, Koch KP. Implantable biomedical microsystems for neural prostheses. IEEE Eng Med Biol Mag 2005;24(5):58–65.
17. Gekeler F, Zrenner E. Status of the subretinal implant project An overview. Ophthalmologe 2005;102(10):941–949.
18. Yamauchi Y, Franco LM, Jackson DJ, et al. Comparison of electrically evoked cortical potential thresholds generated with subretinal or suprachoroidal placement of a microelectrode array in the rabbit. J Neural Eng 2005;2(1):S48–S56.
19. Ameri H, Guven D, Freda R, et al. Surgical implantation of epiretinal prosthesis with spring-mounted electrodes. Invest Ophthalmol Visual Sci 2005;46:1484.
20. Wang K, Fishman HA, Dai H, Harris JS. Fabrication and Characterization of a Carbon Nanotube Microelectrode Array for Retinal Prostheses.
21. Johnson L, Perkins FK, O'Hearn T, et al. Electrical stimulation of isolated retina with microwire glass electrodes. J Neurosci Methods 2004;137(2):265–273.
22. Scribner D, Humayun M, Justus B, et al. Intraocular retinal prosthesis test device, in Proc 23rd Annu Int Conf Ieee Eng Med Biol Soc, Vol. 1–4—Building New Bridges at the Frontiers Eng Med 2001: 3430–3435.
23. Scribner D, Humayun M, Justus B, et al. Towards a retinal prosthesis for the blind; advanced microelectronics combined with a nanochannel glass electrode array, In: Peckerar MC, Postek MT, eds. Nanostructure Sci Metrol Technol 2001:239–244.
24. Scribner D, Johnson L, Klein R, et al. A retinal prosthesis device based on an 80 × 40 hybrid microelectronic-microwire glass array, in Proc Ieee 2003 Custom Integrated Circuits Conf 2003:517–520.

10
Circuit Designs That Model the Properties of the Outer and Inner Retina

Kareem A. Zaghloul, MD, PhD and Kwabena Boahen, PhD

CONTENTS

SILICON MODELS

One goal of understanding neural systems is to develop prosthetic devices that can someday be used to replace lesioned neural tissue. For such prosthesis to be practical, the device must perform these computations as efficiently as, and at a physical scale comparable with the lesioned network, and should adapt its properties over time, independent of external control. The approach to design a successful prosthesis that faithfully replicates the computations performed by a neural circuit is based on a detailed understanding of that circuit's anatomic connections and functional computations.

The retina, one of the best studied neural systems is a complex piece of biological wetware designed to signal the onset or offset of visual stimuli in a sustained or transient fashion *(1)*. To encode its signals into spike patterns for transmission to higher processing centers, the retina has evolved intricate neuronal circuits that capture information, efficiently contained within natural scenes *(2)*. This visual preprocessing, realized by the retina, occurs in two stages, the outer retina and the inner retina. Each local retinal microcircuit plays a specific role in the retina's function, and neurophysiologists have extracted a wealth of data characterizing how its constituent cell types contribute to visual processing. These physiological functions can be replicated in artificial systems by emulating their underlying synaptic interactions.

From: *Ophthalmology Research: Visual Prosthesis and Ophthalmic Devices: New Hope in Sight*
Edited by: J. Tombran-Tink, C. Barnstable, and J. F. Rizzo © Humana Press Inc., Totowa, NJ

Present attempts to engineer a viable retinal prosthesis have focused on the significant problem of efficient electrical stimulation of neurons along the visual pathway *(3,4)*. Microelectrode arrays, implanted epiretinally or subretinally, evoke phosphenes in patients with visual loss (because of outer retinal degeneration) by relying on electrical stimulation of the remaining retinal cells to dictate firing patterns *(5,6)*. Whereas the epiretinal approach relies on an external camera to capture visual information and on an external processor to recreate retinal computation, subretinal devices use photodiodes embedded in the electrode array to locally transduce light into stimulating current. Cortical visual prostheses address disease processes affecting structures postsynaptic to the outer retina *(7,8)*. They are similar to epiretinal prostheses in that they also depend on external devices to capture and process visual information, but they must fully recreate thalamic function as well as retinal function.

Whereas the emphasis on electrical stimulation technology is important in addressing the difficult problem of interfacing with the nervous system, a fully implantable retinal prosthesis would ideally capture all of the functions performed by the mammalian retina in one autonomous device. These neural computations can be performed at an energy efficiency and physical scale comparable with biology by morphing neural circuits into electronic circuits *(9)*. Micron-sized transistors function as excitatory or inhibitory synapses or as gap junctions, thereby recreating the synaptic organization of the retina at a similar physical scale. The time-scale and energy dissipation can be matched as well by operating these transistors in the subthreshold region, where they conduct nanoamperes or even picoamperes, just like small populations of ion channels do. Furthermore, as millions of transistors can be fabricated on a thumb-nail-sized piece of silicon using very large scale integration technology, this neuromorphic approach offers a fully implantable solution for neural prostheses. This implementation efficiency is translated to a higher ability to explore model parameters to further understand the underlying biological system and, by communicating with other neuromorphic chips, a higher ability to replicate more complicated neural systems.

The first effort to morph the retina into silicon, though widely acclaimed, suffered from several shortcomings. First, only outer retina circuitry was morphed: the cones, horizontal cells (HC), and bipolar cells (BC) *(10)*. Second, a logarithmic photoreceptor (cf., cone) was used to capture a wide intensity range, but this degraded the signal-to-noise ratio by attenuating large amplitudes (i.e., signal) whereas leaving small amplitudes (i.e., noise) unchanged. Third, the spatiotemporal average (cf., HC) was subtracted to obtain contrast (cf., bipolar signal) or more precisely, the logarithm of contrast, but this made the signal-to-noise ratio even worse. Subsequent efforts *(11)* overcame these limitations by modulating synaptic strengths locally to control sensitivity, and by including the cone-to-cone gap junctions to attenuate noise. But they still omitted the inner retina, which contains upwards of 44 cell types *(12)*.

The most recent effort to model retinal processing in silicon incorporated outer retina circuitry as well as bipolar and amacrine cell interactions in the inner retina *(13)*. This outer retina circuit took the difference between the photoreceptor signal and its spatiotemporal average, computed by a network of coupled lateral elements HCs, through negative (inhibitory) feedback. Cone-coupling in this model attenuated high-frequency noise to realize a spatial bandpass filter and dynamic range was extended by implementing local automatic gain control. Furthermore, this model used HC activity to boost cone to HC excitation *(13)*, which eliminated the luminance-dependent receptive-field expansion and temporal instability that plagued previous efforts to modulate synaptic strengths locally *(14)*. This gain boosting mechanism has some physiological basis as glutamate release from cones is modulated by HC hemichannels *(15)*, and may be enhanced further by HC autofeedback *(16)*. The outer retina design adopts this approach.

The bipolar and amacrine cell interactions, introduced by this earlier design, represented a model for inner retina processing that much like the circuit discussed here, attempted to capture temporal adaptation through adjustment of amacrine cell feedback inhibition. However, this earlier design did not include the retina's complementary push-pull architecture *(1)*. Hence, at low frequency stimulation, baseline direct current (DC) levels tended to rise, and thus, modulation of the feedback loop gain was not realized as intended *(13)*. Furthermore, this previous implementation did not replicate synaptic interactions at the ganglion cell level, and thus, did not faithfully capture the behavior of the retina's major output pathways.

A silicon retina has been developed modeled on neural circuitry in both the outer and inner retina that addresses the design flaws in earlier designs *(13)* and extends them to include ganglion-cell level synaptic interactions. The model is based on the functional architecture of the mammalian retina and on physiological studies that have characterized the computations performed by the retina. By capturing both outer and inner retina circuitry using single-transistor synapses, the silicon retina, which was built, passes only an intermediate range of frequencies. It attenuates redundant low spatiotemporal frequencies and rejects noisy high frequencies, much like the retina does. And by modulating the strengths of its single-transistor synapses locally, the device adapts to luminance and to contrast. It responds faster, but more transiently as contrast increases, much like the retina does. This contrast gain control arises from a new inner retina circuit design that assigns specific roles to anatomically identified microcircuits in the mammalian retina.

This silicon retina outputs spike trains that capture the behavior of ON- and OFF-center *(17)* versions of wide-field transient and narrow-field sustained ganglion cells (GC) *(18)*, which provide ninety percent of the primate retina's optic nerve fibers *(19)*. And, more significant for a prosthetic application, these are the four major types that project through thalamus, to primary visual cortex. Furthermore, the silicon circuit is constructed at a scale comparable with the human retina and uses under a tenth of a watt, thereby satisfying the requirements of a fully implantable prosthesis.

This chapter describes the silicon implementation of the outer and inner retina circuits. It is organized as follows: In outer retina section, the outer retina model and its silicon implementation are presented. In inner retina section, the novel model for the inner retina and its silicon implementation are presented. In silicon chip layout section, the architecture of the silicon chip layout is presented. In silicon retina response section, the responses of the silicon retina to physiological stimuli are described and are compared with those of the mammalian retina. Finally, conclusion section concludes the article.

OUTER RETINA

The model for outer retinal circuitry is based on identified synaptic interactions and local microcircuits previously described in the literature, obtained using histological and physiological techniques. After constructing this model of the outer retina's synaptic connections, it is used as a blueprint with which the silicon retina is assembled.

Modeling the Outer Retina

The model for the outer retina (Fig. 1A) is designed to realize luminance adaptation by adjusting synaptic strengths locally. Photocurrents from the cone outer segments (CO) in the model drive a network of cone terminals, which subsequently, excite a network of HCs. Because they are coupled through gap junctions, these HCs compute the local average intensity in the model. This signal is used to modulate cone-to-cone coupling strength as well as the cone's membrane conductance (shunting inhibition). Using the local intensity signal to adjust these two synaptic strengths makes the cone terminal's sensitivity inversely proportional to luminance, whereas preventing the changes in spatial frequency tuning that plagued previous attempts at light adaptation *(11)*. To compensate for the resulting signal attenuation at the cone terminal, the HC's local intensity signal is also used to modulate the cone-to-HC synaptic strength. This autofeedback mechanism, whereby the HC regulates its own input, is similar to that found in the retina *(15,16)*.

From the synaptic interactions, the block diagram was derived for the outer retina shown in Fig. 1B by modeling both the cone and HC networks as spatial lowpass filters. The system level equations can be derived that describe how CO activity determines HC and cone terminal activity, represented by i_{hc} and i_{ct}, respectively, from this block diagram. These equations, reproduced from ref. *13* are as follows:

$$i_{hc}(\rho) = \left(\frac{A}{(l_c^2\rho^2 + 1)(l_h^2\rho^2 + 1) + \frac{A}{B}} \right) \frac{i_{co}}{B}$$

$$i_{ct}(\rho) = \left(\frac{(l_h^2\rho^2 + 1)}{(l_c^2\rho^2 + 1)(l_h^2\rho^2 + 1) + \frac{A}{B}} \right) \frac{i_{co}}{B}$$

where i_{co} represents CO activity, excited by photocurrents, l_c and l_h are the cone- and horizontal-network space-constants, respectively, B is the attenuation from the CO to the cone terminal, A is the amplification from cone terminal to HC, and ρ is spatial frequency. HCs have stronger coupling in the model (i.e., l_h is larger than l_c), causing their spatial lowpass filter to attenuate lower spatial frequencies. Thus, HCs lowpass filter the signal whereas cone terminals bandpass filter it, as shown in Fig. 1C, contributing to the GCs' spatial frequency tuning *(20)*.

To realize local automatic gain control in the model, HC activity is set proportional to intensity and this activity is used to modulate CO to cone terminal attenuation,

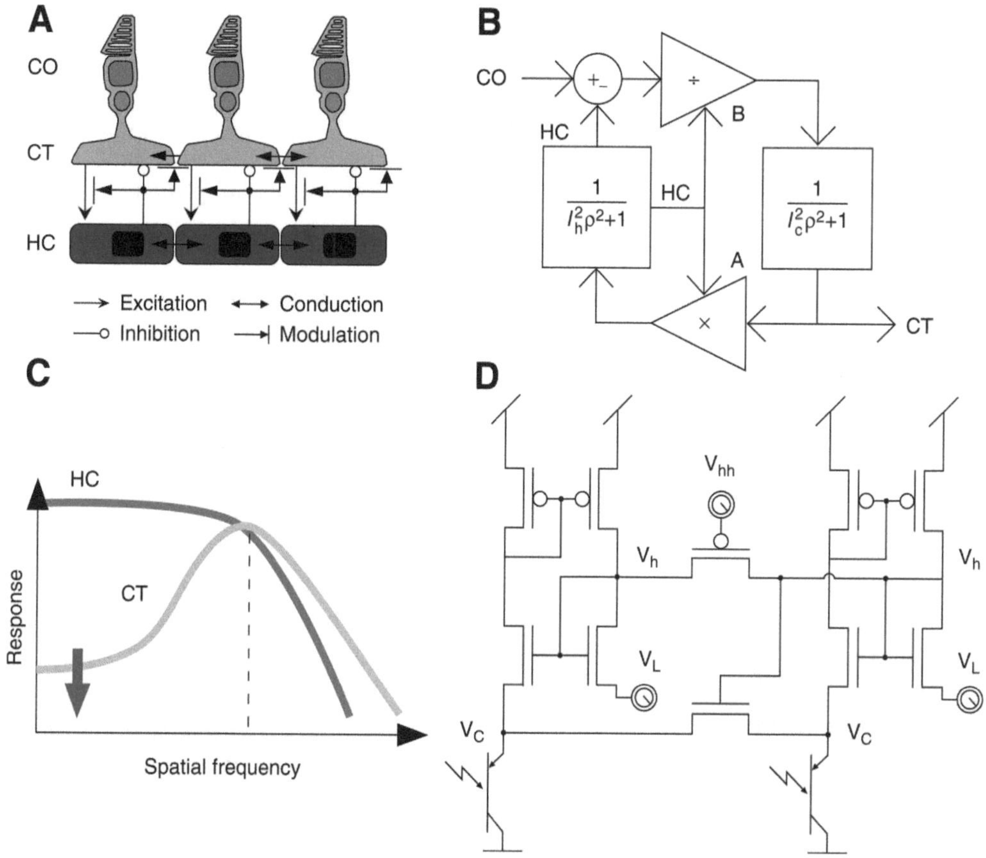

Fig. 1. Modeling the outer retina. **(A)** Neural circuit: cone terminals (CT) receive a signal that is proportional to incident light intensity from the cone outer segment (CO) and from neighboring cones, through gap junctions, and excite horizontal cells (HC). HCs spread their input laterally through gap junctions, provide shunting inhibition onto CT, and modulate cone coupling and cone excitation. **(B)** System diagram: signals travel from the CO to the cone terminal (CT) and on to the HC network, which provides negative feedback. Both cone terminals and HCs form networks, connected through gap junctions that are governed by their respective space constants, l_c and l_h. Excitation of HCs by cone terminals is modulated by the HCs, which also modulate the attenuation from the CO to the cone terminal together with modulation of cone gap junctions to keep l_c constant. These compensatory interactions realize local automatic gain control in the cone terminal and keep receptive field size invariant. **(C)** Frequency responses: both HCs and cone terminals (CT) lowpass filter input signals, but because of the HC network's larger space constant, HC inhibition eliminates low frequency signals, yielding a spatially bandpass response in the cone terminal. **(D)** Outer retina circuitry. A phototransistor draws current through an nMOS transistor whose source is tied to V_c, decreases in which represent increased cone terminal activity, and whose gate is tied to V_h, increases in which represent increased HC activity. This transistor passes a current proportional to the product of cone terminal and HC activity, thus, modeling shunting inhibition from HCs to cones. In addition, this current, mirrored through pMOS transistors, dumps charge on the HC node, V_h, modeling cone terminal excitation of the HCs, and its modulation by HCs. V_L, a global bias set externally sets the mean level of V_c. V_{hh}, another external, bias sets the strength of the HC gap junctions. The strength of those between cones are modulated locally by V_h *(30)*.

B, by changing cone-to-cone conductance. This modulation adapts cone sensitivity to different light intensities *(14)*. Furthermore, to overcome the expansion in receptive-field size that this modulation caused in earlier designs, HC modulation of cone gap-junctions is complemented with HC modulation of cone leakage conductance, through shunting inhibition, making l_c independent of luminance. The change in loop-gain with HC modulation of cone excitation was also compensated by keeping *A*, the amplification from cone terminal to HC, proportional to *B*, thus, fixing the peak spatial frequency of cone terminal response *(21)*.

Cone terminal activity in the model is primarily determined by spatial contrast, saturates when increasing signal fluctuations cause this ratio to become large, and is entirely independent of absolute luminance. From the outer retina model's system equations, how the silicon model's cone terminal activity depends on its cone and horizontal space constants and on contrast *(21)* was derived:

$$i_{\mathrm{CT}} = \frac{2r}{r + 2 - 1/r + 2c} c$$

where i_{CT} represents cone terminal activity and c represents stimulus contrast; $r = l_h/l_c$ is the ratio of the HC space constant to the cone space constant. This behavior is remarkably similar to the mammalian retina. Physiologists have found that cone responses as a function of contrast—intensity of light stimulation relative to a background—are described by a simple equation:

$$V = \frac{IV_m}{I + \sigma}$$

where V is the peak amplitude of the cone response produced by a given level of stimulating light intensity, I. V_m is the maximum response and σ is the background intensity. The response reaches half of the maximum when $I = \sigma$ *(22)*. This adaptive behavior is preserved across five decades of background intensities. Cone responses obtained from the model, where c corresponds to I/σ, behave similarly when l_h is comparable with l_c *(21)*.

Morphing the Outer Retina Into Silicon

The retinal models were morphed into a silicon chip by replacing each synapse or gap-junction in the model with a transistor. One of its terminals is connected to the presynaptic node, another to the postsynaptic node, and a third to the modulatory node. By permuting these assignments, excitation, inhibition, and conduction are realized, all of which are under modulatory control.

Modulation was implemented by exploiting the exponential I(V) relationship of the MOS (metal-oxide-semiconductor) transistor. In the subthreshold regime, the current from the drain terminal to the source terminal is the superposition of a forward component that decreases exponentially with the source voltage (V_s) and a reverse component that decreases similarly with the drain voltage (V_d); both components increase exponentially with the gate voltage (V_g). That is, $I_{\mathrm{ds}} = I_0 e^{\kappa V_g}(e^{-V_s} - e^{-V_d})$ where $\kappa \approx 0.7$ is a non-ideality factor; voltages are in units of $U_T = 25$ mV, at 25°C. This equation describes the n-type device; voltage and current signs are reversed for a p-type *(23)*. Hence, the transistor converts voltage to current exponentially and converts current back to voltage logarithmically. By adding a voltage offset representing the log of the modulation factor, the current can be multiplied by a constant, thereby modulating the synaptic strength. Synaptic strengths are modulated locally, within the chip, except for a small number of biases globally controlled by the user. Current-mirrors are added to reverse the direction of current when necessary.

Morphing the model for the outer retina yielded the electronic circuit shown in Fig. 1D. Omitting adaptation in the phototransduction cascade, photocurrents linearly proportional to luminance discharge the cone terminal node, V_c, which is defined as an increase in cone terminal activity. This drop in V_c produces a current that excites the HC network through an nMOS transistor followed by a pMOS current-mirror. This excitatory current is modulated by HC activity, represented by V_h, and increases as V_h increases to realize autofeedback. But this increased current also releases more charge on to V_c, thereby realizing horizontal-cell inhibition of cone terminal activity. Thus, a single transistor implements two distinct synaptic interactions, one excitatory, the other inhibitory. Cone nodes are electrically coupled to their six nearest neighbors through nMOS transistors whose gates are controlled locally by HCs, implementing the model of cone gap-junction modulation. HCs also communicate with one another, through pMOS transistors, but this coupling is controlled by an externally applied voltage (V_{hh}).

Cone terminals in the model drive two types of BCs that rectify the cone signal into ON and OFF channels, thus, reproducing the complementary signaling scheme found in the mammalian retina *(17)*. the cone signal using push–pull pMOS current mirrors is rectified, not shown here, but described in detail in *(21)*. Briefly, the current generated by cone terminal activity is compared with a reference current set at the mean. Subtracting these currents by mirroring them on to one another removes the baseline; the difference drives the ON bipolar cell if it is positive and drives the OFF bipolar cell if it is negative *(21)*. Thus, the bipolar circuitry divides signals into ON and OFF channels.

INNER RETINA

Adopting the same approach, that was taken with the outer retina, the model for the inner retina is based on identified synaptic interactions and local microcircuits previously described in the literature, obtained using histological and physiological techniques. After constructing this model of the inner retina's synaptic connections, it is used as a blueprint with which the silicon retina is assembled.

Modeling the Inner Retina

The model of the inner retina, which realizes lowpass and highpass temporal filtering, adapts temporal dynamics to input frequency and to contrast, and drives ganglion cell responses, is shown in Fig. 2A. BCs from the outer retina excite amacrine cells with either narrow or wide fields. To ensure that only the ON or OFF channel is active at any time, bipolar and amacrine cells receive inhibition from the complementary channel in the model, similar to vertical inhibition found between the inner-plexiform

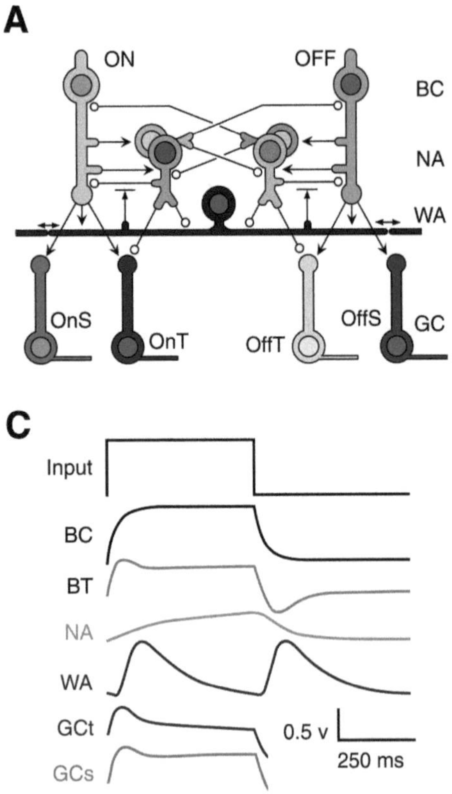

Fig. 2. Inner retina model and circuit. **(A)** Inner retina synaptic interactions: ON and OFF bipolar cells (BC) relay cone signals to ganglion cells (GC), and excite narrow- and wide-field amacrine cells (NA, WA). Narrow-field amacrine cells inhibit BC, wide-field amacrine cells, and transient GCs; their inhibition onto wide-field amacrine cells is shunting. Wide-field amacrine cells modulate narrow-field amacrine cell presynaptic inhibition and spread their signals laterally through gap junctions. BC also excite local interneurons that inhibit complementary bipolar cell and narrow-field amacrine cells. Four types of GCs signal the onset or offset of visual stimuli in a sustained (OnS, OffS) or transient (OnT, OffT) fashion. **(B)** System diagram: narrow-field amacrine cell (NA) signals represent a lowpass filtered version of bipolar terminal (BT) signals and provide negative feedback on to the bipolar cell (BC). The wide-field amacrine cell (WA) network modulates the gain of NA feedback (X). WA receives full-wave rectified excitation from BT and full-wave rectified inhibition from modulated NA (double arrows). BT drives sustained GCs directly whereas the difference between BT and NA drives transient ganglion cells (GCt). **(C)** Numerical solution to inner retina model with a unit step input of 1V. Traces show 1 s of ON cell responses for BC, BT, NA, GCt, and WA. Outer retina time constant, τ_o, is 96 ms; τ_{na} is 1 s. **(D)** Wide field amacrine cell modulation: bipolar terminal (BT) activity (I_{bt}) excites a network of wide-field amacrine cells (WA) through a current mirror; it also excites the narrow-field amacrine cell (NA, excitation circuitry not shown). Wide-field amacrine cell activity modulates the strength of narrow-field amacrine cell feedback inhibition on to the bipolar terminal, subtracting a current wI_{na} from the bipolar cell's excitatory input I_{bc}. The same current is also subtracted from the wide-field amacrine cell's excitatory input, I_{bt}, thereby inhibiting it. V_{bq} controls the quiescent current supplied to the inner retina by the bipolar terminal and V_{aa} controls the extent of gap-junction coupling in the wide-field amacrine cell network *(30)*.

layer's ON or OFF laminae *(24)*. The signal is made at the bipolar cell's terminal more transient (highpass filtered) than its cone input by applying sustained (lowpass filtered) inhibition from the narrow-field amacrine-cell *(25)*. The bipolar cell terminal in the model also excites two types of GCs, which is called transient and sustained. In transient GCs, feedforward inhibition from the narrow-field amacrine cells cancels residual sustained excitation from the bipolar terminal, similar to the synaptic complex found in mammalian retina *(26)*.

Contrast gain control in the inner retina model is determined by the modulatory effects of wide-field amacrine cell activity. Wide-field amacrine cell activity represents the local measure of contrast, weighed on neighboring spatial locations. These cells are excited by both ON and OFF BCs and inhibited by both ON and OFF narrow-field amacrine cells in the model, similar to ON–OFF amacrine cells found in the retina *(27)*, and are coupled together through gap-junctions to form a wide-field amacrine cell network. By modulating their own inhibitory inputs as well as narrow-field amacrine cell inhibition at bipolar terminals, the wide-field amacrine cells compute temporal contrast. That is, their activity reflects the ratio between contrast fluctuations (highpass signal) and average contrast (lowpass signal). As this temporal contrast increases, their modulatory activity increases, the net effect of which is to make the GCs respond more quickly and more transiently. There is also an overall decrease in sensitivity because of the less sustained nature of the response. This adaptation captures properties of contrast gain control in the mammalian retina *(21,28)*.

From the synaptic interactions of Fig. 2A, the block diagram was derived for the inner retina, shown in Fig. 2B, by modeling the narrow-field amacrine cell as a low-pass filter. Responses of the different inner retina cell types in this model to a step input are shown in Fig. 2C. Bipolar cell activity is a low-pass filtered version of light input to the outer retina. Increase in bipolar cell activity causes an increase at the bipolar terminal and slower increase in the narrow-field amacrine cell. The difference between bipolar terminal activity and gain-modulated narrow-field amacrine cell activity determines wide-field amacrine cell activity, which, in turn, sets the gain of narrow-field amacrine cell feedback inhibition on to the bipolar terminal and on to the wide-field amacrine cell. Thus, after a step input, bipolar terminal activity initially rises, but narrow-field amacrine cell inhibition, setting in later, attenuates this rise until bipolar terminal activity is equaled by gain-modulated narrow-field amacrine cell activity. Wide-field amacrine cell activity rises above its baseline value of unity at both onset and offset, driven by full-wave rectified input from ON and OFF bipolar terminals. The bipolar terminal drives the sustained ganglion cell, which responds for the duration of the step, whereas the difference between bipolar terminal and narrow-field amacrine cell activity drives the transient ganglion cell, which decays to zero.

From the block diagram in Fig. 2B, the system level equations can be derived for narrow-field amacrine and bipolar cell acitivity with the help of the Laplace transform:

$$i_{\text{na}} = \frac{g\varepsilon}{\tau_A s+1} i_{\text{bc}}, i_{\text{bt}} = \frac{\tau_A s+\varepsilon}{\tau_A s+1} i_{\text{bc}}$$

where g is the gain of the excitation from the bipolar terminal to the narrow-field amacrine cell, and where

$$\tau_A \equiv \varepsilon \tau_{\text{na}}, \varepsilon \equiv \frac{1}{1 + wg}$$

τ_{na} is the time constant of the narrow-field amacrine cell and w represents wide-field amacrine cell activity, which determines feedback strength. From the equations, it can be seen that the bipolar terminal highpass filters and the narrow-field amacrine cell lowpass filters the bipolar cell signal; they have the same corner frequency, $1/\tau_A$. This closed-loop time-constant, τ_A, depends on the loop gain wg, and therefore on wide-field amacrine cell activity. For example, stimulating the inner retina with a high frequency would provide more bipolar terminal excitation (highpass response) than narrow-field amacrine cell inhibition (lowpass response) to the wide-field amacrine cell network. Wide-field amacrine cell activity, and hence w, would subsequently rise, reducing the closed-loop time-constant τ_A, until the corner frequency $1/\tau_A$ reaches a point where bipolar terminal excitation equaled narrow-field amacrine cell inhibition. This drop in τ_A, accompanied by a similar drop in ε, will also reduce overall sensitivity and advance the phase of the response.

The system behavior governed by these equations is remarkably similar to the contrast gain control model proposed by Victor *(29)*, which accounts for the compression of retinal responses in amplitude and in time with increasing contrast. Victor proposed a model for the inner retina whose highpass filter's time-constant, T_S, is determined by a "neural measure of contrast," c. The governing equation is:

$$T_s = \frac{T_O}{1 + \dfrac{c}{c_{1/2}}}$$

This model's time-constant depends on contrast in the same way that the model's time constant depends on wide-field amacrine cell activity, where Victor's T_O is similar to the τ_{na} and where Victor's ratio $c/c_{1/2}$ is represented by how much wide-field amacrine cell activity increases above its value at DC in the model. As this activity is sensitive to temporal contrast *(21)*, it is proposed that wide-field amacrine cells are the anatomical substrate that computes Victor's neural measure of contrast.

In the model, the loop gain wg is set by the local *temporal contrast*. Wide-field amacrine cell activity reflects inputs from bipolar terminals and narrow-field amacrine cells weighted across spatial locations, so these pooled excitatory and inhibitory inputs should balance when the system is properly adapted:

$$w \updownarrow i_{\text{na}} \updownarrow = \updownarrow i_{\text{bt}} \updownarrow + i_{\text{surr}}$$
$$\Rightarrow w = (\updownarrow i_{\text{bt}} \updownarrow + i_{\text{surr}}) / \updownarrow i_{\text{na}} \updownarrow$$

where i_{surr} is defined as the gap-junction current, resulting from spatial differences in wide-field amacrine cell activity w. $\updownarrow i_{\text{bt}} \updownarrow$ and $\updownarrow i_{\text{na}} \updownarrow$ are full-wave rectified versions of i_{bt} and i_{na}, computed by summing ON and OFF signals. If all different phases are pooled spatially, these full-wave rectified signals will not fluctuate and will be proportional to amplitude. And if i_{surr}, is ignored w will simply be $|i_{bt}|/|i_{na}|$, yielding a temporal measure of contrast, because it is the ratio of a temporal difference (highpass signal, i_{bt}) and a temporal average (lowpass signal, i_{na}). At DC, this ratio is equal to $1/g$; hence the loop gain is unity and the DC gain from the bipolar cell to its terminal is $\varepsilon = 1/2$.

Ganglion cell responses in the inner retina model are derived from the bipolar terminal and narrow-field amacrine cell signals. Specifically, bipolar terminal signals directly excite both sustained and transient types of GCs, but transient cells receive feedforward narrow-field amacrine cell inhibition as well. The system equations determining sustained and transient ganglion cell responses (GCs and GCt), derived from the equations above, as a function of the input to the inner retina, i_{bc}, are as follows:

$$i_{GCs} = \frac{j\tau_A\omega + \varepsilon}{j\tau_A\omega + 1} i_{bc}$$

$$i_{GCt} = \frac{j\tau_A\omega + \varepsilon(1-g)}{j\tau_A\omega + 1} i_{bc}$$

where ω is temporal frequency (and $j = \sqrt{-1}$). When bipolar terminal to narrow-field amacrine cell excitation has unity gain ($g = 1$), feedforward inhibition causes a purely high-pass (transient) response in the transient ganglion cell whereas the sustained ganglion cell retains a low-pass (sustained) component. With a small loop-gain, wg, the DC gain, ε, approaches 1/2 and the bipolar terminal and sustained ganglion cell response becomes all-pass. However, as the loop gain increases, ε decreases and the response becomes highpass. The change in ε with loop gain is matched in both bipolar terminals and narrow-field amacrine cells, and so taking the difference between these two signals always cancels the sustained component in the transient ganglion cell *(21)*. Thus, the transient ganglion cell produces a purely highpass response irrespective of the system's loop gain.

Morphing the Inner Retina Into Silicon

The model for the inner retina is implemented in part by the electronic circuit shown in Fig. 2D. Wide-field amacrine cell modulation of narrow-field amacrine cell inhibition is realized by applying voltages representing wide- and narrow-field amacrine activity to a transistor's source and gate terminals, respectively. This transistor drains current from the node that represents bipolar terminal activity, implementing presynaptic inhibition by the narrow-field amacrine cell. It also sources current onto the node that represents wide-field amacrine activity, charging up that voltage, V_{wa}. This increase corresponds to inhibition of wide-field amacrine activity since, as V_{wa} increases, the strength of narrow-field amacrine inhibition *(w)* decreases. Conversely, as V_{wa} decreases, the strength of this inhibition increases. A pMOS transistor and a current-mirror realize excitation of the wide-field amacrine by the bipolar terminal. Wide-field amacrine cell nodes are coupled to one another through nMOS transistors gated by V_{aa}.

The circuit diagram of Fig. 2D represents only one of the inner retina circuit's complementary halves and does not show bipolar terminal to narrow-field amacrine excitation, lowpass filtering in the narrow-field amacrine cell, or push-pull inhibition. In the complete circuit, ON and OFF signals from bipolar cell circuitry drive either half of the inner retina circuit. Excitatory current from the bipolar terminal excites the narrow-field amacrine cell node on one side of the circuit whereas also inhibiting the narrow-field amacrine cell node on the complementary side, thus, realizing push-pull inhibition. To realize linear lowpass filtering, these inputs are divided by the sum of ON and OFF narrow-field amacrine cell activity, which compensates for the exponential voltage-current relationship in the subthreshold regime *(30)*. ON and OFF bipolar terminal and narrow-field amacrine cell signals converge on to the wide-field amacrine cell network, implementing full-wave rectification. Finally, signals from the bipolar terminal and narrow-field amacrine cells are used to drive ON and OFF transient and sustained GCs.

Because analog signals cannot be relayed over long distances, retinal GCs use spikes to communicate with central structures. Similarly, each ganglion cell in the chip array converts the current it receives from the inner retina circuit to spikes. The spike generating circuit (not shown) exhibits spike-rate adaptation through Ca^{++} activated K^{+} channel analogs, modeled with a current-mirror integrator *(31)*.

SILICON CHIP LAYOUT

In the mammalian retina, cone signals converge on to BCs *(1)*, which makes the receptive field center Gaussian-like *(32)*. To implement signal convergence in the model, chip BCs connect the outputs from a central phototransistor and its six nearest neighbors (hexagonally tiled) to one inner retina circuit, as shown in Fig. 3A. Each of the outer retina circuits actually produces two output currents. A central photoreceptor drives the bipolar cell with both of these outputs while photoreceptors at the six vertices divide these outputs between their two nearest BCs. For symmetry, a similar architecture is implemented for the reference current (*see* ref. *21*).

Transient GCs in the mammalian retina pool their inputs from a larger region than sustained GCs. This difference is maintained in the chip by pooling inner retina signals in the same way that outer retina signals are pooled. Thus, each transient ganglion cell receives input from a central inner retina circuit and its six nearest neighbors, as shown in Fig. 3A. This central circuit excites its ganglion cell with both copies of its transient output whereas those at the six vertices divide their outputs between their two nearest transient GCs. As inner retina circuits are tiled at one-quarter the density of the phototransistors that provide their input, this results in a larger receptive field for each ganglion cell than for each bipolar cell. All GCs tile hexagonally, but the transient ones tile more sparsely than the sustained ones. This architecture gives transient GCs a larger receptive field, as found in cat Y-GCs. It also accounts for transient GCs' nonlinear subunits *(20)* since all the rectified bipolar cell signals can never sum to zero, at any one moment, in response to a sinusoidal grating. Conversely, if the cells were linear, a contrast-reversing grating exactly centered on the cell's receptive field would not modulate the cell's response as decreases in dark regions would exactly cancel out increases in light regions.

The chip design was fabricated in a 0.35 μm minimum feature-size process with its cell mosaics tiled at a scale similar to the mammalian retina (Fig. 3B). Phototransistors are tiled triangularly 40 μm apart; this spacing is only about two and a half times that of human cones at 5 mm nasal eccentricity *(33)*. The phototransistors are only 10 μm

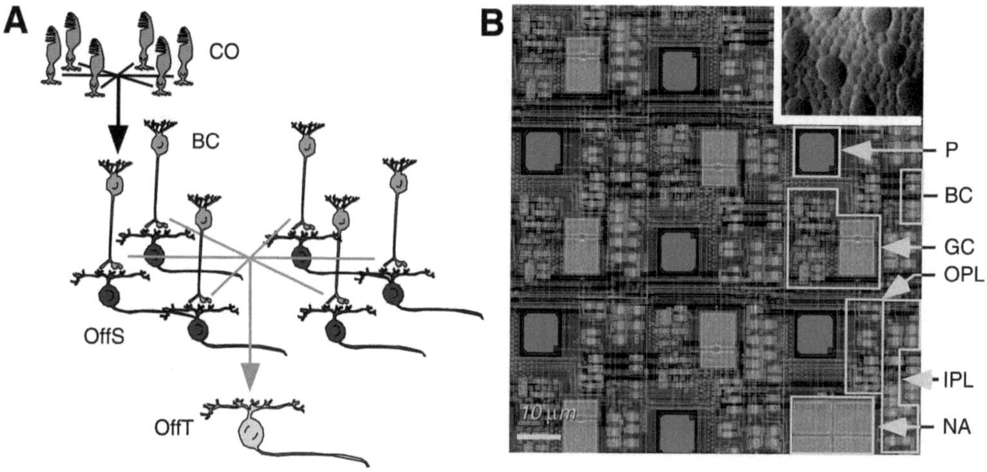

Fig. 3. Chip layout. **(A)** Functional architecture: signals from a central photoreceptor (not shown) and its six neighbors (CO) are pooled to provide synaptic input to each bipolar cell (BC). Each bipolar cell generates a rectified output, either ON or OFF that drives a local inner plexiform layer (IPL) circuit. Sustained GCs (OffS), which have a dendritic field diameter of 80 μm, receive input from a single local IPL circuit. Transient GCs (OffT), however, receive signals from a central IPL circuit (not shown) and its six neighbors, and hence, their dendritic field is 240 μm wide. **(B)** Chip design and human photoreceptor mosaic: each pixel with 38 transistors on average has a phototransistor (P), outer plexiform (synaptic) layer circuitry, bipolar cells, and IPL circuitry. Spike-generating GC are found in five out of eight pixels; the remaining three contain a narrow field amacrine (NA) cell membrane capacitor. Inset: Tangential view of human cone (large) and rod (small) mosaic at 5 mm eccentricity, plotted at the same scale (reproduced from refs. *30, 33*).

on a side, leaving ample space for postsynaptic circuitry, which is interspersed between them. This spacing is necessary because, unlike neural tissue, silicon microfabrication technology cannot produce three-dimensional structures.

One pixel—the basic element that is tiled to create the silicon retina—contains a phototransistor, outer retina circuitry, a pair of BCs, and one-quarter of the inner retina circuit. Hence, 2 × 2- and 4 × 4-pixel blocks are needed to generate a complementary pair of sustained and transient outputs, respectively. Because transient GCs occur at a quarter resolution, not every pixel contains ganglion cell circuitry. Three out of every eight pixels instead contain a large capacitor that gives the narrow-field amacrine cell its long time-constant. A spike generating circuit in the remaining five pixels converts GC inputs into spikes that are sent off chip. The 3.5 × 3.3 mm^2 silicon die has 5760 phototransistors at a density of 722/mm^2 and 3600 GCs at a density of 461/mm^2—tiled in 2 × 48 × 30 and 2 × 24 × 15 mosaics of sustained and transient ON and OFF GCs.

Because of wiring limitations, each ganglion cell output off chip cannot be communicated directly. Instead, an asynchronous, arbitered, multiplexer are used to read spikes out from the silicon neurons *(34)*. Each ganglion cell interfaces with digital circuitry that communicates the occurrence of a spike to the row and each column arbiters. The arbiters select one row and one column at a time, and an encoder outputs their addresses. Row and column addresses are communicated serially off chip. The spike activity of all 3600 GCs can be represented with just seven bits using this address-event representation. By noting the address of each event generated by the chip, Ganglion cell location and type can be decoded in the array. In a future prosthetic application, such a scheme would be unnecessary as each ganglion cell's local spiking circuit could drive a local micro-electrode to stimulate adjacent neural tissue.

SILICON RETINA RESPONSES

To gauge the validity of the outer and inner retina models and to verify their ability to recreate retinal function, the responses of the silicon retina are compared with those of the mammalian retina. Specifically, the silicon retina's responses were recorded to different spatial and temporal frequencies, to different luminances, and to different contrasts and these responses were compared with physiological measurements available in the literature.

Spatiotemporal Filtering

The silicon retina's GCs respond to a restricted band of spatiotemporal frequencies, with transient cells displaying nonlinear spatial summation. In response to a drifting sinusoidal grating, spike trains from GCs of the same type differ significantly because of the cumulative effect of variability between transistors (CV = 20–25% for currents in identically sized and biased transistors). It was possible to obtain results that match physiological data by averaging responses from all cells in a given column, much as physiologists average several trials from the same cell.

The results reveal that at low temporal frequencies, both low and high spatial frequencies are attenuated in our silicon chip. However, sustained cells respond to a higher range of frequencies, as expected from their smaller receptive fields (Fig. 4A,B). Spatial frequency responses of ganglion cell activity (Fig. 4b) were fit with a balanced difference-of-Gaussians, an empirical model for retinal GCs *(35)*. Briefly, the frequency profile of a one-dimensional zero-mean inhibitory Gaussian was computed, with standard deviation σ_{Inh} and unit area, subtracted from a one-dimensional zero-mean excitatory Gaussian, with standard deviation σ_{Exc} and unit area. This frequency response was compared with the data and was optimized σ_{Exc} and σ_{Inh} to give the best fit.

When the phase of a contrast-reversing sinusoidal grating was varied, frequency doubling in transient cells was observed (Fig. 4C,D). This nonlinear summation is the fundamental distinction between narrow- and wide-field mammalian GCs *(20,36,37)* and arises because the bipolar cell signals are rectified before they are summed *(20)*.

Sustained cells in the silicon retina retain bandpass spatial filtering at all temporal frequencies (Fig. 5A,B). This pattern of spatiotemporal filtering matches the mammalian retina, except for a resonance found at very high temporal frequencies *(38)*. In the silicon retina, fast wide-field amacrine-cell modulation augments slow horizontal-cell inhibition to suppress low spatial frequencies, irrespective of whether they are presented at high or low temporal frequencies. And the optics and the cone–cone gap-junctions blur high spatial frequencies, also irrespective of temporal frequency. As a result, the sustained cells pass a restricted band of spatial frequencies at all temporal frequencies.

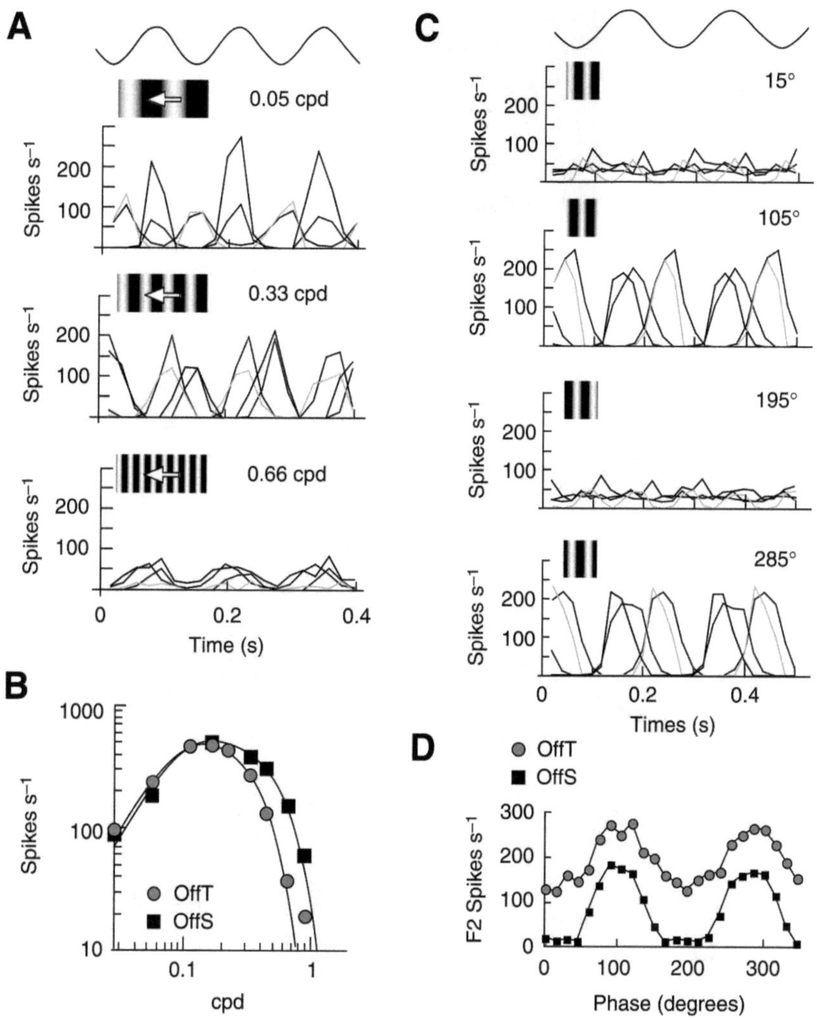

Fig. 4. Spatial filtering and nonlinear summation: (**A**) Varying spatial frequency: responses to 7.5 Hz horizontally drifting sinusoids with three different spatial frequencies. The responses are the strongest at an intermediate frequency, except for OnT cells, which showed an anomalous preference for low frequencies (*see* Fig. 2A for color code). (**B**) Spatial frequency tuning: OffT and OffS amplitudes are plotted for all spatial frequencies tested; they both peaked at 0.164 cpd, but OffS cells pass a higher range of frequencies. Solid lines are the best fit of a balanced difference-of-Gaussian model (OffT: $\sigma_{Exc}/\sigma_{Inh} = 0.20$; OffS: $\sigma_{Exc}/\sigma_{Inh} = 0.15$) *(35)*. (**C**) Varying spatial phase: responses to a 5 Hz 0.33 cpd contrast-reversing grating at four different spatial phases. Transient cells (yellow and blue) show frequency doubled responses at 15° and 195°. (**D**) Null Test: amplitudes of the second Fourier component (F2) of OffT and OffS responses are plotted for all phases tested. The sustained cells' F2 response disappeared at certain phases, but it could not be nulled in the transient cells. Fluctuations in F2 amplitude arise from uneven spatial sampling in the silicon retina *(30)*.

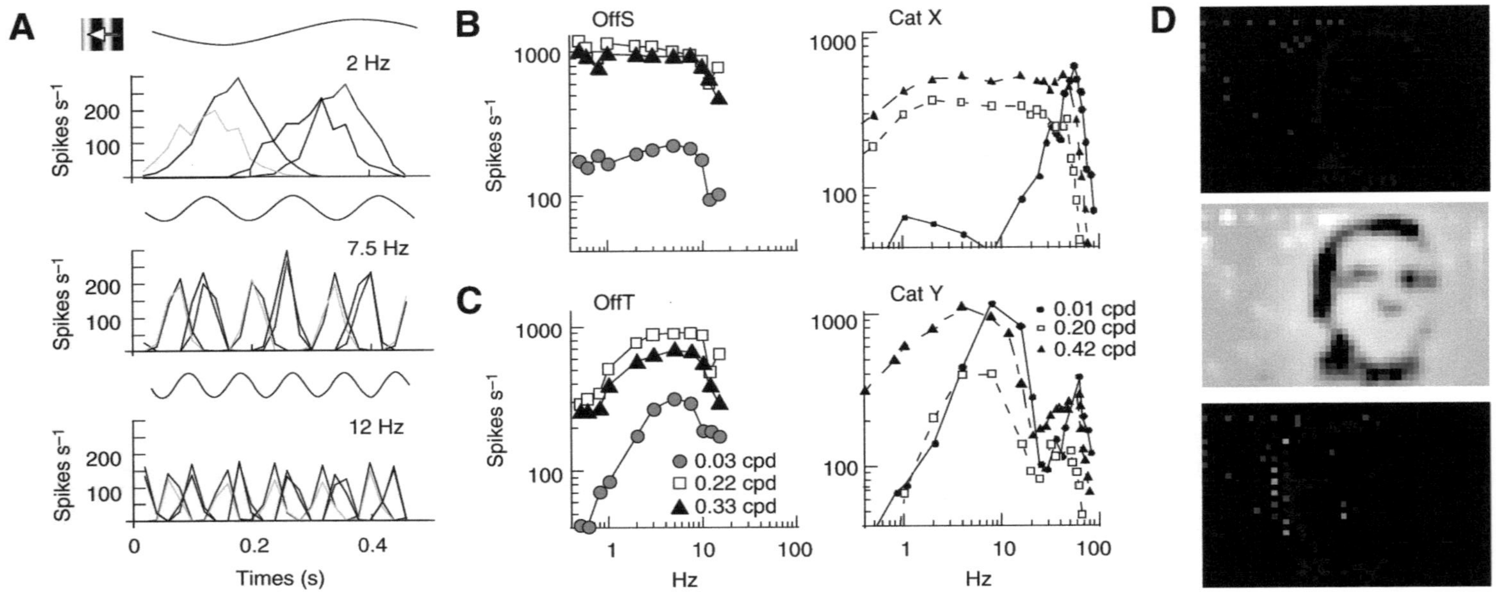

Fig. 5. Temporal filtering: **(A)** Varying temporal frequency: responses to 0.22 cpd horizontally drifting sinusoids at three different temporal frequencies. The response is the strongest at an intermediate frequency for transient cells (yellow and blue), whereas sustained cell (red and green) responses decline monotonically. (**B**) Sustained-cell temporal frequency tuning: responses of OffS and cat ON-center X-cells to low, medium, and high spatial-frequency sinusoidal gratings drifting horizontally at different temporal frequencies (*see* legend in C). Both pass all temporal frequencies less than 10 Hz, except at low spatial frequencies. However, the cat data display a high-frequency resonance (Cat data are reproduced from ref. *38*). The ordinate here and in **c** represents responsivity, which is the amplitude of the fundamental Fourier component divided by the stimulus contrast. **(C)** Transient-cell temporal frequency tuning: same as in B but for OffT cells and cat ON-center Y-cells. Both pass a restricted band of temporal frequencies at all spatial frequencies less than 0.33 cpd. **(D)** Responses to a face: in the static image (top), only sustained GCs respond. Reconstruction of the image from their activity (middle) demonstrates fidelity of retinal encoding. In the moving image (bottom), transient GCs respond as well, highlighting moving edges. The velocity of the image was approx 26.96°/s. The mean spike rate was 19 spikes/cell/s *(30)*.

On the other hand, the silicon retina's transient cells retain bandpass temporal filtering at all spatial frequencies (Fig. 5A,C). This pattern also matches the mammalian retina, except for the high-frequency resonance *(38)*. In the silicon retina, focussed narrow-field amacrine-cell inhibition augments diffuse horizontal-cell inhibition to suppress low temporal frequencies, irrespective of whether they are presented at high or low spatial frequencies. And the cone membrane's capacitance smears high temporal frequencies, also irrespective of spatial frequency. As a result, the transient cells pass a restricted band of temporal frequencies at all spatial frequencies.

The overall effect of spatiotemporal filtering is the best illustrated by natural stimuli (Fig. 5D). Bandpass spatial filtering in sustained GCs enhances edges in the static image. During rapid motion, bandpass temporal filtering in transient GCs captures this information with surprisingly little blurring. To confirm that the chip encodes visual information, the natural stimulus was reconstructed from the sustained ganglion cell spike activity. The visual image was reconstructed from the spikes by convolving ON and OFF sustained ganglion cell spike output with the same difference-of-Gaussian model, whose excitatory and inhibitory standard deviations were determined by the fit to the ganglion cell spatial frequency responses (σ_{Exc} and σ_{Inh}, Fig. 4B). The outputs of this convolution were passed through a temporal low-pass filter with a time constant of 22.7 ms, computing a new frame every 20 ms. The difference between images obtained from ON and OFF spikes was taken, and was displayed on a gray-scale, with ON and OFF activity corresponding to bright and dark pixels, respectively. Activity from transient GCs did not enhance the resolution of the reconstructed image and was not included. Passing spike output through this simple spatiotemporal filter produces an image that is easily recognizable, even with only 30×48 pixels and just 0.4 spikes/cell/frame. This result suggests that cortical structures receiving input from such a visual prostheses can extract useful visual information from the silicon retina's neural code through simple linear filtering.

Light and Contrast Adaptation

The silicon retina's GCs adapt to mean luminance and encode stimulus contrast (Fig. 6). They maintain contrast sensitivity during at least one and a half decades of mean luminance. This intensity range was limited on the low end by leakage currents; the transistors pass a few pico-amps even when their gate voltage is zero. And it was limited on the high end by the projector (could not exceed 200 cd/m^2) in the experimental set-up and by stray photocurrents (light-induced leakage currents) in the silicon chip. To obtain the results presented here, the effect of these photocurrents was compensated by changing two externally applied voltages that would otherwise require no adjustment.

From the outer retina model, discussed in outer retina section, how the silicon model's cone activity depends on contrast was derived. The analytical result is fitted to the measured luminance adaptation curves (Fig. 6C), allowing *r* to increase with decreasing intensities as that data were obtained by exploiting the dependence of contrast sensitivity on horizontal-cell coupling (controlled globally in the circuit by V_{hh}), thereby the effect of stray photocurrents was being compensated. These photocurrents, which set an upper limit for the membrane time-constants that can be realized, reduce the silicon retina's sensitivity by speeding up ganglion-cell spike-frequency adaptation

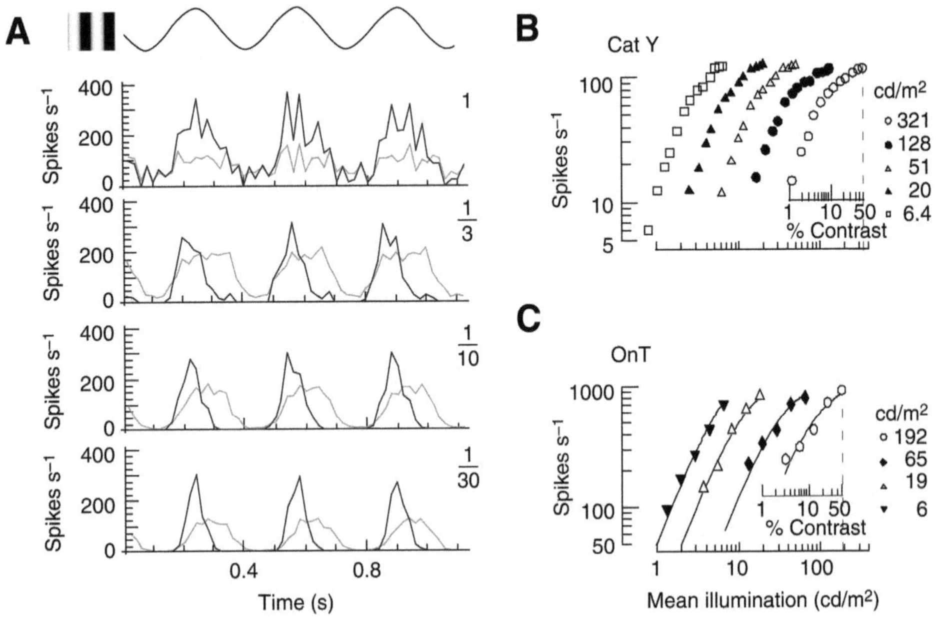

Fig. 6. Luminance adaptation: **(A)** Varying intensity: OnT and OnS responses to a sinusoidal grating (0.22 cpd) whose mean intensity was attenuated by amounts listed, using neutral density filters. Because of increases in sensitivity, the response amplitude hardly changes. The noisier responses at high intensity are due increased background activity, which tends to invoke synchronous firing because of cross-talk (i.e., ephatic interactions) in the silicon chip. **(B)** Cat ON-center Y-cell intensity curves: the sinusoidal grating's (0.2 cpd) contrast varied from 1 to 50% and reversed at 2 Hz, for five mean luminances *(39)*. Here, and in C, response vs contrast (small *x*-axis) curves are shifted to align the 50% contrast response with that particular mean luminance (large *x*-axis). Mean luminance is converted from trolands to cd per m^2 based on a 5 mm diameter pupil (adapted from ref. *39*). **(C)** OnT intensity curves: the sinusoidal grating's (0.22 cpd) contrast varied from 3.25 to 50% and reversed at 3 Hz, for four different mean luminances. Solid lines represent the best fit of the equation given in the text for cone terminal activity; the fit indicates that the ratio of the horizontal cell and cone terminal space constants increased from 0.46 to 0.69, with a corresponding drop in peak spatial frequency from 0.22 to 0.16 cpd *(30)*.

and narrow-field amacrine-cell presynaptic inhibition as intensity increases. The latter effect was compensated by adjusting a second externally applied voltage that sets the narrow-field amacrine-cell's (baseline) membrane leakage.

The analytical fits to the measured luminance adaptation curves support the conclusion that the silicon retina's local modulation of synaptic strengths was largely responsible for adaptation. As mean luminance increased from 6 to 192 cd/m^2, the value assigned to the ratio between the HC and cone space constants, *r*, in the fitted equation decreased monotonically from 0.69 to 0.46 (unexpectedly, the best fits were found with *r* less than one; the corresponding reduction in peak spatial frequency response was from 0.22 to 0.16 cpd). Thus, even though we intervened by manually adjusting V_{hh}, the resulting change in *r*, a drop of only 36%, fell far short of what is required to reduce the sensitivity by a decade and a half. Overall, the silicon retina's ganglion-cell activity remained weakly correlated with absolute light intensity because of the residual effect

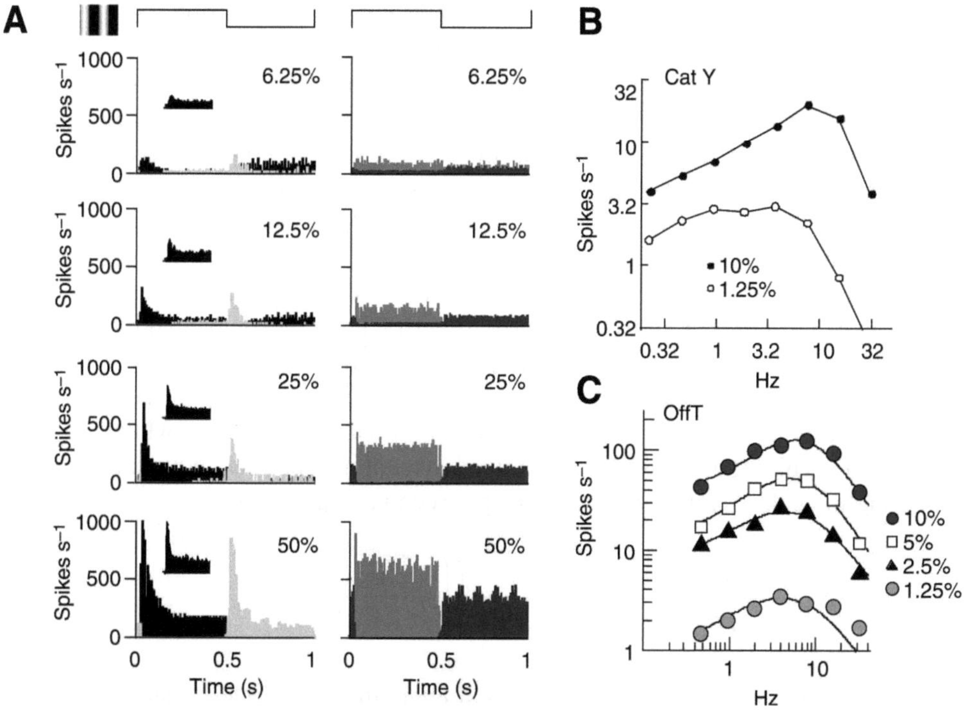

Fig. 7. Contrast gain control: **(A)** Varying contrast: responses to a 1 Hz square-wave contrast reversal of a sinusoidal grating (0.22 cpd) at four different peak stimulus contrasts. Bin width is 4 ms. Responses increase sublinearly and change more rapidly with increasing contrast; these effects are more pronounced in the transient cells. Their responses decayed with a time-constant that decreased from 28 to 22 ms as contrast increased from 6.25 to 50%. Inset: Response of ON-center cat X cell to half a cycle of the same stimulus *(29)*. **(B)** ON-center cat Y cell contrast-dependent temporal filtering: a stationary sinusoidal grating (0.25 cpd) whose contrast was determined by the sum of eight sinusoids was used. All eight sinusoids had the same amplitude whose value relative to the background is stated. The amplitude of the fundamental Fourier component at these eight frequencies was plotted for two different amplitudes (reproduced from ref. *28*). The peak sensitivity shifted from 3.9 to 7.8 Hz as the contrast increased from 1.25 to 10%. **(C)** OffT contrast-dependent temporal filtering: the stimulus was the same except that the spatial frequencywas reduced to 0.14 cpd and four different contrasts were tested. The peak sensitivity shifted from 3.9 to 7.8 Hz as the contrast increased from 1.25 to 10%. Solid lines are the best fit of an analytical model, which indicated that an increase in the strength of narrow-field amacrine-cell feedback inhibition from $w = 1.07$ to $w = 3.76$ could account for this change in temporal dynamics *(30)*.

of the stray photocurrents. Thus, as found in mammalian ganglion cell behavior *(39)*, responses at low contrasts are weaker at lower light intensities.

The silicon retina's GCs also adapt to temporal contrast. When presented with contrast-reversing gratings, the transient GCs respond more quickly, but more transiently with increasing contrast (Fig. 7A). And as the responses became even more transient, the peak-firing rate tended to saturate at the highest contrast levels. This adaptation is similar to the contrast gain control observed in mammalian narrow-field sustained *(29)* and wide-field transient GCs. However, it was not as dramatic in the silicon retina's sustained cells, whose responses did not decay nor saturate as much; they did, however, display a more rapid onset with increasing contrast. This difference between the silicon retina's sustained and transient GCs suggests that narrow-field amacrine cell feed-forward inhibition enhances contrast gain control by making the response more transient.

To better quantify the effect of contrast gain control *(28)*, the silicon retina's temporal frequency was measured, tuning at different contrasts, using a mixture of sinusoids with a flat spectrum. The OFF-Transient cells' peak response shifted to higher frequencies with increasing contrast, moving by an amount similar to that observed in the mammalian retina (Fig. 7B,C). But while this shift in tuning was accompanied by an overall strengthening of responses at all frequencies in the data, it was accompanied by preferential strengthening of high frequency responses in the cat data.

From the inner retina model, discussed in inner retina section, how the silicon model's ganglion cell responses depend on input contrast for a flat spectrum (i.e., white noise) was derived *(21)*. This stimulus was characterized by contrast per unit frequency *d*. The transient ganglion cell response, i_{Gt}, in spikes per second, is given by

$$i_{\mathrm{Gt}} = S\left|\left(d\frac{j\tau_A\omega + \varepsilon(1-g)}{j\tau_A\omega + 1}\right)\left(\frac{1}{j\tau_o\omega + 1}\right)^2\left(\frac{1}{j\tau_P\omega + 1}\right)\right|$$

where $\tau_A = \varepsilon\tau_{\mathrm{na}}$, $\varepsilon = 1/1(1 + wg)$. τ_{na} is the narrow-field amacrine cell's time constant, *g* is the gain from the bipolar terminal to the narrow-field amacrine cell, *w* is the wide-field amacrine cell-modulated strength of narrow-field amacrine cell inhibition onto the bipolar terminal, and *S* is a gain constant. The sustained ganglion cell response can be obtained by setting *g* to zero. The outer retina was approximated, using a lowpass temporal filter with time constant τ_o. The lowpass filtering behavior of the chip's photoreceptors was also included whose time-constant is τ_p. The four data sets are fitted (Fig. 7C) by allowing the system gain term, *S*, and the inhibition modulation *(w)* to vary across different stimulus contrasts, and fixed the remaining parameters.

The best fits of this model to the four input contrast densities (solid lines in Fig. 7C) support the conclusion that the silicon retina's wide-field-amacrine-mediated modulation of presynaptic inhibition was responsible for contrast gain control. The parameter values for these fits were $\tau_p = 33$ ms, $\tau_o = 77$ ms, $\tau_{\mathrm{na}} = 1.0382$ s, and $g = 1.07$. The system gain, *S*, saturated over the four contrasts (352–1358 to 1711–1929), whereas the inhibition strength *(w)* increased exponentially (1.07–1.51 to 2.38–3.76). Thus, the inhibition strength increased with contrast as it was expected, but the system gain was not constant because of a static nonlinearity *(40)*, canceling part of the expected decrease in sensitivity.

CONCLUSIONS

By autonomously extracting the same visual information encoded by the mammalian retina at a similar physical scale and energy efficiency, the silicon retina satisfies the criteria for a fully implantable ocular prosthesis. The device approximates the behavior of the mammalian retina in both linear response and nonlinear adaptation. This success validates the neuromorphic modeling approach for advanced prosthetic applications. In addition, this real-time silicon model may be useful in both further testing specific hypotheses about the retina and in serving as a realistic retinal front-end to other downstream processes like cortical models, other artificial neural systems, or robots.

The approach in constructing this silicon retina was to model synaptic connections found in the mammalian retina and to implement them using transistor primitives. For example, reciprocal inhibition between bipolar and amacrine cells in complementary ON and OFF channels in the model mimics vertical inhibition between ON and OFF laminae *(24)* and serial inhibition found between amacrine cells *(41)* in the mammalian retina. Furthermore, the model proposes an anatomical substrate for computing Victor and Shapley's "neural measure of contrast" *(28,29)*, suggesting that wide-field amacrine cells play this role in the mammalian retina. Whereas this hypothesis remains to be tested experimentally, there are instances in which the model oversimplifies the functional architecture of the mammalian retina. For instance, dopaminergic amacrine cells *(42)*, and light sensing GCs *(43)*, are likely important in modulating mammalian retinal cone and horizontal gap-junction conductance. These dopaminergic cells are not included in the model, and instead the local horizontal-cell signal is relied on to modulate cone coupling.

Capturing the mammalian retina's synaptic organization proved sufficient to recreate its computations in the silicon retina, although there were some specific quantitative differences of note, which arose from technological limitations. The device's sensitivity dropped when luminance was high and its background firing rates increased. These deficiencies can be addressed by reducing photo-induced leakage currents, which tend to speed up inhibition and firing rates in the silicon retina GCs, and by reducing cross-talk between silicon GCs, which produced high background activity when bright or large stimuli were presented. The device's sustained GCs also displayed little or no contrast gain control. A faster photodetector should rectify this situation, which appears to be related to the absence of any significant transient component in the sustained GCs' responses. This speed-up will lead to a transient component at the bipolar terminal, and hence in the sustained GCs.

The goal of designing and fabricating a silicon retina was realized that can operate and adapt independent of external control to a large extent, but there remains some degree of manual intervention necessary to make the device work properly. Whereas the voltages applied to the biases that set mean cone terminal activity, mean bipolar terminal activity, mean ganglion cell activity, and coupling strength in wide-field amacrine cells remained fixed, the voltages applied to the biases had to be manually adjusted that set the coupling strength between HCs and the bias that sets the narrow-field amacrine cell leakage current to compensate for light-dependent leakage currents, a shortcoming of silicon microtechnology at these small length scales. However, this fine tuning was only required for light adaptation. Any bias voltages was not adjusted whatsoever during any of the other experiments. Hence, addressing the leakage issue will make it possible to hard-wire all of the silicon retina's external biases to specific voltages, as required by a final prosthetic application.

The artificial retina satisfies the requirements of a neural prosthesis by matching the biological retina in size and weight and using under a tenth of a watt. Rabbit retina uses 16.2 nW per ganglion cell (82 μmoles of ATP g per min *[44]*, or 88 mW g/min, times 70 mg average weight, divided by 380,000 GCs *[12]*). In contrast, the chip consumes

17 μW per ganglion cell (62.7 mW for the entire chip) at an average spike rate of 45 spikes s per ganglion cell. Although, this energy consumption is one thousand times less efficient than the mammalian retina, it still represents a 100-fold improvement over conventional microprocessors. A 1 GHz Pentium® processor operating at 10 W would dissipate 2.2 mW per ganglion cell to compute the response of a13 × 13 × 13 kernel (X × Y × T) updated at 100 times/s. With an upper limit on a proposed intraocular implant's power dissipation of 100 mW, the Pentium® could thus only compute the responses of under 40 GCs, or a 6 by 6 array, which is too small for functional vision *(5)*. However, with the same 100 mW limit, our neuromophic chip's energy efficiency allows it to compute the responses of 4000 GCs, roughly a 60 × 60 array. This energy efficiency is expected to improve further, together with spatial resolution and dynamic range, as microfabrication technology advances.

In conclusion, based on detailed knowledge of the retina's neuronal specializations, synaptic organization, and functional architecture *(45)*, thirteen neuronal types have been constructed in silicon and linked them together in two synaptic layers on a physical scale comparable with the human retina. Furthermore, a silicon retina has been created that modulates its synaptic strengths locally. The silicon retina realizes luminance adaptation, without using logarithmic compression, and contrast gain control independent of external control, thus, capturing properties of retinal neural adaptation for the first time. The success modeling neural adaptation using single-transistor primitives suggests that a similar approach could be used to morph other neural systems into silicon as well; this may eventually lead to fully implantable neural prostheses *(46,47)* that do not require external interfaces.

REFERENCES

1. Sterling P. Retina. In: Shepherd GM, ed. Synaptic organization of the brain, 4th ed., New York: Oxford University Press, NY, 1998.
2. van Hateren J. A theory of maximizing sensory information. Biol Cybernetics 1992;68:23–29.
3. Rizzo JFr, Wyatt J, Humayun M, et al. Retinal prosthesis: An encouraging first decade with major challenges ahead. Opthalmology 2001;108:13–14.
4 .Margalit E, Maia M, Weiland JD, et al. Retinal prosthesis for the blind. Surv of Ophthalmol 2002;47:335–356.
5. Humayun MS, de Juan EJ, Weiland JD, et al. Pattern electrical stimulation of the human retina. Vision Res 1999;39:2569–2576.
6. Chow AY, Pardue MT, Chow VY, et al. Implantation of silicon chip microphotodiode arrays in the cat subretinal space. IEEE Trans Neural Syst Rehabil Eng 2001;9:86–95.
7. Normann RA, Maynard EM, Rousche PJ, Warren DJ. A neural interface for a cortical vision prosthesis. Vision Res 1999;39:2577–2587.
8. Dobelle WH. Artificial vision for the blind by connecting a television camera to the visual cortex. ASAIO J 1999;46:3–9.
9. Mead CA. Analog VLSI and neural systems, Addison Wesley, Reading, MA, 1989.
10. Mahowald M, Mead C. A silicon model of early visual processing. Neural Networks 1988;1.
11. Baccus SA, Meister M. Fast and slow contrast adaptation in retinal circuitry. Neuron 1997;36:909–919.
12. Masland R. The fundamental plan of the retina. Nature Neurosci 2001;4:877–886.
13. Boahen K. A retinomorphic chip with parallel pathways: Encoding Incresing, On, Decreasing, and Off visual signals. J Analog Integrated Circtuits Signal Processing 2001;30(2).

14. Boahen KA, Andreou AG. A contrast sensitive silicon retina with reciprocal synapses. In: Moody JE, Hanson SJ, Lippmann RP, eds. Advances in neural information processing systems. Vol. 4, Morgan Kaufmann, San Mateo, CA, 764–772.
15. Kamermans M, Fahrenfort I, Schultz K, Janssen-Bienhold U, Sjoerdsma T, Weiler R. Hemichannel-mediated inhibition in the outer retina. Science 2001;292:1178–1180.
16. Kamermans M, Werblin F. GABA-mediated positive autofeedback loop controls horizontal cell kinetics in tiger salamander retina. J Neurosci 1992;12:2451–2463.
17. Kuffler SW. Discharge patterns and functional organization of mammalian retina. J Neurophysiol 1953;16:37–68.
18. Werblin FS, Dowling JE. Organization of the retina of the mudpuppy, Necturus maculosus. II. Intracellular recording. J Neurophyiol 1969;32:339–355.
19. Rodieck RW. The primate retina. Comp Primate Biol 1988;4:203–278.
20. Enroth-Cugell C, Freeman AW. The receptive field spatial structure of cat retinal Y cells. J Physiol 1987;384:49–79.
21. Zaghloul KA, Boahen K. Optic nerve signals in a neuromorphic chip I: Outer and inner retina models. IEEE Trans Biomed Eng 2004;51:657–666.
22. Normann RA, Perlman I. The effect of background illumination on the photoresponses of rod and green cones. J Physiol 1979;286:491–507.
23. Tsividis YP. Operation and modeling of the MOS transistor. New York:McGraw-Hill Book Company, 1987.
24. Roska B, Werblin F. Vertical interactions across ten parallel, stacked representations in the mammalian retina. Nature 2001;410:583–587.
25. Maguire G, Lukasiewicz P. Amacrine cell interactions underlying the response to change in tiger salamander retina. J Neurosci 1989;9:726–735.
26. Kolb H, Nelson R. OFF-alpha and OFF-beta ganglion cells in cat retina: II. Neural circuitry as revealed by electron microscopy of HRP stains. J Comp Neurol 1993;329:85–110.
27. Freed MA, Pflug R, Kolb H, Nelson R. On-Off amacrine cells in cat retina. J Comp Neur 1996;364:556–566.
28. Shapley R, Victor JD. The contrast gain control of the cat retina. Vision Res 1979;19:431–434.
29. Victor JD. The dynamics of cat retinal X cell centre. J Physiol 1987;386:219–246.
30. Zaghloul KA, Boahen K. An On-Off log domain circuit that recreates adaptive filtering in the retina. IEEE Trans Circuits Syst 2005;52:99–107.
31. Boahen KA. The retinomorphic approach: Pixel-parallel adaptive amplification, filtering, and quantization. J Analog Integrated Circtuits Signal Processing 1997;13:53–68.
32. Smith RG. Simulation of an anatomically defined local circuit—The cone-horizontal cell network in cat retina. Visual Neurosci 1995;12:545–561.
33. Curcio CA, Sloan KR, Kalina RE, Hendrickson AE. Human photoreceptor topography. J Comp Neur 1990;292:497–523.
34. Boahen KA. Point-to-point connectivity between neuromorphic chips using address-events. IEEE Trans Circuits Syst 1999;47:100–116.
35. Rodieck R. Quantitative analysis of cat retinal ganglion cell response to visual stimuli. Vision Res 1965;5:583–601.
36. Enroth-Cugell C, Robson JG. The contrast sensitivity of retinal ganglion cells of the cat. J Physiol 1966;187:517–552.
37. Demb JB, Zaghloul KA, Haarsma L, Sterling P. Bipolar cells contribute to nonlinear spatial summation in the brisk-transient (Y) ganglion cell in mammalian retina. J Neurosci 2001; 21:7447–7454.
38. Frishman LJ, Freeman AW, Troy JB, Schweitzer-Tong DE, Enroth-Cugell C. Spatiotemporal frequency responses of cat retinal ganglion cells. J Gen Physiol 1987; 89:599–628.

39. Troy JB, Enroth-Cugell C. X and Y ganglion cells inform the cat's brain about contrast in the retinal image. Exp Brain Res 1993;93:383–390.
40. Zaghloul KA, Boahen K. Optic nerve signals in a neuromorphic chip II: Testing and results. IEEE Trans Biomed Eng 2004;51:667–675.
41. Dowling JE, Boycott BB. Organization of the primate retina: electron microscopy. Proc R Soc Lond B 1966;166:80–111.
42. Jensen RJ, Daw NW. Effects of dopamine and its agonists and antagonists on the receptive field properties of ganglion cells in the rabbit retina. Neuroscience 1986;17:837–855.
43. Berson DM, Dunn FA, Takao M. Phototransduction by retinal ganglion cells that set the circadian clock. Science 2002;295:1070–1073.
44. Ames A, Li YY, Heher EC, Kimble CR. Energy metabolism of rabbit retina as related to function: high cost of Na^+ transport. J Neurosci 1992;12:840–853.
45. Freed MA, Sterling P. The ON-alpha ganglion cell of the cat retina and its presynaptic cell types. J Neurosci 1988;8:2303–2320.
46. Craelius W. The bionic man: Restoring mobility. Science 2002;295:1018–1021.
47. Zrenner E. Will retinal implants restore vision. Science 2002;295:1022–1025.

11
Visual Cortex and Extraocular Retinal Stimulation With Surface Electrode Arrays

John W. Morley, PhD, Vivek Chowdhury, MBBS
and Minas T. Coroneo, MD, FRANZCO

CONTENTS

INTRODUCTION

Severe forms of blindness, resulting in a loss of light perception has a devastating impact on the functional ability *(1)*, physiological state *(2)*, and psychosocial well-being of affected patients *(3)*. Severe visual impairment (acuity 20/200 or worse) affects around 0.7% of the population in Australia older than 49 yr *(4)*, with the population weighted prevalence of profound visual impairment being 1.56/1000. In the United States more than one million people are legally blind with approx 10% having no light perception *(5,6)*. Causes of adult blindness include age-related macular degeneration, glaucoma, end-stage retinal dystrophy, diabetic retinopathy, optic nerve ischaemia, and trauma *(7)*. One common factor with these pathologies is that once damage has occurred to neural tissue it is permanent because of the poor capacity of neural tissue in the central nervous system to regenerate. Recent experimental treatments, such as gene therapy, nerve cell transplantation, and growth factors have not been successful in restoring visual perception to blind patients *(8)*. The only experimental method, which has successfully restored visual sensation to otherwise irreversibly blind patients is electrical stimulation of the retina, optic nerve, or brain with a prosthetic device *(9,10)*. Of these three approaches, a prosthetic device, which acts on the visual area of the brain is the only method that has been investigated extensively in chronic human trials, and has restored visual perceptions, which have increased a blind patient's mobility and independence *(11)*.

From: *Ophthalmology Research: Visual Prosthesis and Ophthalmic Devices: New Hope in Sight*
Edited by: J. Tombran-Tink, C. Barnstable, and J. F. Rizzo © Humana Press Inc., Totowa, NJ

There are several target tissues that are used as the site of electrical stimulation with a visual prosthesis: visual cortex, optic nerve, or retina, and the pathology of the disease causing blindness will determine which of these approaches is appropriate. In cases of cortical blindness from trauma or disease (vascular and neoplastic) of the occipital lobes, there are no prosthetic options to restore visual sensations. If trauma or disease is restricted to the retina and optic nerves, such as retinal detachment or glaucoma, then electrical stimulation of the visual cortex may be able to restore visual sensation. If disease is restricted to a limited part of the retina, such as the photoreceptors in retinitis pigmentosa, then a retinal or optic nerve prosthesis to stimulate remaining retinal ganglion cells or their axons becomes an option. Normal "vision" is a complex sensory phenomenon involving the integration of shapes, spatial orientation, color, movement and so on, through an intricate network of neuronal systems. A visual prosthesis using current technologies will produce visual sensations that consist of a matrix of phosphenes to encode some of these visual features.

VISUAL CORTEX PROSTHESES

Electrical stimulation of the cerebral cortex has a long history. Although, reports of the electrical stimulation of the visual system were made 250 yr ago *(12)*, it was the reports of the German neurosurgeon Foerster in 1929 *(13)* that led to the targeting of the visual regions of the cerebral cortex for electrical stimulation. Foerster electrically stimulated the visual cortex in conscious patients undergoing neurosurgery. The patients reported seeing spots of light (phosphenes) with the position of the phosphene in the visual field dependent on the locus of visual cortex stimulation. Shortly after Kraus and Schum in 1931 *(13)* described the generation of localized phosphenes following electrical stimulation of the visual cortex in a patient who had been blind for 8 yr. However, it was another 30 yr before any real attempt was made to confirm and extend these findings. Button and Putnam *(14)* implanted electrodes in the visual cortex of three blind patients. Two of the patients were able to locate a light source by scanning a room with a photocell, the output of which stimulated the visual cortex through the penetrating electrodes. The first attempt to build a "prosthesis" to stimulate the visual cortex was in 1968 when Brindley and Lewin chronically implanted 80 platinum electrodes (encased in a cap of silicone rubber) on the cortical surface in the calcarine sulcus and neighboring visual cortex of a blind patient *(13)*. Stimulation of individual (surface) electrodes resulted in the patient perceiving phosphenes at a constant position in the visual field with phosphenes produced by stimulation through electrodes 2.4 mm apart spatially resolved. Stimulation through five to eight electrodes simultaneously resulted in the perception of simple patterns, such as a question mark, or a capital L, or V *(13)*.

The work by Brindley and Lewin was continued by Dobelle and colleagues *(15,16)* who also used an implant that consisted of a series of flat electrodes for surface stimulation of the visual cortex. Dobelle constructed an artificial vision system that consisted of a camera mounted on a pair of glasses, connected to a computer, which activated individual or groups of electrodes through a percutaneous connection. During the early 1970s, Dobelle implanted these devices for short period of time in a number of patients. In 1978 he implanted arrays in two patients who still have the implants in place after more than 25 yr. Dobelle reported on the functioning of the arrays in one of the patients in 2000. Activation of the electrodes allowed the patient to perceive enough points of light to independently navigate through, and locate objects in a controlled environment. The implant has led to a significant increase in the patient's mobility, and Dobelle claimed that the prosthesis allowed the (previously) blind patient to recognize 6 in. characters at 5 ft, which corresponds to a visual acuity of approx 20/120 *(11)*. The Dobelle group went into commercial production and implantation of the device and has reported, on their website, that another eight patients have recently been implanted with a prosthetic device, but no details of the implants or function of the device have yet been published.

An alternative to surface stimulation of the cortex is the use of penetrating electrodes to provide intracortical microstimulation. A cortical prostheses consisting of 100 penetrating electrodes, each about 1–1.5 mm long with a tip diameter of 2–3 μm, has recently been developed *(17)*. Intracortical microstimulation has a number of advantages in surface stimulation: lower current thresholds are required for stimulation, increased numbers of electrodes in an array are possible, which may result in higher phosphene resolution. However, there are significant disadvantages associated with an array of penetrating electrodes, such as the difficulty of achieving safe insertion of a large number of penetrating electrodes, the risk of array migration and electrode breakage, and the increased risk of tissue damage because of high charge densities generated at the very fine tips of the electrodes.

Stimulation of the Visual Cortex With a Surface Electrode Array

Electrical stimulation using surface electrode arrays has been demonstrated to be safe and effective for chronic stimulation of the visual cortex *(11,13)*. In recent years, a research program has been carried out to develop a visual cortical prosthesis using a surface electrode array and using transcutaneous stimulation technology developed by Cochlear Ltd. (Sydney, Australia) for the cochlear implant (bionic ear). The aim in the development of a visual prosthetic device is to safely elicit the perception of the largest possible number of distinct and reproducible phosphenes that occupy a stable position in the patient's visual field *(9,10)*. To evaluate the design of electrode arrays and stimulus paradigms a cat model was developed, in which the neural activity resulting from electrical stimulation of the visual cortex using surface electrodes in one cortical hemisphere is assessed by recording evoked cortical responses on the contralateral hemisphere *(18,19)*. The transcallosal pathways provide a direct and localized connection between homotopic points on the cortical surface of the two hemispheres *(20–25)*. The electrode arrays, which have been used in the studies are provided by Cochlear Ltd., and are modified from the electrode array used in the *Nucleus 22 Auditory Brainstem Implant,* which is in clinical use for subdural stimulation of the brainstem *(26)*. The prototype electrode grid is shown in Fig. 1, and consists of 21 active platinum disk electrodes, mounted on a 3 mm by 8.5 mm silicone carrier. Each electrode is 0.7 mm diameter arranged in three rows with 0.95 mm center-to-center spacings.

Both mono- and biphasic bipolar electrical stimulation of visual regions of the cerebral cortex have been used in the anaesthetized cat to investigate the feasibility of the surface electrode array to function as a visual prosthesis. Bipolar monophasic stimulation

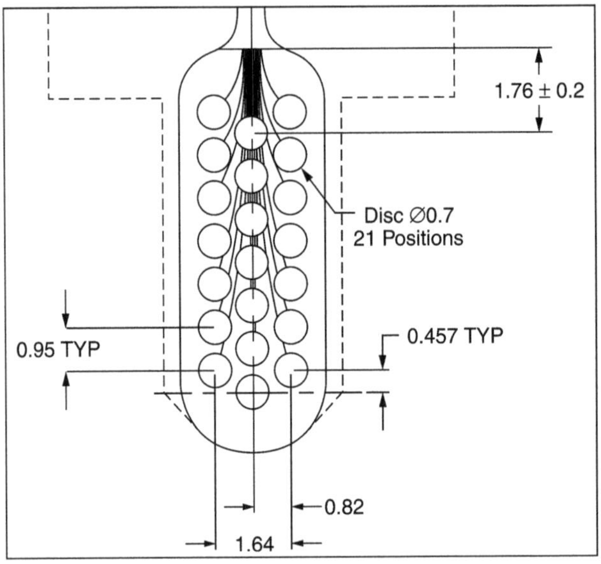

Fig. 1. Diagram of electrode array used in these studies. The array consists of 21 platinum electrodes in a silicon carrier and is modified from the electrode array used in the *Nucleus 22 Auditory Brainstem Implant* manufactured by Cochlear Ltd.

using adjacent electrodes on the stimulating array yielded a transcallosal evoked response (TER) in the opposite hemisphere that was characterized by a positive peak, at a latency of around 7 ms, followed by a negative trough, at a latency of around 25 ms (Fig. 2A), with the amplitude of the initial positive peak of the TER-dependant on both stimulus current intensity and pulse width (Fig. 2B).

The effect of stimulus polarity on evoked cortical activity with electrodes on the array arranged in a monopolar configuration was assessed by comparing TER thresholds for a range of monophasic stimulus pulse widths. Cathodal monopolar stimuli consistently demonstrated lower thresholds for surface cortical stimulation than anodal stimuli. Changes in interelectrode spacing caused marked shifts in the amplitude of the TER. Electrodes placed at a spacing of 0.95 mm evoked lower amplitude responses than electrodes placed at a spacing of 2.85 mm. However, electrodes at a wider spacing of 4.75 mm did not give markedly different responses compared with a spacing of 2.85 mm. The reason for the low-response amplitudes during bipolar stimulation of electrodes at a spacing of 0.95 mm may be because of the current path shunting between electrodes instead of passing through the cortical tissue *(27)*. This has significant implications for the design of electrode arrays for a visual prosthesis, and suggests that electrodes should be arranged further apart to elicit more effective cortical stimulation. If the poor responses at such interelectrode separations correlate with difficulties in eliciting phosphene perception at safe current intensities, it may set a lower limit to the spacing of electrodes that is possible with a visual prosthesis.

The application of charge to a target tissue with a monophasic pulse of one polarity is sufficient to depolarize excitable cells *(27)*. However, for neuroprosthetic applications electrical stimulation commonly uses biphasic pulses, whereby a second pulse of equal but opposite polarity immediately follows the first *(28)*. This prevents tissue

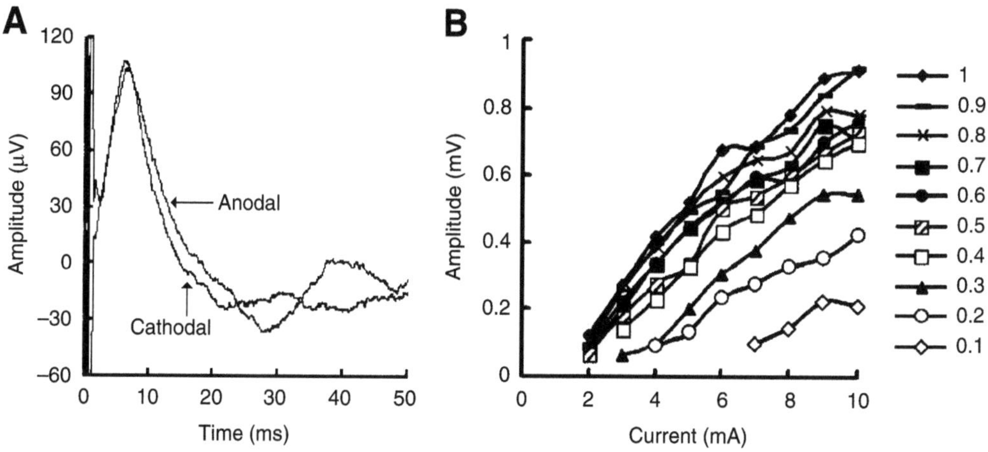

Fig. 2. (A) The transcallosal evoked response is a biphasic wave consisting of a positive peak after the stimulation artifact, followed by a negative peak. Reversing the polarity of a bipolar stimulus inverts the stimulus artifact. The "polarity" of the stimulus refers to the initial deflection of the artifact. **(B)** The change in transcallosal evoked response amplitude with changes in current intensity. The different lines on the graph named 0.1–1 refer to the stimulus pulse width in milliseconds. (Figure modified from ref. *18*.)

injury by ensuring that there is no *net* injection of charge to the tissue being stimulated *(29)*. Charge accumulation in tissue can lead to toxic effects at the electrocatalytic interface, such as the deposition of metal ions in the target tissue *(27,30)*. However, higher current intensities of biphasic stimuli are required to reach threshold (Fig. 3), and therefore present an increased risk of tissue injury from higher charge densities at the electrode-tissue interface *(31)*.

The ideal pulse duration for cortical stimulation with a neural prosthesis depends on both neurophysiological and device engineering considerations. The use of longer pulse durations allows lower stimulus current intensities to be used, but also leads to an increased charge density at the electrode–tissue interface (Fig. 4A). Shorter pulse widths, which decrease threshold charge density, require higher stimulus currents that require a higher voltage to be delivered by the implanted neurostimulator (Fig. 4B). This also causes higher energy delivery to the tissue, which may cause adverse tissue heating effects *(27)*. The ideal pulse width may depend on aspects of electrode configuration, such as the interelectrode separation (Fig. 4). The use of different pulse widths for different stimulus configurations may further improve the efficiency of a neuroprosthetic device.

Chronic stimulation of the visual cortex using surface electrodes has been shown to be safe and effective. Further development of visual prostheses based on surface stimulation will require developing and evaluating new electrode configurations and stimulus paradigms that effectively activate different retinotopic areas of the visual cortex, and generate resolvable visual phosphenes from a limited number of implanted electrodes. The preliminary studies into cortical stimulation with multielectrode arrays in the cat provide the basis for investigating new stimulus paradigms. These physiological results in the anaesthetized cat will need to be correlated with the visual sensations associated with cortical stimulation in psychophysical studies in human patients.

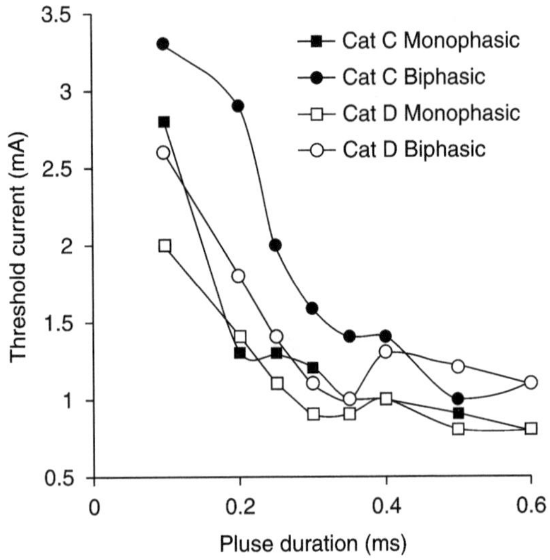

Fig. 3. The current intensity required to elicit a threshold TER (more than 30 µV peak-to-peak amplitude) for monopolar monophasic cathodal stimuli and monopolar biphasic cathodal-first stimuli. (Figure from ref. *19.*)

RETINAL PROSTHESES

The development of a retinal prosthesis has been the most active area of visual prosthesis research in recent times *(32)*. A number of technical issues have been identified in the development of intraocular retinal prostheses *(33)*. The use of epiretinal/subretinal electrodes *(32)* or indeed suprachoroidal–transretinal stimulation with a return electrode placed in the vitreous chamber *(34,35)* requires invasive intraocular surgery, which may further damage the already diseased eye. The implanted components may cause mechanical damage to the intraocular tissues, and generate foreign body reactions. If the stimulation electronics are implanted wholly within the globe, this will generate challenges with ensuring hermetic encapsulation of the device, puts the retina at risk of damage from heat effects *(36)*, and makes replacement of damaged devices a difficult procedure. The alternative use of a perscleral *(37)* connection to extraocular electronics creates a defect in the integrity of the globe, which may be a pathway for infection. It has proved difficult to safely and stably implant electrode arrays in the subretinal space or on the epiretinal surface. Subretinal arrays require a transretinal *(38)* or transchoroidal *(39)* approach for implantation, both of which may damage the retina or impair its blood supply from the choroidal circulation, while epiretinal arrays need to be attached with tacks *(40)* or adhesives *(41)*. Both the arrays and the attachment methods may cause mechanical damage to the retina, and the electrodes may be dislodged because of the high velocities generated during saccadic movements of the eye.

Extraocular Retinal Stimulation

An alternative approach to an intraretinal implant for electrical stimulation of the underlying retina is an extraocular retinal prosthesis (ERP), which uses electrodes sutured to the scleral surface of the globe *(42–44)*. Such a device would avoid many

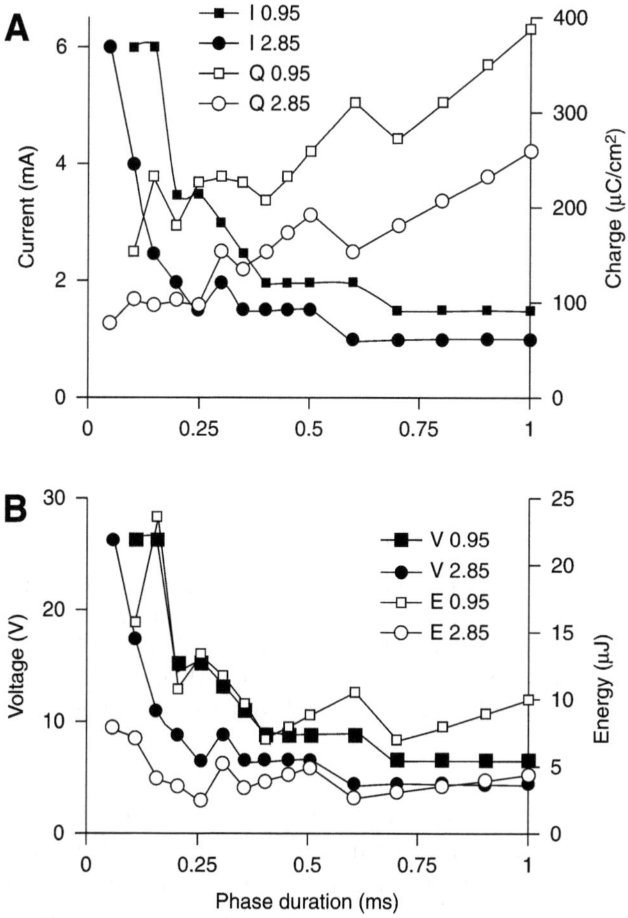

Fig. 4. **(A)** Threshold current and threshold charge density for eliciting a 100 μV transcallosal evoked response for a bipolar biphasic stimulus with phase durations between 100 μs and 1000 μs, for interelectrode spacings of 0.95 mm and 2.85 mm. **(B)** Calculations of voltage and energy at current intensities for eliciting a threshold TER of 100 μV for interelectrode spacings of 0.95 mm and 2.85 mm. (Figure from ref. *19.*)

of the difficulties associated with positioning electrodes within the cavity of the eye, and could be a simple way of restoring basic visual sensations to blind and severely visually impaired patients. However, the greater distance from the retina of the ERP would require larger electrodes and higher stimulus currents *(45)*, which will limit the number and resolution of phosphenes that can be generated by an ERP, a distinct disadvantage compared with the resolution proposed for intraocular prostheses. Despite a lower phosphene resolution, the generation of even a few localized visual sensations may be of considerable benefit to blind patients *(6)*, and at the very least may assist in restoring disturbed circadian rhythms *(46)*.

Recently, a series of experiments was begun to investigate the feasibility of developing a low-resolution ERP for blind patients. As far as there is awareness, the only report of extraocular retinal stimulation is anecdotal *(47)* and no systematic animal or human study using a multielectrode ERP has been performed *(32)*. To demonstrate the feasibility of an

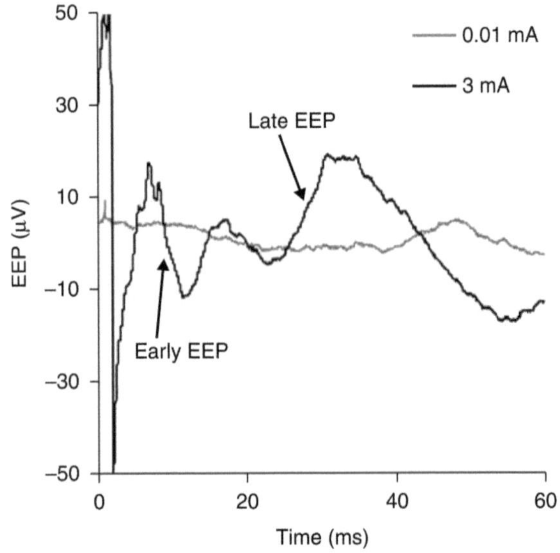

Fig. 5. Electrical evoked potential (EEP) recorded from the ipsilateral cortex from bipolar stimulation between electrodes A and B. Single biphasic stimuli with a phase width of 1 ms (resulting in 2 ms pulses) were delivered with electrode A cathodal in the first phase. A suprathreshold stimulus of 3 mA elicits a characteristic EEP with early positive–negative and late negative–positive wave components. No EEP is elicited with a subthreshold stimulus of 0.01 mA. (Figure from ref. *43*.)

ERP, and evaluate stimulus and electrode configurations, we needed to develop a suitable electrode array, demonstrate its method of attachment to the eye, show that stimulation through the sclera can elicit retinal excitation and that the charge density for retinal stimulation is at a safe level for chronic neural stimulation. In order to evaluate whether useful sensations can be produced we also needed to show that an ERP causes localized retinal stimulation, which can be expected to produce localized phosphenes in the visual field for blind patients. We also needed to obtain information about the effects of electrode spacing and orientation, which will allow to design prosthesis suitable for human implantation.

The electrode array used in the ERP studies is the same array that was used in the studies of a cortical visual prosthesis (Fig. 1). The multielectrode array is easily attached to the scleral surface of the eye by suturing the silicon carrier directly to the sclera, with the long axis of the array parallel to the horizontal meridian of the eye *(44)*. Bipolar biphasic stimulation of the underlying retina was performed between pairs of electrodes on the array, and its effectiveness was assessed by recording cortical evoked activity (electrical evoked potential [EEP]) using a low-impedance subdural electrode positioned on the cortical surface over the primary visual cortex (area 17). Bipolar stimulation with the array is effective in exciting the retina and eliciting an EEP at the visual cortex (Fig. 5). The EEP waveform is similar to those described in previous studies of retinal stimulation in cats *(35,48,49)*.

Biphasic pulses with a current intensity of 500 μA and phase duration of 500 μs were able to elicit an EEP with amplitude 1.5 times the average noise amplitude. The

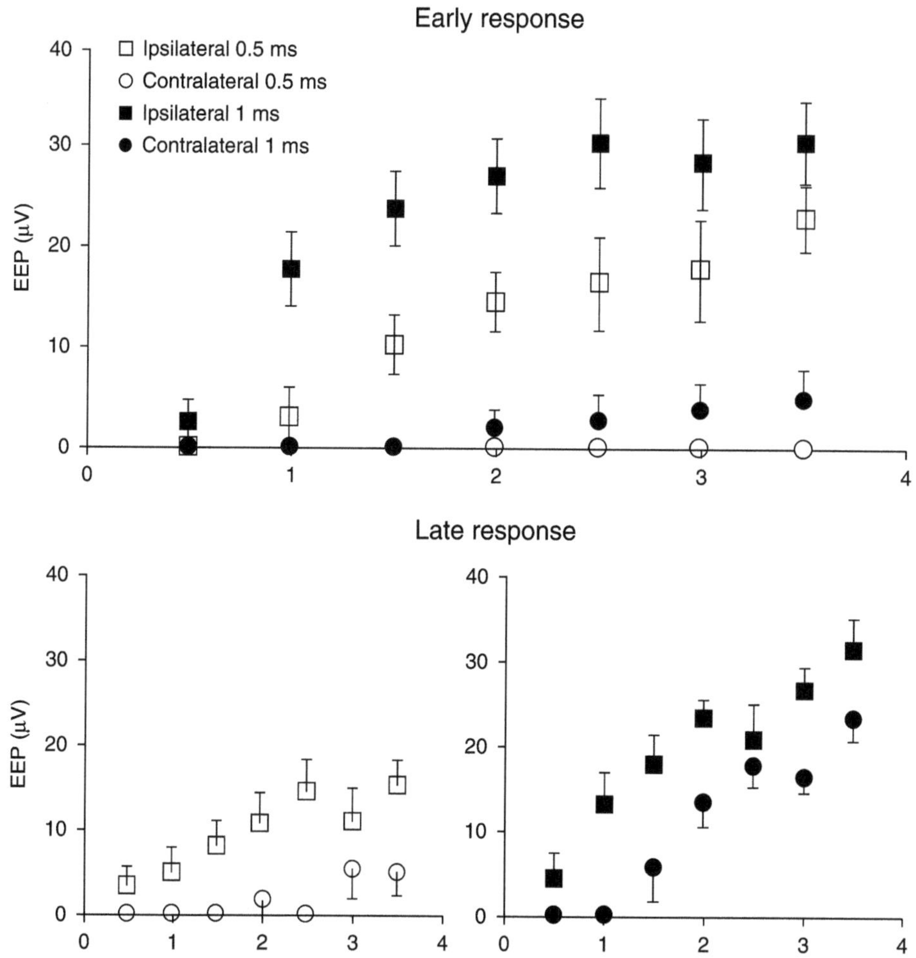

Fig. 6. Amplitudes of the early components of the electrical evoked potential (*Early Response* – Upper graph), and the late components of the electrical evoked potential (*Late Response* – Lower graph). Bipolar biphasic stimulation between adjacent electrodes A (cathodal first phase) and B (anodal first phase) was carried out with 500 µs (open points) and 1000 µs (solid points) pulses. Data points are mean +/– SE. Responses recorded from the ipsilateral cortex are shown as squares, and responses recorded from the contralateral cortex are shown as circles. For the early response (upper graph) responses for both stimulus pulse durations are shown in the same figure. For the late response (lower graph), responses from 500 µs pulses are shown in the left

diameter of electrodes on the array was 700 µm, giving a surface area of 3.85×10^{-3} cm^2, resulting in a threshold charge density of 64.96 µC/cm^2, which is lower than the safe limits for charge injection with platinum electrodes *(10,50)*. In the cat as in the human, the lateral retina projects to the ipsilateral primary visual cortex and the fibers from the medial retina cross at the optic chiasm to pass to the contralateral primary visual cortex. The ERP in the experiments was laterally placed on the globe of the eye, and if localized stimulation of the retina was occurring then stimulation should produce an

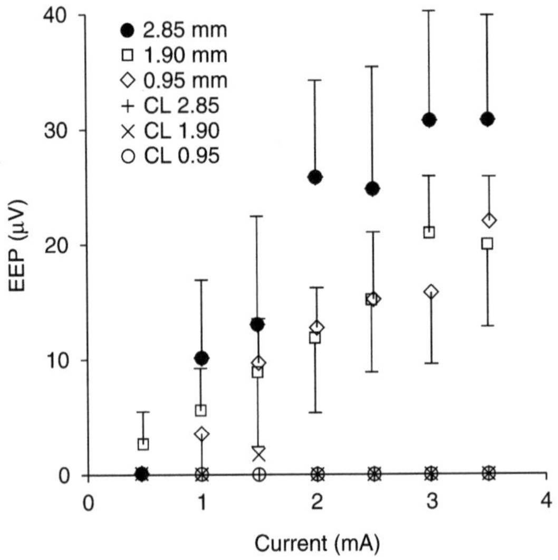

Fig. 7. Average ipsilateral and contralateral electrical evoked potential responses for extraocular stimulation with 500 μs pulses through electrode pairs at center-to-center interelectrode separations of 0.95 mm, 1.90 mm, and 2.85 mm. Ipsilateral responses are labeled according to interelectrode separation. Contralateral responses are labeled with "CL" before spacing distance. Data points are mean +/– SE. (Figure from ref. *43.*)

EEP that was localized to the ipsilateral primary visual cortex. This was clearly the case with the ERP (Fig. 6), and with one hemisphere activated, any phosphenes generated would be expected to be localized to 50% of the visual field *(51)*. Further estimation of phosphene localization using an ERP will require more detailed retinal stimulation and mapping of the visual cortical responses in animal models and psychophysical studies with human patients.

As was the case with cortical stimulation using this multielectrode array, retinal stimulation with electrodes at a spacing of 2.85 mm elicited the highest EEP responses. This may be because more area of the retina is activated, or because it decreases the shunting of current that may occur between closely spaced electrodes *(27)*. Even with the largest electrode spacing tested EEPs were still restricted to the ipsilateral hemisphere (Fig. 7).

By placing electrodes on the external scleral surface of the eye, the need for invasive intraocular surgery, placement of a foreign body within the cavity of the eye, and miniaturization of stimulation electronics is avoided, as is the possibility of direct injury to the retina. Although, an ERP would only function as a low-resolution prosthesis, the results of the experiments suggest that it may be a feasible approach to develop a retinal prosthesis. Further experiments on animal models with retinal degeneration, and more detailed cortical mapping will allow to assess more fully the localization of retinal stimulation, and evaluate thresholds for stimulation of the diseased retina. Human studies will be needed to evaluate the psychophysical properties of the phosphene sensations produced by extraocular retinal stimulation.

CONCLUSIONS

The aim of the research is to develop visual prostheses for blind and severely visually impaired patients that could be adapted from the neurostimulation and electrode technology that is already available in other neuroprosthetic devices, such as cochlear implants, auditory brainstem implants, and devices for functional neuromuscular stimulation. This would avoid the need to develop new neural stimulation technology, and allow to use devices and materials that already had a good safety and biocompatibility profile in human trials. A visual prosthesis needs to be safe and effective in activating the visual system to produce phosphene sensations and characteristics that are achievable using surface electrode arrays. Such a device could lead to the development of a clinically useful visual prosthesis in the short-term, and allow more detailed studies of phosphene perception to occur in chronic human trials.

ACKNOWLEDGMENTS

The authors' research described in this chapter has been supported by Ophthalmic Research Institute of Australia, The University of New South Wales, Retina Australia, the Brain Foundation, and the National Health & Medical Research Council, Australia. Some equipment used in this research were provided at no expense by Cochlear Ltd.

REFERENCES

1. Weih L, McCarty CA, Taylor HR. Functional implications of vision impairment. Clin Exp Ophthalmol 2000;28(3):153–155.
2. Sack RL, Lewy AJ, Blood ML, Keith LD, Nakagawa H. Circadian rhythm abnormalities in totally blind people: incidence and clinical significance. J Clin Endocrinol Metab 1992;75(1):127–134.
3. DeLeo D, Hickey PA, Meneghel G, Cantor CH. Blindness, fear of sight loss, and suicide. Psychosomatics 1999;40(4):339–344.
4. Attebo K, Mitchell P, Smith W. Visual acuity and the causes of visual loss in Australia. The Blue Mountains Eye Study. Ophthalmology 1996;103(3):357–364.
5. Chiang YP, Bassi LJ, Javitt JC. Federal budgetary costs of blindness. Milbank Q 1992; 70(2):319–340.
6. Ross RD. Is perception of light useful to the blind patient? (comment). Arch Ophthalmol 1998;116(2):236–238.
7. Krumpaszky G, Klauss V. Epidemiology of blindness and eye disease. Ophthalmologica 1996;210(1):1–84.
8. Sharma RK, Ehinger B. Management of hereditary retinal degenerations: present status and future directions. Surv Ophthalmol 1999;43(5):427–444.
9. Maynard EM. Visual prostheses. Ann Rev Biomed Eng 2001;3:145–168.
10. Margalit E, Maia M, Weiland JD, et al. Retinal prosthesis for the blind. Surv Ophthalmol 2002;47(4):335–356.
11. Dobelle WH. Artificial vision for the blind by connecting a television camera to the visual cortex. ASAIO J 2000;46(1):3–9.
12. Le Roy M. (Mémoire) Ou l'on rende compte de quelques tentatives que l'on a faites pour guérir plusieurs maladies par l'électricité. Mém Mat Phys Acad Roy Sci Paris 1775;60:98.
13. Brindley GS, Lewin WS. The sensations produced by electrical stimulation of the visual cortex. J Physiol (Lond) 1968;196(2):479–493.

14. Button J, Putnam T. Visual responses to cortical stimulation in the blind. J Iowa State Med Soc 1962;52:17–21.
15. Dobelle WH, Mladejovsky MG. Phosphenes produced by electrical stimulation of human occipital cortex, and their application to the development of a prosthesis for the blind. J Physiol (Lond) 1974;243(2):553–576.
16. Dobelle WH, Quest DO, Antunes JL, Roberts TS, Girvin JP. Artificial vision for the blind by electrical stimulation of the visual cortex. Neurosurgery 1979;5(4):521–527.
17. Normann RA, Maynard EM, Rousche PJ, Warren DJ. A neural interface for a cortical vision prosthesis. Vision Res 1999;39(15):2577–2587.
18. Chowdhury V, Morley JW, Coroneo MT. An in-vivo paradigm for the evaluation of stimulating electrodes for use with a visual prosthesis. Aust N Z J Surg 2004;74(5):372–378.
19. Chowdhury V, Morley JW, Coroneo MT. Surface stimulation of the brain with a prototype array for a visual cortex prosthesis. J Clin Neurosci 2004;11(7):750–755.
20. Curtis HJ. Intercortical connections of corpus callosum as indicated by evoked potentials. J Neurophysiol 1940;3:405–413.
21. Curtis HJ. An analysis of cortical potentials mediated by the corpus callosum. J Neurophysiol 1940;3:414–421.
22. Chang HT. Cortical response to activity of callosal neurons. J Neurophysiol 1953;16: 117–131.
23. Peacock SM. Activity of anterior suprasylvian gyrus in response to transcallosal afferent volleys. J Neurophysiol 1957;20:140–155.
24. Kawamura K. Corticocortical fiber connections of the cat cerebrum. 2. The parietal region. Brain Res 1973;51:23–40.
25. Testerman RL. The cortical response to callosal stimulation: a model for determining safe and efficient stimulus parameters. Ann Biomed Eng 1978;6(4):438–452.
26. Sollmann WP, Laszig R, Marangos N. Surgical experiences in 58 cases using the Nucleus 22 multichannel auditory brainstem implant. J Laryngol Otol Suppl 2000;27:23–26.
27. Jayakar P. Physiological principles of electrical stimulation. Adv Neurol 1993;63:17–27.
28. Finn WE, LoPresti PG. Handbook of neuroprosthetic methods. Boca Raton: CRC Press, 2003.
29. Yeomans JS. Principles of Brain Stimulation. New York: Oxford University Press, 1990.
30. McCreery D, Agnew WF. Neuronal and axonal injury during functional electrical stimulation. Ann Int Conf IEEE Eng Med Biol Soc 1990;12(4):1488–1489.
31. McCreery DB, Agnew WF, Yuen TG, Bullara L. Charge density and charge per phase as cofactors in neural injury induced by electrical stimulation. IEEE Trans Biomed Eng 1990;37(10):996–1001.
32. Weiland J, Liu W, Humayun MS. Retinal prosthesis. Annu Rev Biomed Eng 2005;7: 361–401.
33. Rizzo JF, Wyatt J, Humayun M, et al. Retinal prosthesis: an encouraging first decade with major challenges ahead. Ophthalmology 2001;108(1):13–14.
34. Kanda H, Morimoto T, Fujikado T, Tano Y, Fukuda Y, Sawai H. Electrophysiological studies of the feasibility of suprachoroidal-transretinal stimulation for artificial vision in normal and RCS rats. Invest Ophthalmol Vis Sci 2004;45(2):560–566.
35. Sakaguchi H, Fujikado T, Fang X, et al. Transretinal electrical stimulation with a suprachoroidal multichannel electrode in rabbit eyes. Jpn J Ophthalmol 2004;48(5):515.
36. Piyathaisere DV, Margalit E, Chen SJ, et al. Heat effects on the retina. Ophthalmic Surg Lasers Imaging 2003;34(2):114–120.
37. Humayun MS, Weiland JD, Fujii GY, et al. Visual perception in a blind subject with a chronic microelectronic retinal prosthesis. Vision Res 2003;43(24):2573–2581.
38. Chow AY, Pardue MT, Chow VY, et al. Implantation of silicon chip microphotodiode arrays into the cat subretinal space. IEEE Trans Neural Syst Rehabil Eng 2001;9(1):86–95.

39. Shinoda K, Gekeler F, Eckert E, Zrenner E, Gabel V-P, Kobuch K. Externo-Implantation, -explantation and postoperative follow up of subretinal electronic devices. ARVO Meeting Abstracts 2002;43(12):4471.
40. Majji AB, Humayun MS, Weiland JD, Suzuki S, D'Anna SA, de Juan E Jr. Long-term histological and electrophysiological results of an inactive epiretinal electrode array implantation in dogs. Invest Ophthalmol Vis Sci 2002;40(9):2073–2081.
41. Margalit E, Fujii GY, Lai JC, et al. Bioadhesives for intraocular use. Retina 2000;20(5): 469–477.
42. Chowdhury V, Morley JW, Coroneo MT. Evaluation of extraocular electrodes for a retinal prosthesis from evoked potentials in cat visual cortex. J Clin Neurosci 2005a;12(5): 574–579.
43. Chowdhury V, Morley JW, Coroneo MT. Stimulation of the retina with a multielectrode extraocular visual prosthesis. Aust N Z J Surg 2005;75:697–704.
44. Chowdhury V, Morley JW, Coroneo MT. Feasibility of Extraocular Stimulation for a Retinal Prosthesis. Can J Ophthalmol 2005c;40:563–572.
45. Ranck JB. Extracellular Stimulation. In: Patterson MM, Kesner RP, eds. Electrical stimulation research techniques. New York: Academic Press, 1981:1–36.
46. Chiquet C, Dkhissi-Benyahya O, Cooper HM. Is the study of blind patients useful for understanding light perception? Arch Ophthalmol 1999;117(6):848.
47. Humayun MS, Prince M, de Juan E Jr., et al. Morphometric analysis of the extramacular retina from postmortem eyes with retinitis pigmentosa. Invest Ophthalmol Vis Sci 1999;40(1): 143–148.
48. Dawson WW, Radtke ND. The electrical stimulation of the retina by indwelling electrodes. Invest Ophthalmol Vis Sci 1977;6(3):249–252.
49. Schanze T, Wilms M, Eger M, Hesse L, Eckhorn R. Activation zones in cat visual cortex evoked by electrical retina stimulation. Graefes Arch Clin Exp Ophthalmol 2002;240(11): 947–954.
50. Robblee LS, Cogan SF. Metals for medical electrodes. In: Bever MB, ed. Encyclopedia of materials science and engineering. Oxford Pergamon Press, 1986; pp. 276–281.
51. Lamme VA, Super H, Landman R, Roelfsema PR, Spekreijse H. The role of primary visual cortex (V1) in visual awareness. Vision Res 2000;40(10–12):1507–1521.

12
Glaucoma Drainage Devices
Advances in Design and Surgical Techniques

Cheryl L. Cullen, DVM, MVetSc

Contents

INTRODUCTION

Glaucoma is a group of optic neuropathies that share a slowly progressive degeneration of the retinal ganglion cells and their axons, resulting in a distinct appearance of the optic disc and a concurrent pattern of vision loss *(1)*. Glaucoma is the second leading cause of blindness in the world with estimates that it affects 66.8 million individuals worldwide *(2)*; at least 6.7 million of these affected people suffer from complete blindness *(2)*. The vision loss associated with this disease is irreversible, yet the biological basis of glaucoma and the factors contributing to its progression have not been completely elucidated *(1)*. Intraocular pressure (IOP) is the only proven treatable risk factor in glaucoma *(1)*. As such, glaucoma is theoretically defined as a progressive optic neuropathy as a result of elevation of IOP above the physiological level of individuals *(3)*. The upper limit of "normal" IOP, based on a large number of subjects, is internationally accepted as being approx 21 mmHg as a standard in the clinical diagnosis of glaucoma *(3)*.

There are numerous medical and surgical strategies used in the management of glaucoma. The goal of this chapter is to briefly review the physiology of aqueous humor production and drainage, retinal, and optic nerve anatomy pertinent to glaucoma, and the pathophysiology of glaucoma. Further, the main emphasis of this chapter will be on glaucoma drainage devices (GDDs). In particular, historical GDD designs will be reviewed, and the remainder of the chapter will address current designs and surgical strategies used with GDDs for the management of glaucoma, the complications following GDD implantation, and the challenges encountered in developing and using these devices in the eye.

From: *Ophthalmology Research: Visual Prosthesis and Ophthalmic Devices: New Hope in Sight*
Edited by: J. Tombran-Tink, C. Barnstable, and J. F. Rizzo © Humana Press Inc., Totowa, NJ

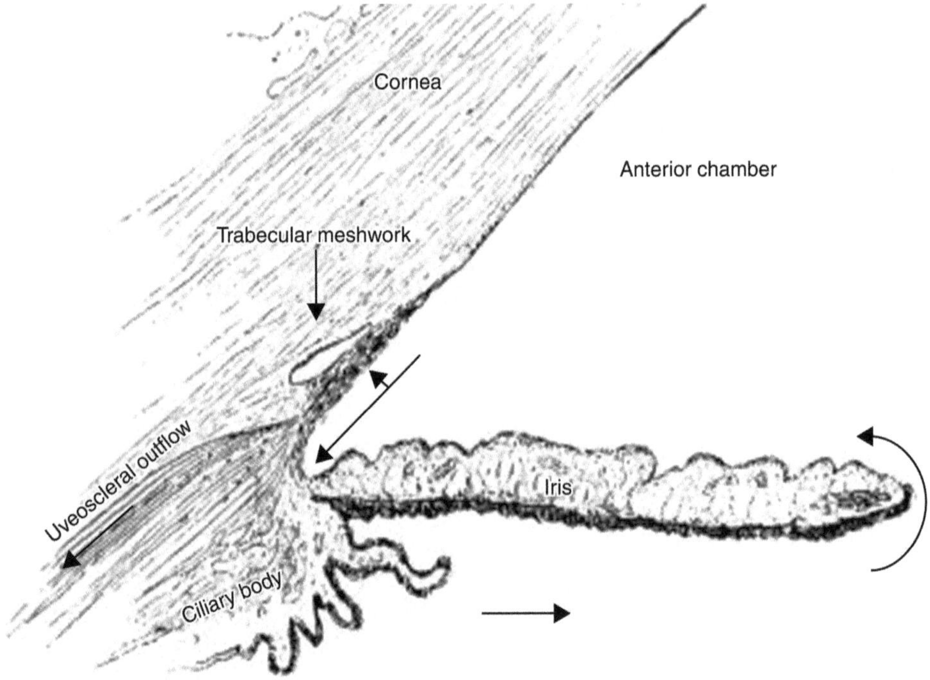

Fig. 1. Diagram illustrating aqueous humor dynamics. Arrows indicate the direction of aqueous humor flow from the ciliary body (located in the posterior chamber) through the pupil, into the anterior chamber, and exiting the eye through the trabecular meshwork (conventional outflow pathway) or uvea (uveoscleral/unconventional outflow pathway) (Reprinted with permission from Elsevier *[1]*).

Ocular Anatomy and Physiology

Aqueous Humor Dynamics

IOP is regulated by a balance between the secretion and drainage of aqueous humor from the eye (Fig. 1). Aqueous humor is secreted posterior to the iris by the nonpigmented ciliary epithelium of the ciliary body and this fluid then flows anteriorly through the pupil to the anterior chamber. Aqueous humor exits the eye into the venous circulation through the trabecular meshwork (conventional outflow pathway) and independently through the uveoscleral pathway (unconventional outflow pathway) (Fig. 1).

Retinal Ganglion Cells and Optic Nerve

Axons from the retinal ganglion cells consist of the innermost layer of the retina, the nerve fiber layer (Fig. 2B). These axons converge on the optic disc and form the optic nerve, which contains a central depression called the cup. Most optic nerves have a visible physiological cup, which is surrounded by a neuroretinal rim (Fig. 2A). The human optic nerve contains approx 1 million nerve fibers (Fig. 2C), which exit the eye after passing through the lamina cribrosa, a series of perforated connective tissue sheets, and synapse in the lateral geniculate nucleus of the brain *(1)*. Trophic factors are transported both retrogradely from the axonal terminals of the retinal ganglion cells to their cell bodies in the inner retina as well as anterogradely from the retinal ganglion cell

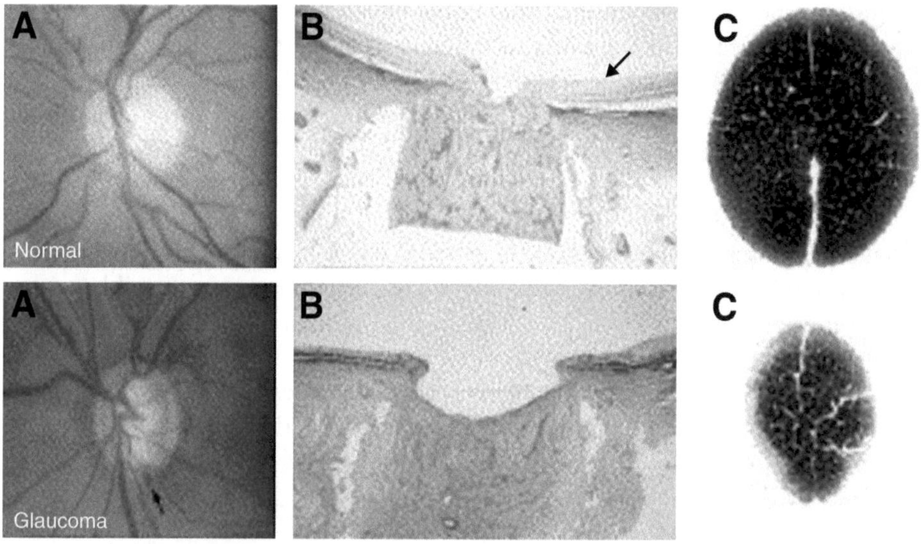

Fig. 2. (A) Fundic photograph from nonglaucomatous (top) and glaucomatous (bottom) human eyes. Note the enlarged optic cup, thinned neuroretinal rim, and focal optic disc hemorrhage in the glaucomatous compared with the nonglaucomatous optic disc. **(B)** Photomicrograph of optic nerve and peripapillary retina of nonglaucomatous (top) and glaucomatous (bottom) eyes. The arrow depicts the innermost nerve fiber layer. Note the optic disc cupping and the inner retinal degeneration in the glaucomatous globe. **(C)** Transverse section of a nonglaucomatous (top) and glaucomatous (bottom) optic nerve. Note the reduction in the diameter of the glaucomatous optic nerve compared with the nonglaucomatous optic nerve (Reprinted with permission from Elsevier *[1]*).

axons to the lateral geniculate nucleus *(1)*. Survival of neuronal cells in these regions depends on the transport of these trophic factors.

Pathophysiology of Glaucoma

Glaucoma is a neurodegenerative disease characterized by a slowly progressive loss of retinal ganglion cells. The pathophysiology of neurodegeneration in glaucoma is not completely understood. However, nearly all of the glaucomas in humans and other animals develop from impairment of aqueous humor drainage rather than from increased rates of aqueous humor secretion *(4)*. The resistance to the outflow of aqueous humor is generated mainly by the trabecular meshwork (Fig. 1) and/or the juxta canalicular connective tissues *(4)*. IOP becomes elevated when this outflow resistance increases thereby causing damage to the retinal ganglion cells and optic nerve. In particular, when IOP is elevated above physiological levels, the pressure gradient across the lamina cribrosa increases, thereby causing mechanical stress across the lamina cribrosa and retinal ganglion cells *(5)*. Compression of the axons of the retinal ganglion cells can result in impaired transport of trophic factors and subsequent cell death because of trophic insuffficency *(1)*. Other factors including retinal ischemia–hypoxia *(6)*, excessive excitatory amino acid glutamate *(7)*, inflammatory cytokines *(8)*, oxidative stress, and formation of free radicals *(9)*, and aberrant immunity *(10)* may also contribute to glaucomatous neurodegeneration. When the optic nerve develops a reduction in the number of nerve fibers, a decrease in the width of the neuroretinal rim with concomitant enlargement of the cup of the optic disc is seen clinically (Fig. 2A) *(1)*. In addition to glaucomatous neuronal cell death of the retinal ganglion cells, neurons in the lateral geniculate nucleus and the visual cortex are also affected *(11)*.

GLAUCOMA FILTRATION SURGERY

The goal of glaucoma filtration surgery (GFS) is to lower IOP by increasing aqueous humor outflow. Open-angle glaucoma (OAG), the second leading cause of irreversible blindness in people in the United States *(12)*, is one of many forms of glaucoma for which GFS has been used. In particular, GFS has been shown to be more effective at arresting disease progression than other primary therapeutic modalities in OAG *(13)*. If poor aqueous humor flow control and its complications could be avoided, primary GFS would likely be offered more widely *(13)*.

Trabeculectomy

The most effective GFS, and hence, the procedure of choice in conventional GFS is trabeculectomy *(13)*. In this procedure, a guarded channel is created, allowing aqueous humor to flow from the anterior chamber into the sub-Tenon's and subconjunctival space, thereby bypassing the blocked trabecular meshwork *(14)*. A drainage bleb (aqueous humor under Tenon's capsule and the bulbar conjunctiva) is commonly detectable under the upper eyelid *(14)*. The most common cause for failure of trabeculectomy is episcleral fibrosis, resulting in blockage of aqueous humor drainage from the eye. The introduction of antimetabolite agents, such as mitomycin C and 5-fluorouracil applied intra- or postoperatively, has aided wound healing at the trabeculectomy site and has reduced the risk of filtration failure secondary to fibrosis *(14–16)*. However, flow control remains inexact and postoperative hypotony as a result of overfiltration may arise as a complication despite implementing various suture adjustment techniques *(13)*.

Glaucoma Drainage Devices

GDDs create alternate pathways for the diversion of aqueous humor by channeling aqueous from the anterior chamber through a long tube to various explant devices, the most common of which utilize an equatorial plate that promotes bleb formation. GDDs have potential advantages over conventional GFS (trabeculectomy), including the potential to consistently regulate aqueous humor outflow thereby avoiding hypotony *(13)*, and with the advent of newer designs and surgical techniques, the possibility of avoiding filtration failure secondary to fibrosis. However, despite these potential advantages, GDDs have generally been reserved for use in the management of complicated glaucomas that do not respond to medications and/or carry a high risk of failure from or do not respond to conventional GFS. These forms of complicated glaucoma include neovascular glaucoma, and glaucomas associated with pseudophakia or aphakia, uveitis, trauma, epithelial downgrowth, iridocorneal endothelial syndrome, vitreoretinal disorders, and penetrating keratoplasty *(1,17,18)*, and primary congenital glaucoma *(19)*.

The designs and materials used for GDDs have changed significantly during the past few decades. However, the surgical drainage sites to which aqueous humor has been diverted have not tended to vary substantially during that time with the most common drainage site being the subconjunctival space. Recent developments in GDDs are exciting given the advent of new biomaterials and flow control mechanisms, and the use of novel drainage sites. In addition, the occurrence of late-onset blebitis with or without endophthalmitis in up to 2.6% of eyes that have undergone mitomycin C trabeculectomy *(20)* provides increased incentive for continued interest in GDDs. As such, GDDs may become more commonly used in GFS.

Historical Perspectives

In 1906, horse hair was placed through a corneal paracentesis in an attempt to drain hypopyon out of the anterior chamber to the external cornea near the limbus *(21)*. The same technique was used in 1907 to treat two patients with glaucoma *(22)*. Since that time, several attempts to drain aqueous humor out of the anterior chamber and externally have been made with little success until 1969 *(18)*. These attempts have included the use of: (1) various materials, such as silk thread *(23)*, gold *(24)*, platinum *(25,26)*, magnesium *(27)*, lacrimal canaliculus *(28)*, tantalum *(27,29)*, glass rod *(30)*, supramid *(31)*, polyethylene *(32)*, and polyvinyl *(13,33)*, and/or (2) a variety of unconventional drainage sites, such as the vortex vein *(34)*, and the nasolacrimal duct *(35)*. Similar to contemporary GDDs, the most common drainage site used in which to divert aqueous humor was the subconjunctival space. Common complications encountered included lack of flow control and subsequent postoperative hypotony, chronic inflammation (i.e., uveitis and conjunctivitis) as a consequence of intracameral and subconjunctival foreign material, and early filtration failure.

Several decades later, GDD filtration failure was theorized to be the result of fibrosis in the bleb wall (subconjunctival fibrosis) *(36,37)*. Hence, Molteno introduced the concept of a large surface area required to divert aqueous humor. He developed a GDD utilizing a short acrylic tube placed in the anterior chamber and attached to a thin acrylic plate placed subconjunctivally, curved to fit the sclera, and sutured to the sclera. The tube was to guarantee the patency of the fistula, whereas the plate was to form the floor of the bleb, thereby preventing its shrinkage to an area less than that of the plate despite fibrosis *(36,37)*. However, most of these surgeries failed within 6 mo as a result of erosion of the plate through the conjunctiva, tube exposure, and subconjunctival fibrosis. Molteno eventually developed a GDD in 1973 that gained acceptance and remains the "gold standard" commercially available device against which other GDDs are compared. This new concept of the Molteno implant consisted of a long silicone tube attached to a large explant plate placed 9–10 mm posterior to the limbus *(38)*. This plate device was to act as a filtration reservoir despite continued subconjunctival fibrosis. However, the Molteno GDD provided no resistance to aqueous humor outflow, and severe postoperative hypotony, resulting in anterior chamber flattening, hyphema, and choroidal effusions, was a consequence.

Additional modifications arose to GDDs subsequent to the introduction of the Molteno implant including: (1) the use of a unidirectional valve to increase outflow resistance and prevent postoperative hypotony (e.g., slit valve of Krupin in 1976 *[39]*; Ahmed glaucoma valve in 1993 *[40]*); and (2) increased surface area of the explant plate (e.g., double plate implant by Molteno in 1981 *[41]*) to further lower IOP *(41–43)*.

Contemporary GDDs

Design

Currently, GDDs with tube and plate designs remain the predominant commercially available implants used for managing complicated glaucomas as an alternative to other forms of GFS, such as trabeculectomy. Key examples of these GDDs, including the year each was introduced and followed by the manufacturing company, are the Molteno (1979; Molteno Ophthalmic Ltd., Dunedin, New Zealand) (Fig. 3A), Krupin (1990; E. Benson Hood Laboratories, Inc., Pembroke, MA), Baerveldt (1990; Advanced Medical Optics, Inc., Santa Ana, CA), and Ahmed (1993; New World Medical, Inc., Rancho Cucamonga, CA). Contemporary designs have used the concept introduced earlier by Molteno with a silicone tube that opens onto an explant plate that is positioned posteriorly away from the limbus, and placed beneath Tenon's capsule, to avoid filtration failure associated with subconjunctival fibrosis and to prevent implant extrusion and erosion through the conjunctiva. The GDDs differ from one another in plate design and presence or absence of resistance mechanism to control outflow of aqueous humor. In particular, some of these devices offer no internal resistance mechanisms (e.g., Molteno and Baerveldt implants), others depend on tissue apposition to provide resistance to outflow (e.g., Molteno dual chamber single plate device [Fig. 3B,C] *[44]* and Baerveldt glaucoma implant with bioseal *[45]*), whereas some have nonadjustable resistance mechanisms, such as valves, which provide some resistance to outflow (e.g., Ahmed glaucoma valve [Fig. 4A,B] and Krupin slit valve).

Several modifications have been proposed for Molteno and similar GDDs, which offer no internal resistance mechanism and hence, result in postoperative hypotony. These adjustments in surgical technique include: (1) a two-stage procedure in which the device is implanted and allowed to encapsulate before intracameral tube insertion at a second procedure following 2–6 wk, and (2) modifications of a single-stage procedure including temporary ligature using absorbable suture material placed around the silicone tube *(46)*, internal occlusion of the tube with suture that is removed postoperatively *(47)*, or a combination of internal suture occlusion with external ligature placed around the tube *(48)*. These single stage methods provide transient occlusion of the implant to permit partial encapsulation of the explant plate. Laser suture lysis has been used to removed these suture ligatures *(46)*. Although, both the two- and single-stage modifications have successfully reduced complications with hypotony, the single stage procedures are preferred for GDDs with no internal resistance mechanism *(13)*.

Other contemporary GDDs used less commonly include: the Schocket tube implant in which the explant plate is an encircling silicone band, which is placed under the extraocular muscles around the globe at the equator and sutured to the sclera *(49)*; White glaucoma pump implant (Tamcenan Corp., Sioux Falls, SD) consisted of a reservoir sutured to the sclera posterior to the equator and a centrally located pump, which shunts aqueous from the anterior chamber to the reservoir by eyelid action and digital massage *(50,51)*; the Joseph implant, which is similar to the Krupin and Schocket GDDs in its mechanism of valve control *(52,53)*; and the Susanna drainage device (Tecprosil Ltd., Brazil), a nonvalved device, which combines features of both Molteno and Baerveldt implants *(54)*.

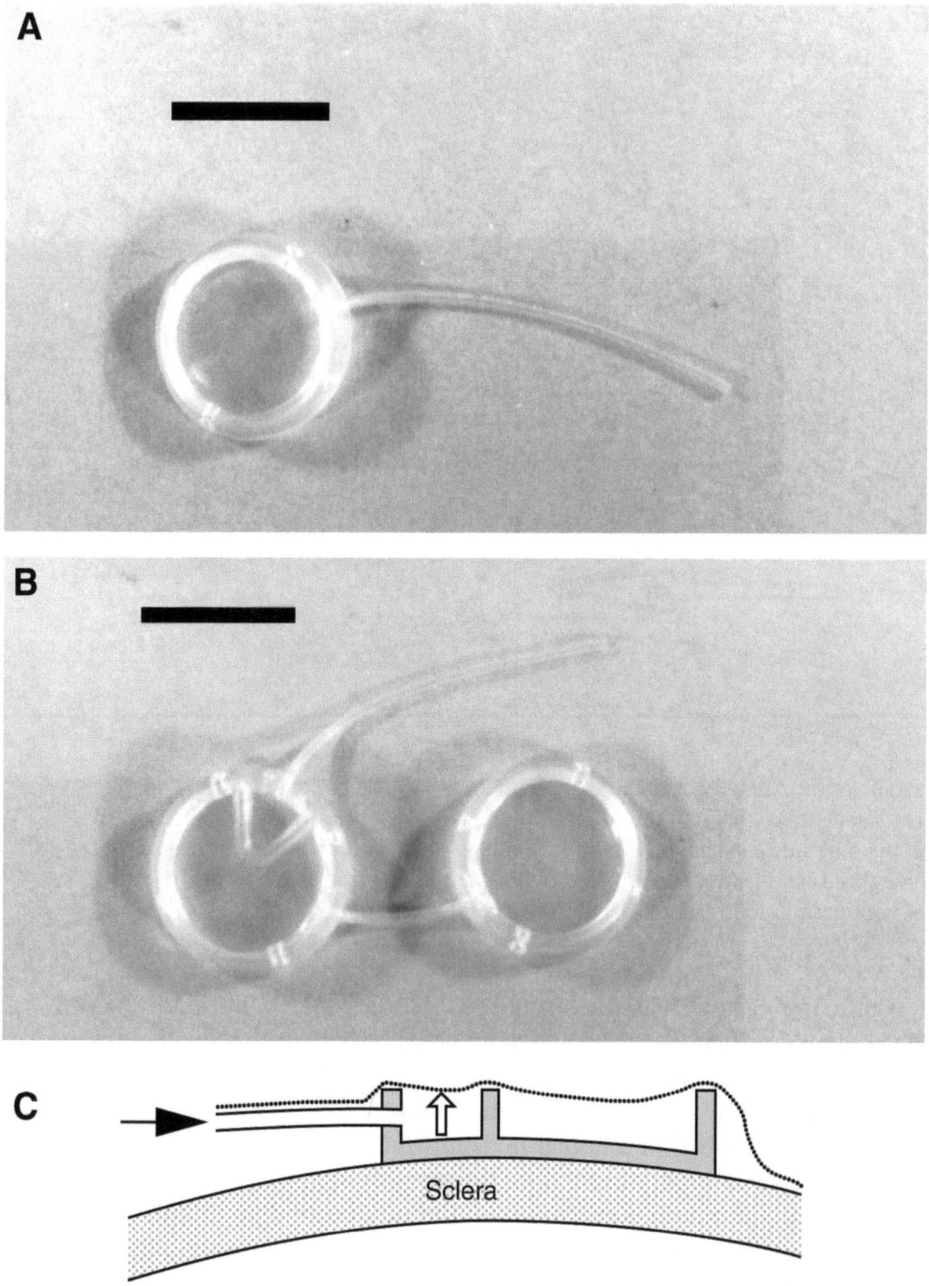

Fig. 3. Photographs of single **(A)** and double **(B)** plate Molteno implants (Scale bars = 1 cm). **(C)** Diagram of resistance mechanism of a dual chamber single plate Molteno implant. Black arrow indicates flow of aqueous humor into smaller proximal chamber. Once adequate pressure is achieved within the chamber to cause elevation of overlying conjunctiva (open block arrow) and subsequent drainage of aqueous humor (Reprinted with permission from BMJ Publishing Group *[13]*).

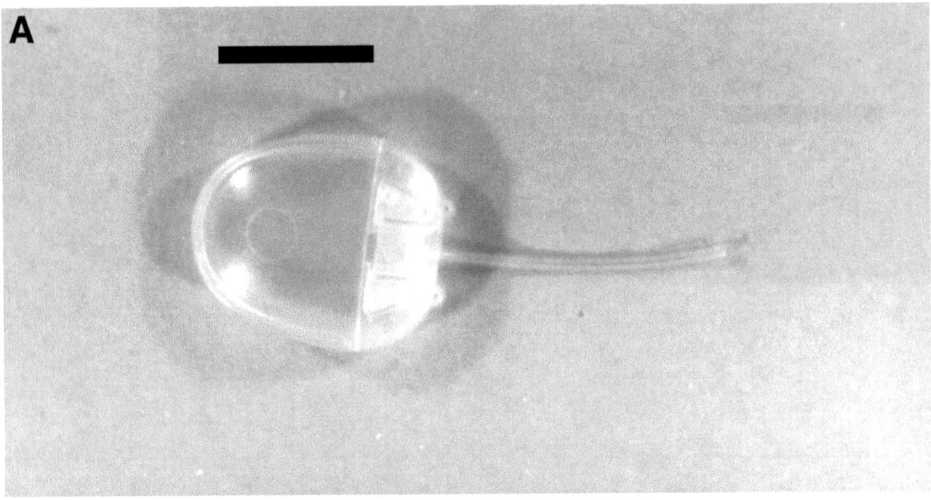

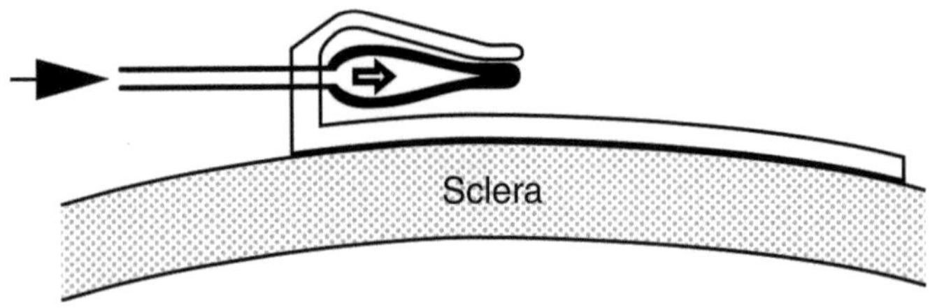

Fig. 4. Photograph of the Ahmed glaucoma valve implant (**A**) (Scale bar = 1 cm). (**B**) Diagram illustrating the resistance mechanism of the Ahmed valve. The black arrow indicates the flow of aqueous through the tube into a chamber within the explant plate. This chamber is created by a silicone membrane (bold black line) that is folded over such that it's free edges form a one-way valve (Reprinted with permission from BMJ Publishing Group *[13]*).

A recent design using current manufacturing technologies for GDDs is the Ex-PRESS (Optonol Ltd., Neve Illan, Israel), a miniature nonvalved glaucoma implant for draining aqueous humor into the subconjunctival space *(55)*. Short-term efficacy of the Ex-PRESS has been reported in a preliminary clinical study *(56),* and implantation of this GDD into rabbit eyes resulted in minimal peri-implant inflammation or fibrosis histologically *(55)*. However, a separate retrospective study evaluating the Ex-PRESS miniature glaucoma implant placed directly under the conjunctiva in 11 cases of advanced glaucoma revealed an unacceptably high incidence of complications including failure (unacceptably high IOP requiring revision or explantation) in 4/11 eyes, choroidal detachment in 3/11 eyes, and suprachoroidal hemorrhage in 2/11 eyes *(57)*. Subsequently, a recent prospective multicenter clinical study evaluating the long-term effectiveness of this miniature GDD for the management of primary OAG and cataract when combined with phacoemulsification, revealed an overall success rate (IOP < 21 mmHg at the last visit with or without antiglaucoma medications) of 76.9% (mean follow-up = 23.9 [+/–10.4 SD] months) *(58)*.

MATERIALS

The most commonly used material in contemporary GDDs continues to be elastomeric silicone (polydimethylsiloxane) *(13)*. The tubes of most of these GDDs including the Molteno, Krupin, Baerveldt, and Ahmed are all made up of silicone *(13)*. The explant plates in the Ahmed glaucoma valve and the double plate Molteno implant are made of polypropylene, whereas those of the Baerveldt and Krupin implants are made of silicone *(59)*. However, despite the similarities in materials across these GDDs, there are differences in the texture and flexibility of these implants possibly because of differences in the medical grades of: (1) polypropylene in the Molteno implant compared with the Ahmed glaucoma valve (2) silicone in the Baerveldt compared with the Krupin implant *(59)*. Unlike other more commonly used GDDs, the Ex-PRESS miniature GDD is made of stainless steel *(55)*.

SURGICAL DRAINAGE SITES

During the past several decades, the subconjunctival space has become the standard explant site to which aqueous humor is diverted using GDDs. Unfortunately, despite several advances in implant design including posterior placement and enlargement of the explant plate, and improvements in the biocompatibilty of materials used in the manufacturing of these devices, early filtration failure as a result of fibrosis at the explant site continues to be a common complication of GDDs in humans and other animals *(60–62)*. In part, this may be because of the location into which aqueous humor is being shunted. Cytokines including transforming growth factors β1 and β2 and fibroblastic growth factor have been demonstrated in the aqueous humor of humans and dogs with chronic ocular disease including primary glaucoma *(63–67)*. The subconjunctival space is made up of mesenchymal tissues with mitogenic activity. This, combined with these cytokines from the draining aqueous humor, further contribute to fibrosis at the site of the filtration bleb and reduced absorptive area for aqueous humor from this scarring. As such, this prompted recent research in dogs in which aqueous humor was diverted from the anterior chamber to the frontal sinus, an epithelium-lined, air-filled space, using a nonvalved silicone implant *(68)*. Results of this pilot study in clinically normal dogs indicated that frontal sinus shunting of aqueous humor was a safe and effective means of extraorbital aqueous diversion with potential applicability in the management of glaucoma *(68)*.

Subsequently, studies have been conducted using a professionally manufactured, valved GDD (Cullen canine frontal sinus valved glaucoma shunt; EagleVision, Memphis, TN (Fig. 5) implanted in glaucomatous eyes of dogs *(69,70)*. Results from both the preliminary clinical and multicenter studies, with follow-up of up to 9 mo, revealed that frontal sinus shunting of aqueous humor is a safe and effective means of controlling glaucoma without the need for additional glaucoma medications in many cases. Intracameral fibrin formation with or without blood was the most common postoperative complication occurring in 15/19 eyes *(70)*. Intracameral fibrinous shunt occlusion occurred in 7/19 eyes, and typically, resolved following injection of intracameral tissue plasminogen activator (tPA) *(70)*. One globe developed endophthalmitis following GDD implantation *(70)*; however, this eye was from a diabetic dog that had previously

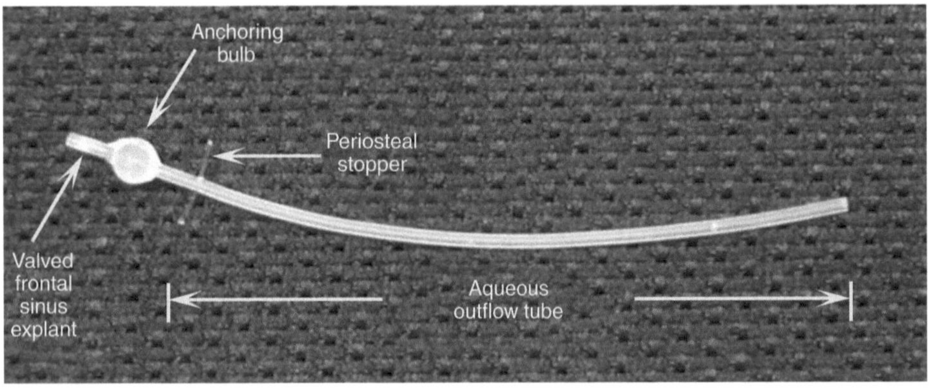

Fig. 5. Photograph of the Cullen Canine Frontal Sinus Valved Glaucoma Shunt. Note the silicone anchoring bulb and periosteal stopper that together, prevent migration of the shunt into and out of the frontal sinus. The valved end of the shunt is modelled after the Krupin valve. The valve is designed to open with IOPs of >18–20 mmHg (*Note:* high normal IOP in the dog is >25 mmHg) (Reprinted with permission from Blackwell Publishing *[69]*).

undergone phacoemulsification and intraocular lens implantation, and the globe had received three intracameral tPA injections at various time-points following the GDD procedure.

Subsequent to canine frontal sinus shunting of aqueous, Dohlman et al. *(71)* reported on the feasability of diverting aqueous humor from a glaucoma shunt extended to the lacrimal sac or ethmoid sinus in humans with keratoprothesis. A longer term study ensued in which modified Ahmed shunts (Fig. 6) were used to direct aqueous to the maxillary or ethmoid sinus or lacrimal sac in 19 patients with keratoprothesis as a result of severe ocular surface disease and with intractable glaucoma *(72)*. After follow-up times ranging from 1 to 30 mo (mean = 14.3 mo; SD = 9.8 mo), IOP was well-controlled without glaucoma medications in two-thirds of patients, whereas none of the patients developed endophthalmitis *(72)*.

Complications Following Placement of GDDs

Early and late postoperative complications may arise following GDD implantation. Early complications of GDD include hypotony, inflammation, and tube occlusion *(73)*, whereas late complications include increased IOP, hypotony, implant erosion through the conjunctiva, tube migration, cataract, corneal decompensation, calcification of the drainage implant, tube occlusion, endophthalmitis, and strabismus *(17,74,75)*. Aqueous misdirection, an uncommon complication of GDD procedures can develop days to months postoperatively *(76)*.

The most common complications include those associated with hypotony as a result of poor flow control including anterior chamber flattening, serous choroidal detachment, suprachoroidal hemorrhage, hypotony maculopathy, and corneal decompensation *(13,17)*. One retrospective study reported that delayed suprachoroidal hemorrhage occurred in 2.8% of valved and 7.1% of nonvalved GDD implantations *(77)*. Hypotony may be transient because of overfiltration in eyes with normal aqueous humor production. However,

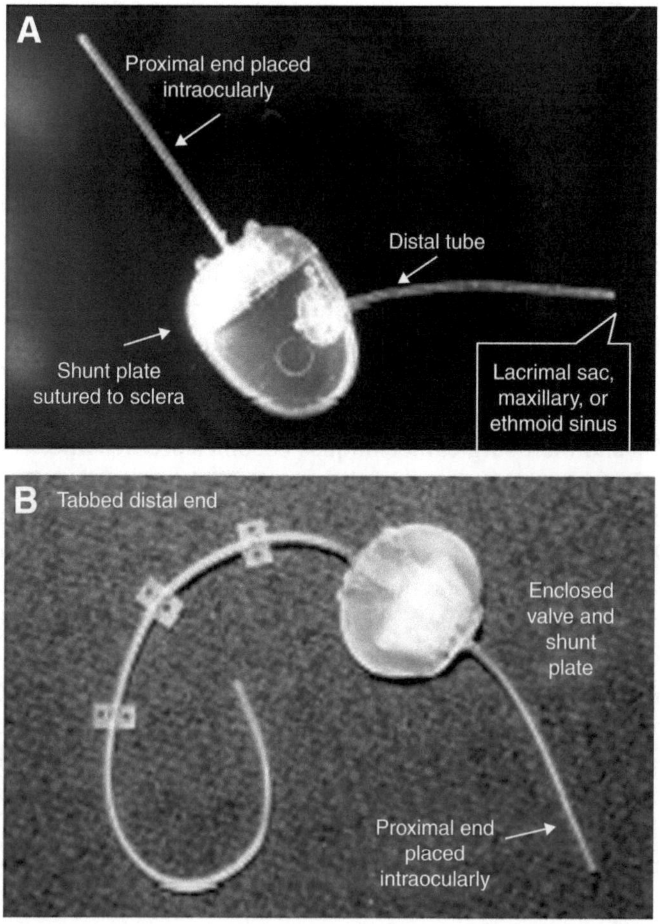

Fig. 6. Photographs of modified Ahmed open shunt (**A**) and closed shunt (**B**) for diversion of aqueous humor to the lacrimal sac, or the maxillary, or ethmoid sinuses in human patients with refractory glaucoma and keratoprosthesis (Reprinted with permission from Lippincott Williams and Wilkins *[72]*).

chronic hypotony may arise in eyes with reduced aqueous humor production secondary to ciliary body dysfunction including uveitic glaucoma, neovascular glaucoma, or eyes having prior cycloablative surgeries *(17)*.

Other less frequent complications of GDD implantation include a transient sterile uveitis following tube ligature absorption or polypropylene stent release, which is typically responsive to topical steroids, and late-onset sterile endophthalmitis with intracameral fibrin *(17)*. Intracameral tPA has been reported to successfully clear or prevent tube occlusion by fibrin/blood clots in 88.9% of affected eyes implanted with GDDs with the mean number of injections required to achieve successful outcomes being 1.6 *(78)*. Although rare, late-onset culture-positive endophthalmitis has also been reported following GDD procedures with incidences ranging from 0.9 to 6.3% *(75)*. Exposure of the GDD tube appears to be a major risk factor for these infections *(75)*.

Orbital complications, although rare, have also been described following GDD implantation including orbital cellulitis *(73,79)*, ocular proptosis *(80)*, myositis *(81)*, and silicone oil migration *(82)*.

Challenges Encountered in Developing GDDs

One of the most significant challenges encountered in developing GDDs is the difficulty in accurately evaluating the long-term efficacy of currently used devices in direct comparison with one another in the management of glaucoma. This difficulty arises, in part, because of variability in the clinical data published and the limitations of retrospective case series. Variables across many of these studies, including the investigator, the stage and type of glaucoma, the number of eyes, duration of follow-up, definition of success, and the use of antiglaucoma medications, make accurate evaluations of the currently available GDDs challenging. As such, determination of viable modifications to the contemporary GDDs and their surgical techniques becomes problematic.

Two of the main complications encountered followng GDD implantation include hypotony and the development of filtration failure because of fibrosis at the explant site. Developing a GDD to successfully avoid these complications is proving to be challenging, to say the least. Both valved and nonvalved GDDs with or without the use of various suture adjustment techniques have been used in attempts to address both acute and chronic hypotony. Various materials have been used to enhance tissue biocompatibility and thereby attempt to prevent the inflammation, which promotes fibrosis and tube occlusion. However, there is little published data available to substantiate the manufacturers' claims regarding flow performance or biocompatibility of most GDDs *(13)*. As well, antimetabolites, such as mitomycin C and 5-fluorouracil have been used successfully to decrease bleb fibrosis following trabeculectomy. However, the findings in support of using antimetabolites adjunctively with GDDs remain mixed *(18,83)*. A retrospective study assessing modified Baerveldt implants with varied, nonrandomized exposure to antimetabolites (mitomycin C or 5-fluorouracil) found the use of these agents did not appear to influence early postoperative IOP control or the development of complications *(84)*. One study investigating the use of controlled release, biodegradable collagen plugs of 5-fluorouracil placed within silicone tubes of modified Baerveldt GDDs, demonsrated improved IOP-lowering function, and reduced bleb wall thickness in eyes implanted with antimetabolite-containing tubes in comparison with control eyes *(85)*. Regardless, despite these improvements in GDDs and techniques used during their implantation during the past several decades, most GDDs have a functional lifespan of less than 5 yr before failure through fibrous encapsulation arises *(13)*.

Future Directions

Current designs and surgical techniques of GDDs (for a current comprehensive review of GDDs and the existing literature *see* Hong et al., 2005 *[18]*) must address the failure rate through fibrosis. Manners in which this complication may be prevented long-term will likely depend on a multifactorial, innovative strategy involving modifications to improve flow control, the use of new polymers with enhanced tissue biocompatibility, critical evaluation of novel drainage sites including epithelium-lined sinuses, and the use of current and newer antimetabolites at various delivery routes and time-points (for a comprehensive review of the use of antimetabolite/antifibrotic agents adjunctive to glaucoma surgery, recently published studies that address current use of these agents, and new methods of wound modulation refer to Yoon and Singh, 2004 *[83]*).

ANIMAL MODELS

Increased utilization of animal models of glaucoma will become crucial for novel GDDs to be more critically evaluated in comparison with existing implants. For a comprehensive review of spontaneously occurring and experimental animal models of glaucoma refer to Gelatt et al. *(4,86)*. Such animal models are useful when studying, not only the effects of new GDD designs, but also the effects of new drainage sites for the diversion of aqueous humor *(69,87)* and antimetabolites to reduce the occurrence of postoperative fibrosis.

FLOW CONTROL: NEW VALVE DESIGNS

The Ahmed Glaucoma Valve (New World Medical, Inc., Rancho Cucamonga, CA) was invented by Mateen Ahmed, and was approved by the US Food and Drug Administration in November 1993 *(88)*. The Ahmed valve design offers a unidirectional, pressure-sensitive valve created of a silicone, elastomer membrane that resides in a trapezoidal chamber. The membranes of this valve have an opening IOP of 8 mmHg and a Venturi effect is created to assist with the flow of aqueous through the chamber *(88)*. The Ahmed valve design has been reported to reduce the complications associated with overfiltration and subsequent hypotony in the immediate postoperative period compared with other drainage devices, although the long-term results are similar to other GDDs *(89)*. As such, new Ahmed valve designs have subsequently been developed and studied, and different designs of Ahmed GDDs are available varying in size, shape, number of explant plates, and materials used *(88)*.

Future studies continuing to investigate novel designs for GDD valve mechanisms, building on the significant information gained from current valve mechanisms may help to prevent acute and chronic hypotony associated with GDDs.

TISSUE BIOCOMPATIBILITY: ALTERNATIVE MATERIALS

Inflammation associated with the biomaterials used in the construction of GDDs, and their explant plates in particular may contrbute to bleb failure secondary to fibrosis *(59)*. Consequently, choosing a biomaterial with limited potential to cause inflammation may improve the success rate of GDDs *(59)*.

Experimental GDDs made of new polymers or polymer modifications to help reduce the risks of chronic inflammation and subconjunctival fibrosis have been described. The alternative materials used in these experimental studies with potential application in GDDs used the management of glaucoma in humans, include: GDDs constructed of expanded polytetrafluoroethylene *(90)*; hydroxylapatite reservoir tube implants *(91)*; and the potential to use the same material used to manufacture Acrosof, a new acrylic (polymethylmethacrylate), foldable intraocular lens implant, for producing explant plates of GDDs *(59)*. In addition, successful development of a new group of bioinert materials synthesized from monomers based on phosphorylcholine appears promising regarding improving the biocompatibility of medical devices, such as GDDs *(92)*. One study reported significantly less protein (fibrinogen and fibrin) and cellular (macrophages and scleral fibroblasts) adhesion to phosphorylcholine-coated polymethylmethacrylate discs compared with levels of adhesion to materials used in contemporary GDDs including uncoated phosphorylcholine-coated polymethylmethacrylate, PTFE, polypropylene, and silicone *(92)*.

New Drainage Sites: Epithelium-Lined Extraorbital Sites

Postoperative fibrosis at the explant site may be prevented with GDDs, which divert aqueous humor to extraorbital sites with epithelial lining. Recipient sites in the periorbital region including the lacrimal drainage apparatus and paranasal sinuses have been implanted with GDDs in patients with keratoprosthesis and severe glaucoma *(71,72)*. Based on studies conducted not only in humans *(71,72)*, but also in dogs *(69,70)*, additional studies investigating these and similar innovative surgical procedures for shunting aqueous humor to these epithelium-lined sites in different forms of glaucoma are warranted. In particular, because the explant plate placed subconjunctivally is the main site of the fibrosis, designing shunts for extraorbital diversion without explant plates, as is performed in dogs *(69,70)*, may be beneficial.

CONCLUSIONS

In order to effectively and critically evaluate such modifications in GDDs, future research should be directed at prospective studies using animal models of glaucoma with systematic investigation of variables including biocompatible materials, novel drainage sites, and current and newer antimetabolites, and standard guidelines regarding success with short- and long-term follow-up.

REFERENCES

1. Weinreb RN, Khaw PT. Primary open-angle glaucoma. Lancet 2004;363:1711–1720.
2. Quigley HA. Number of people with glaucoma worldwide. Br J Ophthalmol 1996;80:389–393.
3. Shiose Y. Intraocular pressure: new perspectives. Surv Ophthalmol 1990;34:413–435.
4. Gelatt KN, Brooks DE, Samuelson DA. Comparative glaucomatology. I: The spontaneous glaucomas. J Glaucoma 1998;7:187–201.
5. Bellezza AJ, Rintalan CJ, Thompson HW, Downs JC, Hart RT, Burgoyne CF. Deformation of the lamina cribrosa and anterior scleral canal wall in early experimental glaucoma. Invest Ophthalmol Vis Sci 2003;44:623–637.
6. Stefansson E, Pedersen DB, Jensen PK, et al. Optic nerve oxygenation. Prog Retin Eye Res 2005;24:307–332.
7. Dreyer EB, Zurakowski D, Schumer RA, Podos SM, Lipton SA. Elevated glutamate levels in the vitreous body of humans and monkeys with glaucoma. Arch Ophthalmol 1996;114: 299–305.
8. Yan X, Tezel G, Wax MB, Edward DP. Matrix metalloproteinases and tumor necrosis factor alpha in glaucomatous optic nerve head. Arch Ophthalmol 2000;118:666–673.
9. Sacca S. Nitric oxide as a mediator of glaucoma pathogenesis. Med Sci Monit 2002;8: LE33–LE34.
10. Tezel G, Edward DP, Wax MB. Serum autoantibodies to optic nerve head glycosaminoglycans in patients with glaucoma. Arch Ophthalmol 1999;117:917–924.
11. Yucel YH, Zhang Q, Weinreb RN, Kaufman PL, Gupta N. Effects of retinal ganglion cell loss on magno-, parvo-, koniocellular pathways in the lateral geniculate nucleus and visual cortex in glaucoma. Prog Retin Eye Res 2003;22:465–481.
12. Tielsch JM, Sommer A, Witt K, Katz J, Royall RM. Blindness and visual impairment in an American urban population. The Baltimore Eye Survey Arch Ophthalmol 1990;108:286–290.

13. Lim KS, Allan BDS, Lloyd AW, Muir A, Khaw PT. Glaucoma drainage devices; past, present, and future. Br J Ophthalmol 1998;82:1083–1089.
14. Khaw PT, Shah P, Elkington AR. Glaucoma—2: treatment. Br Med J 2004;328:156–158.
15. Chen CW, Huang HT, Bair JS, Lee CC. Trabeculectomy with simultaneous topical application of mitomycin-C in refractory glaucoma. J Ocul Pharmacol 1990;6:175–182.
16. Smith MF, Sherwood MB, Doyle JW, Khaw PT. Results of intraoperative 5-fluorouracil supplementation on trabeculectomy for open-angle glaucoma. Am J Ophthalmol 1992;114:737–741.
17. Assaad MH, Baerveldt G, Rockwood EJ. Glaucoma drainage devices: pros and cons. Curr Opin Ophthalmol 1999;10:147–153.
18. Hong CH, Arosemena A, Zurakowski D, Ayyala RS. Glaucoma drainage devices: a systematic literature review and current controversies. Surv Ophthalmol 2005;50:48–60.
19. Rodrigues AM, Corpa MV, Mello PA, de Moura CR. Results of the Susanna implant in patients with refractory primary congenital glaucoma. J AAPOS 2004;8:576–579.
20. Muckley ED, Lehrer RA. Late-onset blebitis/endophthalmitis: incidence and outcomes with mitomycin C. Optom Vis Sci 2004;81:499–504.
21. Rollett M, Moreau M. Traitement de hypopyon par le drainage capillary de la chambre anterieure. Rev Gen Ophtalmol 1906;25:481–489.
22. Rollett M, Moreau M. Le drainage au crin de la chambre anterieure contre l'hypertonie et la douleur. Rev Gen Ophtalmol 1907;26:289–292.
23. Zorab A. The reduction of tension in chronic glaucoma. Ophthalmoscope 1912;10:258–261.
24. Stefansson J. An operation for glaucoma. Am J Ophthalmol 1925;8:681–693.
25. Muldoon WE, Ripple PH, Eilder HC. Platinum implant in glaucoma surgery. Arch Ophthalmol 1951;45:666–672.
26. Row H. Operation to control glaucoma:preliminary report. Arch Ophthalmol 1934;12:325–329.
27. Trancosco MU. Use of tantalum implants for inducing a premanent hypotony in rabbits' eyes. Am J Ophthalmol 1949;32:499–508.
28. Gibson G. Transcleral lacrimal canaliculus transplants. Trans Am Ophthalmol Soc 1942;40:499–515.
29. Bick MW. Use of tantalum for ocular drainage. Arch Ophthalmol 1949;42:373–388.
30. Bock RH. Subconjunctival drainage of the anterior chamber by a glass seton. Am J Ophthalmol 1950;33:929.
31. Losche W. Vorschlage zur Verbesserung der Zyklodialyse. Klin Monatsbl Augenheilkd 1952;121:715–716.
32. Bietti GB. The present state of the use of plastics in eye surgery. Acta Ophthalmol (Copenh) 1955;33:337–370.
33. La Rocca V. Int. Cong. Ophthalmol (Brussels) 1958.
34. Lee PF, Wong WT. Aqueous-venous shunt for glaucoma: report on 15 cases. Ann Ophthalmol 1974;6:1083–1088.
35. Mascati NT. A new surgical approach for the control of a class of glaucomas. Int Surg 1967;47:10–15.
36. Molteno AC. New implant for drainage in glaucoma. Clinical trial Br J Ophthalmol 1969;53:606–615.
37. Molteno AC. New implant for drainage in glaucoma. Animal trial Br J Ophthalmol 1969;53:161–168.
38. Molteno AC, Straughan JL, Ancker E. Long tube implants in the management of glaucoma. S Afr Med J 1976;50:1062–1066.
39. Krupin T, Podos SM, Becker B, Newkirk JB. Valve implants in filtering surgery. Am J Ophthalmol 1976;81:232–235.

40. Ayyala RS, Zurakowski D, Smith JA, et al. A clinical study of the Ahmed glaucoma valve implant in advanced glaucoma. Ophthalmology 1998;105:1968–1976.
41. Molteno AC. The optimal design of drainage implants for glaucoma. Trans Ophthalmol Soc NZ 1981;33:39–41.
42. Britt MT, LaBree LD, Lloyd MA, et al. Randomized clinical trial of the 350-mm2 versus the 500-mm2 Baerveldt implant: longer term results: is bigger better? Ophthalmology 1999;106:2312–2318.
43. Smith MF, Sherwood MB, McGorray SP. Comparison of the double-plate Molteno drainage implant with the Schocket procedure. Arch Ophthalmol 1992;110:1246–1250.
44. Freedman J. Clinical experience with the Molteno dual-chamber single-plate implant. Ophthalmic Surg 1992;23:238–241.
45. Baerveldt G, Chou JS, Longren B. Comparison of the Baerveldt glaucoma implant with bioseal to the Baerveldt implant in rabbits. Invest Ophthalmol Vis Sci 1997;38:52.
46. Price FW Jr., Whitson WE. Polypropylene ligatures as a means of controlling intraocular pressure with Molteno implants. Ophthalmic Surg 1989;20:781–783.
47. Egbert PR, Lieberman MF. Internal suture occlusion of the Molteno glaucoma implant for the prevention of postoperative hypotony. Ophthalmic Surg 1989;20:53–56.
48. Latina MA. Single stage Molteno implant with combination internal occlusion and external ligature. Ophthalmic Surg 1990;21:444–446.
49. Schocket SS, Lakhanpal V, Richards RD. Anterior chamber tube shunt to an encircling band in the treatment of neovascular glaucoma. Ophthalmology 1982;89:1188–1194.
50. White TC. Clinical-Results of Glaucoma Surgery Using the White Glaucoma Pump Shunt. Ann Ophthalmol 1992;24:365–373.
51. White TC. A new implantable ocular pressure relief device. A preliminary rep Glaucoma 1985;7:289–294.
52. Hitchings RA, Joseph NH, Sherwood MB, Lattimer J, Miller M. Use of one-piece valved tube and variable surface area explant for glaucoma drainage surgery. Ophthalmology 1987; 94:1079–1084.
53. Joseph NH, Sherwood MB, Trantas G, Hitchings RA, Lattimer L. A one-piece drainage system for glaucoma surgery. Trans Ophthalmol Soc UK 1986;105(Pt 6):657–664.
54. Susanna R Jr. Modifications of the Molteno implant and implant procedure. Ophthalmic Surg 1991;22:611–613.
55. Nyska A, Glovinsky Y, Belkin M, Epstein Y. Biocompatibility of the Ex-PRESS miniature glaucoma drainage implant. J Glaucoma 2003;12:275–280.
56. Kaplan-Messas A, Traverso CE, Glovinski Y, et al. The ExPRESS miniature glaucoma implant: Intermediate results of a prospective multi-center study. Invest Ophthalmol Vis Sci 2001;42:S552.
57. Wamsley S, Moster MR, Rai S, Alvim HS, Fontanarosa J. Results of the use of the Ex-PRESS miniature glaucoma implant in technically challenging, advanced glaucoma cases: a clinical pilot study. Am J Ophthalmol 2004;138:1049–1051.
58. Traverso CE, De Feo F, Messas-Kaplan A, et al. Long term effect on IOP of a stainless steel glaucoma drainage implant (Ex-PRESS) in combined surgery with phacoemulsification. Br J Ophthalmol 2005;89:425–429.
59. Ayyala RS, Michelini-Norris B, Flores A, Haller E, Margo CE. Comparison of different biomaterials for glaucoma drainage devices: part 2. Arch Ophthalmol 2000;118: 1081–1084.
60. Bentley E, Nasisse MP, Glover T, Nelms S. Implantation of filtering devices in dogs with glaucoma: preliminary results in 13 eyes. Vet Comp Ophthalmol 1996;6:243–246.
61. Glover TL, Nasisse MP, Davidson MG. Effects of topically applied mitomycin-C on intraocular pressure, facility of outflow, and fibrosis after glaucoma filtration surgery in clinically normal dogs. Am J Vet Res 1995;56:936–940.

62. Tahery MM, Lee DA. Review: pharmacologic control of wound healing in glaucoma filtration surgery. J Ocul Pharmacol 1989;5:155–179.
63. Pasquale LR, Dorman-Pease ME, Lutty GA, Quigley HA, Jampel HD. Immunolocalization of TGF-beta 1, TGF-beta 2, and TGF-beta 3 in the anterior segment of the human eye. Invest Ophthalmol Vis Sci 1993;34:23–30.
64. Kostick AM, Grahn BH, Murphy PH, Romanchuk KG. Comparison of aqueous humor TGF-B-2 from cataract and glaucoma patients. Invest Ophthalmol Vis Sci 1996;37:717.
65. Kostick AM, Grahn BH, Murphy PH, Romanchuk KG. Analysis of Growth-Factors in Aqueous-Humor from Glaucoma and Cataract Patients. Invest Ophthalmol Vis Sci 1995;36:S726.
66. Tripathi RC, Li J, Chan WF, Tripathi BJ. Aqueous humor in glaucomatous eyes contains an increased level of TGF-beta 2. Exp Eye Res 1994;59:723–727.
67. Grahn BH, Wilcock B, Hayes MA. Demonstration of mediators of fibrous and fibrovascular reactions in the canine, feline and equine eye. Vet Pathol 1992;25:216.
68. Cullen CL, Allen AL, Grahn BH. Anterior chamber to frontal sinus shunt for the diversion of aqueous humor: a pilot study in four normal dogs. Vet Ophthalmol 1998;1:31–39.
69. Cullen CL. Cullen frontal sinus valved glaucoma shunt: preliminary findings in dogs with primary glaucoma. Vet Ophthalmol 2004;7:311–318.
70. Cullen CL, Corcoran KA, Bartoe J, et al. Preliminary findings from the multicenter clinical study group evaluating the Cullen canine frontal sinus valved glaucoma shunt. Vet Ophthalmol 2003;6:356.
71. Dohlman CH, Grosskreutz CL, Dudenhoefer EJ, Rubin PA. Can a glaucoma shunt tube be safely extended to the lacrimal sac or the ethmoid sinus in keratoprosthesis patients? Digital J of Ophthalmol 2002;9:http://www.djo.harvard.edu/site.php?url/physicians/oa/361&PHPSESSIDba3e304919c94c495374a52f318ef18d.
72. Rubin PA, Chang E, Bernardino CR, Hatton MP, Dohlman CH. Oculoplastic technique of connecting a glaucoma valve shunt to extraorbital locations in cases of severe glaucoma. Ophthal Plast Reconstr Surg 2004;20:362–367.
73. Marcet MM, Woog JJ, Bellows AR, Mandeville JT, Maltzman JS, Khan J. Orbital complications after aqueous drainage device procedures. Ophthal Plast Reconstr Surg 2005;21: 67–69.
74. Joos KM, Lavina AM, Tawansy KA, Agarwal A. Posterior repositioning of glaucoma implants for anterior segment complications. Ophthalmology 2001;108:279–284.
75. Gedde SJ, Scott IU, Tabandeh H, et al. Late endophthalmitis associated with glaucoma drainage implants. Ophthalmology 2001;108:1323–1327.
76. Greenfield DS, Tello C, Budenz DL, Liebmann JM, Ritch R. Aqueous misdirection after glaucoma drainage device implantation. Ophthalmology 1999;106:1035–1040.
77. Tuli SS, WuDunn D, Ciulla TA, Cantor LB. Delayed suprachoroidal hemorrhage after glaucoma filtration procedures. Ophthalmology 2001;108:1808–1811.
78. Zalta AH, Sweeney CP, Zalta AK, Kaufman AH. Intracameral tissue plasminogen activator use in a large series of eyes with valved glaucoma drainage implants. Arch Ophthalmol 2002;120:1487–1493.
79. Lavina AM, Creasy JL, Tsai JC. Orbital cellulitis as a late complication of glaucoma shunt implantation. Arch Ophthalmol 2002;120:849–851.
80. Danesh-Meyer HV, Spaeth GL, Maus M. Cosmetically significant proptosis following a tube shunt procedure. Arch Ophthalmol 2002;120:846–847.
81. Oh KT, Alward WL, Kardon RH. Myositis associated with a Baerveldt glaucoma implant. Am J Ophthalmol 1999;128:375–376.
82. Nazemi PP, Chong LP, Varma R, Burnstine MA. Migration of intraocular silicone oil into the subconjunctival space and orbit through an Ahmed glaucoma valve. Am J Ophthalmol 2001;132:929–931.

83. Yoon PS, Singh K. Update on antifibrotic use in glaucoma surgery, including use in trabeculectomy and glaucoma drainage implants and combined cataract and glaucoma surgery. Curr Opin Ophthalmol 2004;15:141–146.
84. Trible JR, Brown DB. Occlusive ligature and standardized fenestration of a Baerveldt tube with and without antimetabolites for early postoperative intraocular pressure control. Ophthalmology 1998;105:2243–2250.
85. Jacob JT, LaCour OJ, Burgoyne CF. Slow release of the antimetabolite 5-fluorouracil (5-FU) from modified Baerveldt glaucoma drains to prolong drain function. Biomaterials 2001;22:3329–3335.
86. Gelatt KN, Brooks DE, Samuelson DA. Comparative glaucomatology. II: The experimental glaucomas. J Glaucoma 1998;7:282–294.
87. Hakanson NW. Extraorbital diversion of aqueous humour in the treatment of glaucoma in the dog: a pilot study including two recipient sites. Vet Comp Ophthalmol 1996;6:82–88, 90.
88. Bhatia LS, Chen TC. New Ahmed valve designs. Int Ophthalmol Clin 2004;44:123–138.
89. Coleman AL, Hill R, Wilson MR, et al. Initial clinical experience with the Ahmed Glaucoma Valve implant. Am J Ophthalmol 1995;120:23–31.
90. Boswell CA, Noecker RJ, Mac M, Snyder RW, Williams SK. Evaluation of an aqueous drainage glaucoma device constructed of ePTFE. J Biomed Mater Res 1999;48:591–595.
91. Pandya AD, Rich C, Eifrig DE, Hanker J, Peiffer RL. Experimental evaluation of a hydroxylapatite reservoir tube shunt in rabbits. Ophthalmic Surg Lasers 1996;27:308–314.
92. Lim KS. Cell and protein adhesion studies in glaucoma drainage device development. The AGFID project team. Br J Ophthalmol 1999;83:1168–1171.

13
The Efficacy and Safety of Glaucoma Drainage Devices

Jenn-Chyuan Wang, MB BCH BAO, MRCSED, MMED, FRCS (OPHTH), FAMS **and Paul Chew,** MBBS, FRCOPHTH, FRCSED, FAMS

CONTENTS

INTRODUCTION

The glaucoma drainage devices (GDD) are increasingly popular as the primary procedure over trabeculectomy and cyclo-destructive in the management of refractory glaucoma. Interest in this field is warranted because studies have found significant rates of desirable intraocular pressure (IOP) control as well as maintenance of visual acuity in these difficult cases. GDD have evolved from the first crude attempt, in 1907, using horsehair to drain aqueous from the anterior chamber into the subtenon/conjunctival space at the limbus *(1)*. Since then, various materials, such as silk, gold, platinum, tantalum, glass rod, and polythene tube have been used near the limbus and have failed because of conjunctival erosion, implant migration, extrusion as well as scarring. To overcome these problems, current implants were based on the Molteno implant, which has a long tube attached to an explant placed 9–10 mm posterior to the limbus. Current GDD are made from various biomaterials, such as acrylic, silicone, and polypropylene, however, their influence on efficacy and safety has not yet been determined.

Significant progresses have been made to the design of GDD. Postoperative hypotony, flat anterior chambers, and choroidal effusion were very common observations in nonvalved implants. These complications resulted in the introduction of several modifications, such as temporary ligature of the tube. An important advancement was the addition of unidirectional valves with fixed opening IOP to the design in order to increase the resistance to aqueous outflow. Another concern was to lower IOP to a level that reduces the risk of glaucoma progression. In order to achieve this, Molteno

From: *Ophthalmology Research: Visual Prosthesis and Ophthalmic Devices: New Hope in Sight*
Edited by: J. Tombran-Tink, C. Barnstable, and J. F. Rizzo © Humana Press Inc., Totowa, NJ

Table 1
Types of Implants

Implant	*Valve*	*Material*	*Surface area/mm²*
Ahmed	Yes	Polypropylene Silicone	96/184/364
Baerveldt	No	Barium-impregnated silicone plate	250/350/425/500
Molteno	No	Polypropylene	270

Table 2
Indications for Glaucoma Drainage Device Implantation

Neovascular glaucoma
Failed trabeculectomy
Glaucoma following Retinal detachment surgery
Glaucoma following previous ocular surgery with extensive conjunctival scarring
PKP with glaucoma
Congenital glaucoma
Uveitic glaucoma
Traumatic glaucoma
Silicone oil glaucoma
Irido-corneal endothelial syndrome
Sturge–Weber syndrome
Glaucoma with chronic topical medication
Aphakic glaucoma

increased the surface area of his GDD by introducing the double plate implant in 1981. Surgeons are now faced with many choices of both valved and nonvalved implants (Table 1) in the management of refractory glaucoma (Table 2). The following chapter would attempt to summarize the current knowledge on the efficacy and safety of these implants.

EFFICACY IN IOP CONTROL

In the literature, GDD surgical success is usually defined as IOP less than 22 mmHg, but more than 4 mmHg and these eyes should not become totally blind or undergo phthisis. However, comparisons between studies are difficult because they differ in terms of surgical indications, surgical techniques, postoperative management, criteria for defining success as well as duration of follow-up. Furthermore, the numbers are usually small and lacked the power to elucidate risk factors for failures. The success rates were impressive in most reports considering the fact that GDD are usually implanted in refractory glaucoma where conventional surgery would otherwise have failed. They were quoted in the region of 58–98% for the Caucasian population *(2–5)*. Long-term follow-up of Molteno tubes over a median of 44 mo had success rates of 57% *(6)*.

Similar short-term results were echoed in Asian patients. Aung et al. *(7)* reported success in 73.5% and qualified success in 12% for Molteno and Baerveldt tubes in East Asians with average follow-up of 13.41 mo. While Lai et al. *(8)* had 73.8% success at 12 mo for Ahmed tubes in Hong Kong Chinese patients. Success rates of 82.83% at 24 mo were seen in Indian patients *(9)*. Complete success experienced for Baerveldt was 83.3% and 66.7% for Ahmed implants and qualified success were 4.2 and 16.7%, respectively with a mean follow-up of 22.8 mo *(10)*. As for the intermediate term, Seah et al. *(11)* had 54% complete success and 22% qualified success for Baerveldt GDD. Future randomized comparative trials with long-term follow-up may shed light on the impact of race on GDD outcomes.

Plate Area

Historically, it has been assumed that the plate size determined the size of the encapsulated bleb and therefore, has a positive influence on surgical success *(12)*. With twice the surface area, the double-plate Molteno device achieved higher success rates and lower IOP than single-plate Molteno GDD *(13)*. Although, most surgeons considered these findings as proof of theory, Camras *(14)* pointed to the fact that the success rates and IOP reduction had not been twice as good. Further evidence from Smith et al. *(15)* and Wilson et al. *(16)* showed that the smaller double-plate Molteno produced similar or lower IOP when compared with the Schocket encircling device. Studies comparing similar devices, i.e., the Baerveldt 250 mm^2 with 350 mm^2 in Asians *(11)* and 350 mm^2 with 500 mm^2 in Western population did not find significant differences in final IOP control *(17)*. Evidently, IOP decrease is not proportional to the plate area used as the effective surface area may be quite different from the total surface area of the GDD endplate. For example, a significant proportion of the Baerveldt endplate is tucked under both the vertical and the horizontal recti muscles and the extension of bleb under these muscles has been demonstrated echogenically *(18)*. However, the plate-muscle adhesions and pseudocapsule here may not be as permeable to aqueous, therefore, the resultant effective surface area may be significantly less.

Molteno described the best plate-size for the development of optimal bleb surface area. He mentioned that the circumferential distance between recti muscle delimits the size and shape of the episcleral plate. Plates larger than an ocular quadrant with the involvement of recti muscles may not necessarily increase functioning encapsulation. The posterior ciliary artery limits the posterior dimension. Finally, encapsulation within the orbit as opposed to being near the conjunctiva reduces permeability to aqueous humor *(19)*.

Second Tube Implant

In the event of suboptimal IOP control or failure of surgery, Management options include shunt revision, additional tube implant, and supplemental cyclophotocoagulation. Burgoyne et al. examined a group of patients with second tube implant and found that corneal morbidity is a common complication, whereas other rates of complications were not higher than expected. Pressure control was achieved with fewer hypotensive agents and stable visual acuity could be maintained *(20)*. Shah et al. compared additional tube shunt with shunt revision and found more patients (62% vs 42%) achieved at least qualified success (no statistical significance). Moreover, the sample size was very small with overrepresentation of neovascular glaucoma in the shunt revision group *(21)*.

Semchyshyn et al. treated a small group of tube failures with supplemental *trans*-scleral diode laser cyclophotocoagulation. Twenty-one eyes were treated and followed up during a mean period of 26.9 mo. The average IOP was significantly reduced from a preoperative level of 35.7 +/– 14.7 (SD) mmHg to a postoperative level of 13.6 +/– 7.1 (SD) mmHg. Therefore, in cases of tube failure, adjunctive *trans*-scleral diode cyclophotocoagulation treatment(s) is a viable option to lower IOP *(22)*.

Comparison With Other Surgical Modalities

GDD do not offer any advantage in terms of improving IOP control or reducing complication rates when compared with trabeculectomy alone in uncomplicated glaucoma *(23)*. However, for the Asian patients who have high risk of trabeculectomy failure *(24)*, GDD may have more to offer as the primary procedure *(7,10)*. Pars plana Baerveldt GDD surgery is superior to neodymium:YAG cyclophotocoagulation in terms of IOP control, less hypotony and with higher preservation of visual acuity. As compared with 5.6% in the Baerveldt group 26.3% of Nd:YAG failed at 6 mo. The neodymium:yttrium aluminum garnet (Nd:YAG) group lost light perception in 23.3% vs 5.6% in the Baerveldt group *(25)*.

When Lima et al. compared endoscopic cyclophotocoagulation (ECP) with Ahmed GDD in the treatment of refractory glaucoma, efficacy was similar, but there were more complications with Ahmed GDD, specifically, choroidal detachment (Ahmed 17.64%, ECP 2.94%), shallow anterior chamber (Ahmed 17.64%, ECP 0.0%), and hyphema (Ahmed 14.7%, ECP 17.64%) *(26)*. As such, ECP may be promising as a treatment modality for refractory glaucoma.

Augmentation With Mitomycin C

Antimitotic agents have not been used in any of the patients mainly because of the lack of evidence for their efficacy. The issue of using antimitotic agents to improve the outcome of GDD remained unresolved as the few comparative studies produced contradicting results. Kook et al. *(27)* reported increased success rates with Ahmed GDI without increasing complication rates whereas Costa et al. *(28)* did not observe any significant improvement in the IOP lowering effect. Meanwhile Irak et al. *(29)* reported no differences with the Baerveldt GDD in a noncomparative study. Likewise, Molteno GDD did not show significant differences in terms of survival rates, IOP control, and antiglaucoma medication use *(30)*. Conversely, complications related to mitomycin C use, such as conjunctival melts *(31)*, infections, and hypotony have been reported *(32,33)*.

Combined Cataract and GDD Surgery

Combined phacoemulsification and lens implantation with GDD surgery has been met with good outcomes. Chung et al. *(34)* reported a series of 32 patients who underwent such surgery with both Ahmed and Baerveldt GDD, and they benefited from visual rehabilitation without increased complications and glaucoma surgery failure. Similarly, excellent results for IOP control were also reported for Molteno implants *(35)*. However, cataract surgery performed in the presence of a functioning tube may precipitate corneal edema and failure of IOP control *(36)*. Therefore, screening of corneal health with specular microscopy and patient counseling on the risk of GDD failure should be considered before cataract surgery.

Corneal Transplant

Corneal complication rates of 11–50% were observed *(5,37,38)* where corneal decompensation and bullous keratopathy occurred most frequently *(39,40)*. Glaucoma control with GDD in penetrating keratoplasty (PKP) is similar to the results for other indications. Several studies have reported glaucoma controlled at 74–92% at 1 yr and 82–86% at 3 yr. On the other hand, the percentages of clear graft survival were quite disappointing at 58.5–92% and 25.8–55% at 1 and 2 yr, respectively *(39–41)*. The majority of those failures resulted from immunological graft rejection and tube endothelial touch *(39)*. The sequence of PKP in relation to GDD implantation may also influence the outcome of the graft where tube-first surgery is 3.8 and 4.7 times more likely to have graft failure than simultaneous and PKP-first surgery *(41)*. The reason for this is still unclear.

Modifiable aspects of tube surgery to reduce tube-endothelial contact include decreasing the length of the tube to 3–4 mm in the anterior chamber, and in pseudophakic patients, redirecting the tip more posteriorly toward the pupil plane during insertion. Direct contact between the tube and endothelium during the early part of the postoperative period when anterior chamber can be unstable would be less likely. Clinically, it has also been found that firm digital pressure on the anterior edge of the plate during examination may also cause tube-endothelial contact. Second, temporary increase in filtration may also shallow the anterior chamber. Hence, it is important to advise the patients against rubbing or messaging the eye after tube implantation. Using an alternative insertion site, i.e., the pars planar *(42)* may be achieved with pneumatically stented Baerveldt GDD modified with a Hofmann elbow to facilitate pars plana insertion *(43)*. Indeed, increased success rate of 83% vs 48% graft survival against anterior chamber placement, without compromise of IOP control, makes a convincing case for pars plana tube placement *(44)*. In the event that anterior segment complications, such as corneal decompensation, shallow anterior chamber, or recurrent tube erosion occurred after anterior chamber placement of the tube, a surgical revision to reposition the tube into the vitreal cavity may be attempted successfully *(42)*. When Sidoti et al. reported a noncomparative series of PKP with pars plana Baerveldt GDD with 12- and 24-mo success rates for corneal graft clarity at 64% and 41%, respectively. However, they encountered higher incidence of posterior segment complications, which included retinal detachment *(45)*, vitreous blockage of tube, and kinking of tube at the scleral entry site *(46)*. The placement of the tube through the ciliary sulcus offers a theoretical advantage in reducing tube-corneal touch and endothelial decompensation minus the retinal complications *(47)*. However, these advantages have not been confirmed by current published literature. Pediatric PKP success rates were reported at 85, 44, and 33% for

IOP control and 85, 43, and 17% graft success at 2, 24, and 48 mo *(48)*. The pars plana insertion and its results have not been reported for this group.

Neovascular Glaucoma

Luttrull and Avery reported excellent IOP reduction as well as preservation of vision in a series of pars plana implantation and vitrectomy for neovascular glaucoma. At 16 mo follow-up, vision improved in 86%, IOP ranged between 9 and 21 mmHg while on a mean of 0.7 medications. One patient required second implant and unfortunately, two patients were complicated with retinal detachment where one patient lost light perception *(49)*.

Sidoti et al. reviewed 18 patients implanted with Baerveldt 350 mm^2 and 16 eyes with 500 mm^2 for neovascular glaucoma. The 12- and 18-mo life table success rates were similar at 79 and 56%, respectively, although the 500 mm^2 required less medication. There were also no differences in terms of IOP reduction and complication rates. Visual acuity remained stable or improved in 31% and a similar number progressed to no light perception. They were able to determine that better preoperative visual acuity and increasing patient age correlate with successful outcome *(50)*.

Comparison of GDD surgery with neodymium:YAG cyclophotocoagulation in a case matched comparative group study reported IOP control in 66.7% vs 37.5% over a mean follow-up of 15.2 +/– 11.8 mo and 16.9 +/– 14.6 mo, respectively. In the tube-shunt group, the cumulative proportions of failure were 12.5, 29.2, and 43.3% at 6 mo, 1 yr, and 3 yr respectively. The cumulative proportion of failure in the Nd:YAG group was 20.8, 35.4, and 71.2%, respectively. More patients in the Nd:YAG group lost light perception (45.8%) then in the tube-shunt group (16.7%) *(51)*.

Uveitic Glaucoma

The success rates achieved with tubes in uncontrolled uveitic glaucoma are excellent. Ahmed, Baerveldt, and Molteno GDD showed comparable results up to the intermediate term. Cebellos et al. followed up a group Baerveldt GDD in uveitic glaucoma for a mean of 20.8 mo. Cumulative life-table success rates were 95.8% at 3 mo and 91.7% at 6, 12, and 24 mo. Total success at 24 mo was 58.3%. Best-corrected visual acuity improved or remained within two lines of preoperative visual acuity in 19 (79.2%) *(52)*. The Otago glaucoma surgery outcome study group reported that insertion of a Molteno implant was effective in controlling the IOP at 21 mmHg or less with a probability of 0.87 and 0.77 at 5, 10, or more years after surgery. The mean visual acuity improved from 20/100 to 20/70. However, this value declined to 20/130 at 5 and 10 yr and then improved slightly to 20/120 at 15 yr. The Kaplan-Meier *(53)* estimated the probability of retaining useful vision (visual acuity >20/400; visual field >5° radius) at 0.75 and 0.71 at 5, 7, or more years after surgery. Gil-Carrasco *(54)* reported follow-up for Ahmed GDD from 11 to 40 mo (mean 22.6 mo). Success was achieved in eight of 14 eyes (57.14%) with 0.71 +/– 0.99 antiglaucoma medications. A group of juvenile rheumatoid arthritis patients with uveitic glaucoma and Molteno implants were followed up for a mean of 40 mo and total success was achieved in 89%. Life-table analysis success rates were 95% after 27 mo and 90% after 52 mo of follow-up. Eighty-five percent had Snellen visual acuity within one line of the preoperative level or improved *(55)*.

Pediatric Glaucoma

Refractory pediatric glaucoma usually presents for tube surgery after two to three previous glaucoma procedures. Indications for surgery may include aphakia, persistent hyperplastic primary vitreous, primary infantile, juvenile secondary, Peter syndrome, and Lowe syndrome. Success rates with or without medication for IOP control with Ahmed GDD were reported at 93% at 12 mo and 45% at 48 mo, respectively *(56)*. Whereas Coleman et al. *(57)* observed success at 77.9 and 60.6% at 12 and 24 mo, respectively. Only 34% of IOP were controlled with initial Molteno implant after 6–59 mo follow-up *(58)*. Englert et al. *(59)* showed that tube surgery is safe for use in eyes with previous cycloablation, and early results showed lower postoperative IOP and fewer medication requirements up to 18 mo after surgery when compared with those without cycloablation. In high risk eyes, such as Sturge–Weber syndrome, two-staged Baerveldt implants appeared to be safe without suprachoroidal hemorrhage *(60)*. Complications included retinal detachment, corneal-tube contact, corneal decompensation, corneal graft rejection, dislocated tubes, and recurrent uveitis. Failures were considered if the child loses useful vision, incurred phthisis, or undergo enucleation. Analyses of risk factors suggested that the diagnosis, number of previous glaucoma procedures, and surgeon's experience are related to GDD survival *(61)*.

Maintenance of Visual Acuity

Improvement or maintenance of visual acuity within one Snellen line may be seen between 50 and 82% of patients *(5,37,62,63)*. It is believed that poor IOP control in patients with advanced glaucoma despite maximal medication after surgery may benefit from further intervention to prevent progressive visual loss. Employment of diode *trans*-scleral cyclophotocoagulation either sequentially or simultaneously with GDD may be useful in these instances. IOP should be monitored closely at around 1 mo postsurgery because spikes in IOP or hypertensive phase (HP) may lead to severe loss of visual acuity.

SAFETY

Complications are common in GDD surgery although most of them are self-limiting. The complications of GDD may be categorized according to three stages, namely, intraoperative, early postoperative (<1 mo), and late postoperative (>1 mo). Intraoperative complications consist of iris and lens trauma, tube misdirection, and scleral perforation. Early postoperative complication includes hypotony and flat anterior chamber, tube obstructions, choroidal effusions, hyphema, and suprachoroidal hemorrhage. Late postoperative complications consist of HP, corneal decompensation, tube migration, tube or plate extrusion, motility disturbance, infection, and phthisis.

Infection

Late infections associated with conjunctival defects *(64)* and tube exposure. These may occur as late as 4 yr after surgery and ultimately may require the removal of the GDD *(65–67)*. Infective organisms commonly implicated are *Propionibacterium acnes* and *Pseudomonas aeruginosa*. Not surprisingly, *P. aeruginosa* endophthalmitis are devastating even with prompt administration of intravitreal antibiotics to which the organisms are sensitive *(68)*. *P. acnes* is an ocular surface flora that probably gained access to the prosthesis through defects in the conjunctiva. Visual acuity worsened in approx 50% of the patients despite treatment and implants removal. Once infection is eradicated, subsequent IOP control may be achieved with either medication or GDD replacement. In view of the risk of endophthalmitis and poor visual outcome, urgent revision of exposed tubes with patch grafts are recommended *(67)*. Bacterial endophthalmitis has been reported 4 mo after cataract surgery in the presence of functioning tube *(65)* as well as after repositioning of tube *(69)*. In both instances, tube removal was performed before inflammation could be controlled.

Blockage

Common causes of tube blockage include blood in hyphema, vitreous in aphakics, and with pars plana tube placement, silicone oil, and iris knuckle. Vitreous incarceration can still occur after anterior vitrectomy, causing raised IOP as well as combined tractional-rhegmatogenous retinal detachment. Nd:YAG laser may be used to reopen blocked tubes and maintain their patency over time. Although, the initial success rates may be as high as 80%, reblockage is frequent. Complications arising from this procedure included moderate anterior chamber reaction, corneal edema, and hyphema *(70)*. Desatnik et al. *(71)* found that YAG laser vitreolysis was only 25% successful as compared with 100% for pars plana vitrectomy in relieving the obstruction.

In vitrectomized eyes, the internal orifices of the tubes may be obstructed by silicone oil. In author's experience, anterior chamber washouts are frequently needed and the intraocular IOP control may be difficult. Furthermore, silicone oil may migrate through the tube into the orbit to complicate management *(72)*. Patients who were aphakic or had significant amounts of silicone oil in the anterior chamber (fish egg sign or reverse hypopyon sign) are probably at risk of tubal obstruction. In some pseudophakic patients, small adherent droplets of silicone oil have been observed on the silicone tubes. Therefore, it is believed that in the absence of contraindications, silicone oil should be removed before tube implantation. When this is not possible, diode *trans*-scleral cyclophotocoagulation may be used as an alternative treatment *(73)*. Al-Jazzaf et al. *(74)* have recommended inferior quadrant placement of tubes, which can prevent tube orifice-silicone oil contact in the upright position. However, the tube orifice may still remain exposed in the supine position. The inferior fornix may be disadvantageous because it is shallower and GDD implantation may result in conspicuous bulging of the lower eyelid. Finally, inferiorly located blebs are more exposed, thereby increasing the likelihood of scarring, wound breakdown, and loss of IOP control.

Aqueous Misdirection

Aqueous misdirection is rare in tube surgery. Greenfield reported a small series in Baerveldt GDD, occurring at a median time of 33.5 d (range 1–343 d). Only 1 of 10 eyes responded to aqueous suppressants and cycloplegia. Four eye required Nd:YAG laser hyloidotomy, two eyes required vitrectomy alone, one eye had vitrectomy and lensectomy, and two eyes received vitrectomy and intraocular lens explantation. In essence, response to medical therapy is poor and anterior chamber depth normalization required aggressive laser and surgical therapy *(75)*.

Hypotony

Hypotony is a major problem in GDD surgery resulting in increased risk of suprachoroidal hemorrhage. Shallowing of the anterior chamber may also lead to tube-corneal-lens and iridocorneal touch that result in peripheral synechiae, cataracts, and corneal decompensation. Hypotony was first seen in single staged nonvalved tubes, where early overfiltration occurs before the bleb lining could mature with a thin layer of dense fibrous tissue.

Current noncomparative reports of ligated nonvalved Baerveldt and valved Ahmed GDD have shown that rates of hypotony are rather similar at 5–37.5% *(2,8,10,31, 62,63,76,77)*. The prevalence was 37.5% in Baerveldt GDD and 11.1% in Ahmed GDD *(10)*. Strategies to overcome postoperative hypotony in nonvalved GDD include complete *(78)* or partial tube occlusion as well as two-staged procedures *(60)*. The disadvantage of complete ligation was uncontrollably high IOP in some eyes. The addition of a venting stab (Sherwood slits) on the anterior chamber side of the occlusive ligation reduces the risk of high IOP, but moderately increase the frequency of ocular hypotony. This strategy is the best compromise to avoid short-term ocular hypotony while achieving long-term pressure control *(79)*. Some authors also believed that complete ligation with Sherwood slits is a superior technique as it facilitates capsular formation before the flow of aqueous humor through the tube *(80)*. Franks and Hitchings described the perioperative use of sodium hyaluronate, sulfur hexafluoride, or perfluoropropane in addition to tube ligation. They found less frequent hypotony with perfluoropropane vs sodium hyaluronate, but the mean IOP was significantly higher for the first 4 d after surgery *(81)*.

Although spontaneous recovery is the rule in most cases, the odd one may require anterior chamber reformation with either viscoelastic (in either GDD) or gas (in ligated GDD). Choroidal effusion usually resolves as the eye stabilized in terms of IOP over time. The surgeon has to bear in mind that peritubular filtration can be a cause of hypotony *(82)*. To prevent this complication, a scleral flap up to 1 mm is usually created from the limbus, a sclerostomy with long passage up to 2 mm beyond the limbus was made with a 23-gauge needle directed toward the pupil. If the sclerostomy is too long, the tip of the tube may sit too near to the endothelial surface. A 23-gauge needle gives a tighter wound, although the 21-gauge needle allows easier insertion. In cases, where previous limbal surgery, such as extracapsular surgery had been performed, thinning of tissue can make peritubular leakage more likely. Even with modern GDD, hypotony continues to be a significant postoperative morbidity that demands further improvement of implant design and evolution of surgical techniques.

Suprachoroidal Hemorrhage

Suprachoroidal hemorrhage is a dreaded complication in tube surgery and reported rates were between 2 and 11% *(62,80,83)*. In a study of Baerveldt GDD complications, Nguyen et al. *(80)* observed a bimodal fashion of occurrence in these ligated tubes and related that to IOP fluctuation. Peritubular leaks immediately after surgery or suturelysis at 1 mo precipitate significant drop in IOP, which can result in hypotony and choroidal effusion. Surgical experience with the Molteno GDD and trabeculectomy with 5-fluorouracil have indicated that rapid and large fall in IOP are risk factors for supraclioroidal hemorrhage *(84,85)*. Other risk factors included aphakic and intraoperative vitrectomy *(86)*, previous ocular surgery *(85)*, hypertension, and artherosclerosis *(80)*. The pathogenesis is thought to be mechanical stress on the posterior ciliary arteries by choroidal effusion during hypotony *(86–88)*. Therefore, preventive measures include prevention as well as treatment of hypotony.

Hemorrhage

Hyphema is a common complication frequently seen in rubeotic cases. Bad prognostic signs include concurrent high IOP in the presence of an eight-ball hyphema. These cases should be treated urgently with anterior chamber washout. Neither ligated nor valve tubes offer additional advantage in terms of prevention *(10)*.

Scleral Erosion/Tube Exposure

Tube erosions are not frequently encountered and when it occurs, the site of occurrence is usually near the limbus. Previous surgery at the limbus and consequent scleral thinning is probably a risk factor if a scleral trapdoor is to be raised. Uveitis was found to be a significant risk factor in children *(56)*. Various materials, such as donor sclera, dura, and cadaveric pericardium had been used for tube coverage. Smith et al. *(89)* reported similar efficacy among these materials and the tube erosion rates were low. No tube erosion were reported in either pericardial grafts *(90)* or fascia lata grafts *(91)*.

A broad and long (approx 5×5 mm^2), partial thickness (2/3) scleral flap is created for tube coverage. A thin and long flap may be predisposed to melting when blood supply is compromised after diathermy and surgical dissection. The flap has to be long enough such that it drapes over the tube to come in contact with the dissected scleral edge and may subsequently regain some vascular supply. In instances, where passive contact has not been possible, minimal diathermy over the flap as well as suturing the free flap edges to the scleral bed would be advantageous. Although, thinning of the sclera flaps was noticed in many patients, conjunctival erosion by the silicone tube remained rare in the series. Currently, Tutopatch® (Tutogen Medical Inc., Alachua, FL) has been used to patch the tube at the limbus. Tutopatch is a bovine pericardium collagen framework that allows scar tissue to be incorporated for reinforced support. The long-term results are still not available at this stage.

Diplopia

GDD with large plates are at risk of diplopia. The plate of the Baerveldt tube that extends under the recti muscles forms plate-muscle adhesions that potentially interfere with ocular motility. As such, high rates of heterotropia (77%), restriction of gaze in the quadrant of implant and diplopia in the primary gaze (65%) surfaced during early clinical experience *(3)*. Later studies reported intermittent symptomatic diplopia of smaller incidences at 6–8.7% *(2,92)*. A study comparing the 350 and 500 mm^2 Baerveldt implant reported strabismus rates of 16 and 19%, respectively. It is associated with both nasal and temporal implants, and usually, after flowthrough the tube has been established. Although, some strabismus did resolve in some eyes, persistent cases may require muscle surgery *(17)*. Minoz et al. *(93)* described characteristic patterns of either exotropia or hypertropia, or both where the deviation increased with gaze opposite to the action of ipsilateral superior and lateral muscles and decreased in the field of action of the involved muscles. All patients in their series demonstrated exotropia with convergence insufficiency.

The muscle dysfunction was probably originated from a mechanical cause because of the large blebs in the orbit. These blebs, which also extend under the muscles may potentiate the functional impairment. Fenestrations on the plates can encourage fibrous ingrowths that connect the roof and floor of the blebs, thereby limiting the bleb height *(17)*. Coats et al. described a case of acquired pseudo-Brown's syndrome immediately following Ahmed valve glaucoma implant in the superonasal quadrant. Their patient experienced both vertical and horizontal diplopia. Apparently, the implant became wedged between the globe and orbit superonasally and was confirmed during forced duction testing. As such, forced duction testing during implantation was recommended *(94)*. Christmann and Wilson *(95)* proposed that motility disorders in Molteno GDD are because of posterior fixation or Farden effect. True Tenon cyst, albeit, a rare cause of motility disturbance has been described *(96)*.

Bleb Fibrosis

The modeling of bleb around the endplate of GDD is influenced by the timing of aqueous introduction, IOP during the healing phase, material in the endplate, handling of the endplate, and usage of anti-inflammatory agents. In the Otago Glaucoma Surgery Outcome Study, Molteno *(97)* described the histopathology of Molteno implant capsules in correlation to the surgical techniques and outcomes with Molteno GDD. They studied 75 autopsy eyes obtained between 4 d and 23 yr after the insertion of Molteno GDD. The results of their study are summarized in the Table 3.

The interaction between IOP, aqueous, and the bleb is interesting. Histological studies have demonstrated that the high IOP stimulates fibrovascular proliferation *(98)* whereas aqueous from glaucomatous eye with controlled IOP stimulated connective tissue degeneration *(99)*. Further studies from Molteno implants showed that increased IOP exerts a temporary inflammatory and fibrosing effect on episcleral tissues lasting 4–6 wk, followed by a noninflammatory degenerative action *(100,101)*. The noninflammatory degenerative process is dependent on sufficient increase in IOP in order for the aqueous to displace the interstitial fluid from the deep layers of the capsule *(98)*. In practice, the two-stage procedure was associated with an initial HP, especially after suture lysis, followed by normalization of IOP. Two-stage procedure developed thinner capsules compared with the single-stage or three-staged procedure. However, a review of literature comparing postoperative IOP level and surgical success rates between the single stage and two-stage technique failed to show any difference *(102)*.

Intraoperative and postoperative factors may also contribute to bleb outcome. Rabbit studies have shown that surgical trauma that roughens the endplate may enhance fibrous ingrowths onto the plate and increase the risk of late failure. The capsule thickness may be further modulated by the use of systemic colchicines or topical anti-inflammatory agents *(103)*.

The histopathological implication of having a valved GDD as opposed to two-staged nonvalved GDD on bleb fibrosis is unknown as valves are designed to open at predetermined IOP. Hence, further studies examining the relationship between IOP and bleb

Table 3
Staged Insertion of Molteno

	Single stage	*Two stage*	*Three stage*
Technique	Nonligatured implant with immediate aqueous flow	Two-staged procedure where the ligature is released at 4–6 wk after surgery	Externalization of aqueous flow onto the conjunctival space and IOP not exceeding 12 mmHg
Capsule	Thick capsules 300–600 μm Inner layer of fibrodegenerative and outer fibrovascular layer of equal thickness	Stage 1: Thin and avascular collagenous layer (20–60 μm). Stage 2: Thin permeable capsules (190–250 μm) with fewer fibrovascular than fibrodegenerative component	Stage 1: Plate implanted and aqueous drain through accessory tube through the Tenon's capsule into the subcon junctival space Stage 2: 6–8 wk, tube inserted into the anterior chamber. Free end of secondary tube exposed in the fornix to allow aqueous to escape to maintain IOP at 6–12 mmHg Stage 3: 8–11 wk, secondary tube is cut and allowed to retract into the fornix Thick capsule (375–700 μm) heavily fibrosed and impermeable, which has dense fibrovascular tissue, but no fibrodegenerative layer
IOP		Temporary IOP spike to 25–35 mmHg	High IOP at 2–3 wk, procedure was abandoned subsequently

thickness or composition using histopathological as well as imaging techniques may provide more evidence on the impact of IOP on bleb modeling and consequent surgical success. Careful control of postoperative IOP to achieve a certain desired bleb may play an increasingly important role during postoperative management.

Hypertensive Phase

Molteno and Dempster first observed HP in the single-plate nonvalved Molteno implant. After surgery, there was a short-lived hypotensive phase lasting 7–10 d. IOP then rise gradually, accompanied by formation of a well-circumscribed bleb. This phase was characterized histologically by the disappearance of edema and the emergence of a dense layer of fibrous tissue over the plate. During the first few weeks of HP, intense congestion of the bleb was noted with IOP rising up to 50 mmHg. Concurrent to reduction in IOP and subsequent stabilization, there is decrease in congestion, inflammation, and bleb thickness *(104)*. HP is a curious phenomenon of raise in IOP usually during the first 2–4 wk and last for 4–6 mo *(102)*. Some authors have defined it as an IOP higher than 21 mmHg during the first 6 postoperative mo in the presence of an encapsulated bleb *(31,105)*. Nouri-Mahdavi and Caprioli defined HP in their study as IOP higher than 21 mmHg during the first 3 mo after surgery after an initial reduction of IOP to less than 22 mmHg, in the presence of a functioning device. They believed that rise in IOP after 3 mo postoperatively may be demonstrating poor IOP control rather than HP *(106)*.

They further defined the resolution of HP as IOP < 22 mmHg along with (1) a reduction of IOP by 3 mmHg or more with the same number of medications or less or (2) reduction of at least one medication with a change of less than 3 mmHg. Indeed, identifying the resolution of HP is difficult because (1) some patients continued to be on medication while IOP stabilizes, thereby masking the end point and (2) the arbitrary duration of treatment after which the implant should be considered an absolute failure or qualified success rather than still in HP. This is probably in part because of the fact that HP may share the same pathology as bleb fibrosis and failure of IOP control. Further studies are needed to verify the histopathology.

Incidences of HP based on arbitrary definitions from various authors were as follows for Ahmed GDD (56.4–82%) *(31,106)*, Baerveldt (0–30%) *(2,105)*, double-plate Molteno (43.5%) *(103)*. HP is an IOP manifestation of bleb remodeling and fibrosis with the sequential histological events of congestion, edema, and fibrosis. The behavior of bleb fibrosis around the plates may be influenced by surgical technique and postoperative management and these postulates are listed in the following Table 4.

The bleb remodeling and fibrosis is probably a common pathway shared by both HP and late bleb failure from encapsulation. In the practice, patients who entered HP may be dependent on medication for IOP control beyond 4–6 mo. The suspicions were confirmed when Nouri-Mahdavi and Caprioli *(106)* followed up their patients with Ahmed GDD up to 1 yr and they found that resolution occurred in only a minority of the of the eyes whereas a majority of eyes continued to require the same number of glaucoma medications as they did during the HP. The more flexible silicone plate Ahmed FP7 and FP8 GDD has been recently introduced. There is theoretical advantages of less microtrauma with flexible plate and use of biocompatible silicone. A lower incidence of HP in these new devices would point to biomaterial and design as the main risk factors.

Table 4
Risks Factors for Hypertensive Phase

Surgical technique	
	Immediate filtration in valved GDD as opposed to delayed filtration with ligated nonvalved GDD *(105)*. Delayed filtration elicit less fibrotic reaction and therefore lower risk of encapsulation
Aqueous proinflammatory factors	
	Aqueous humor containing certain factors that stimulate a fibrotic response in the subconjunctival space *(99)*. The nature of these proinflammatory factors remained unknown
Plate size	
	Higher incidence of HP in Ahmed GDD could be due to its smaller plate (185 mm^2) compared with Molteno (270 mm^2) or Baerveldt (350 mm^2)
Shape and consistency	
	The rigid and longer plate of the Ahmed GDD show greater micro motion resulting in proinflammatory trauma as compared to the softer and disk-shaped Molteno implants that may be more stable on the sclera
Design	
	Ridged plates prevent direct growth of the capsule onto the plate as compared with smooth plates, which attracts white blood cells and fibroblasts
Biomaterial	
	Silicone may incite less inflammation on the implant endplates as compared with polypropylene *(109,110)*

The Latin American Glaucoma Society Investigators compared mitomycin C augmentation with or without partial Tenon's capsule resection during surgery. This technique had significant, but small effect on HP (40% vs 46.8%). However, there were no advantages conferred in terms of IOP control and complication rates *(107)*. Lai et al. reported an incidence of 26.4% of encapsulation in their series of Ahmed GDD, but not HP. The authors attributed this to intervention with either antiglaucoma medication or needling with mitomycin C *(8)*. Surgical excision of the bleb has been recommended in cases that do not respond to needling. Hwang and Kee tried to use pericardial membrane (preclude) to expand the surface area of the endplate to 300 mm^2. Two (20%) of 10 eyes in the pericardial membrane + Ahmed GDD group, as opposed to 8(80%) of 10 eyes in Ahmed GDD group, developed HP *(108)*. However, these numbers are small and would require further studies to ascertain these results.

FUTURE

GDD await design refinement for the control of postoperative hypotony, HP, corneal-tube touch, tube erosion, and encapsulation. The definition of surgical success at IOP of 21 mmHg may require future adjustments when long-term data on disease progression and field loss becomes available. In Asians where the high success rates of these GDD in refractory glaucoma contrasted sharply with the lower success rate of trabeculectomy *(24)* for uncomplicated glaucoma warrants randomized studies to address its role as a primary surgical option in uncomplicated glaucoma.

REFERENCES

1. Rollet M. Le drainage au irin de la chambre anterieure contre l'hypertonie et al. douleur. Rev Gen Ophthalmol 1906;25:481.
2. Siegner SW, Netland PA, Urban RC Jr, et al. Clinical experience with the Baerveldt glaucoma drainage implant. Ophthalmology. 1995;102(9):1298–1307.
3. Smith SL, Starita RJ, Fellman RL, Lynn JR. Early clinical experience with the Baerveldt 350-mm^2 glaucoma implant and associated extraocular muscle imbalance. Ophthalmology 1993;100(6):914–918.
4. Lloyd MA, Baerveldt G, Heuer DK, et al. Initial clinical experience with the baerveldt implant in complicated glaucomas. Ophthalmology 1994;101:640–650.
5. Britt MT, LaBree LD, Lloyd MA, et al. Randomized clinical trial of the 350-mm^2 versus the 500-mm^2 Baerveldt implant: longer term results: is bigger better? Ophthalmology 1999;106:2312–2318.
6. Mills RP, Reynolds A, Emond MJ, Barlow WE, Leen MM. Long-term survival of Molteno glaucoma drainage devices. Ophthalmology 1996;103(2):299–305
7. Aung T, Seah SK. Glaucoma drainage implants in Asian eyes. Ophthalmology 1998;105: 2117–2122.
8. Lai JS, Poon AS, Chua JK, et al. Efficacy and safety of the Ahmed glaucoma valve implant in Chinese eyes with complicated glaucoma. Br J Ophthalmol 2000;84:718–721.
9. Das JC, Chaudhuri Z, Sharma P, Bhomaj S. The Ahmed Glaucoma Valve in refractory glaucoma: experiences in Indian eyes. Eye 2000;19(2):183–190.
10. Wang JC, See JL, Chew PT. Experience with the use of Baerveldt and Ahmed glaucoma drainage implants in an Asian population. Ophthalmology 2004;111(7):1383–1388.
11. Seah SK, Gazzard G, Aung T. Intermediate-term outcome of Baerveldt glaucoma implants in Asian eyes. Ophthalmology 2003;110(5):888–894.
12. Minckler DS, Shammas A, Wilcox M, Ogden TE. Experimental studies of aqueous filtration using the Molteno implant. Trans Am Ophthalmol Soc 1987;85:368–392.
13. Heuer DK, Lloyd MA, Abrams DA, et al. Which is better? One or two? A randomized clinical trial of single-plate versus double-plate Molteno implantation for glaucomas in aphakia and pseudophakia. Ophthalmology 1992;99(10):1512–1519.
14. Camras CB. Discussion Of: Wilson RP, Cantor L, Katz LJ, et al. Aqueous shunts. Molteno versus Schocket. Ophthalmology 1992;99:672–678.
15. Smith MF, Sherwood MB, McGorray SP. Comparison of the double-plate Molteno drainage implant with the Schocket procedure. Arch Ophthalmol 1992;110(9):1246–1250.
16. Wilson RP, Cantor L, Katz LJ, Schmidt CM, Steinmann WC, Allee S. Aqueous shunts. Molteno versus Schocket. Ophthalmology 1992;99(5):672–676; discussion 676–678.
17. Lloyd MA, Baerveldt G, Fellenbaum PS, et al. Intermediate-term results of a randomized clinical trial of the 350- versus the 500-mm^2 Baerveldt implant. Ophthalmology 1994; 101(8):1456–1463; discussion 1463–1464.
18. Lloyd MA, Minckler DS, Heuer DK, Baerveldt G, Green RL. Echographic evaluation of glaucoma shunts. Ophthalmology. 1993;100(6):919–927.
19. Molteno AC. The optimal design of drainage implants for glaucoma. Trans Ophthalmol Soc N Z 1981;33:39–41.
20. Burgoyne JK, WuDunn D, Lakhani V, Cantor LB. Outcomes of sequential tube shunts in complicated glaucoma. Ophthalmology 2000;107(2):309–314.
21. Shah AA, WuDunn D, Cantor LB. Shunt revision versus additional tube shunt implantation after failed tube shunt surgery in refractory glaucoma. Am J Ophthalmol 2000;129(4):455–460.
22. Semchyshyn TM, Tsai JC, Joos KM. Supplemental trans-scleral diode laser cyclophotocoagulation after aqueous shunt placement in refractory glaucoma. Ophthalmology 2002;109(6): 1078–1084.

23. Wilson MR, Mendis U, Smith SD, Paliwal A. Ahmed glaucoma valve implant vs. trabeculectomy in the surgical treatment of glaucoma: a randomized clinical trial. Am J Ophthalmol 2000;130(3):267–273.
24. Tan C, Chew PT, Lum WL, Chee C. Trabeculectomy—success rates in a Singapore hospital. Singapore Med J 1996;37(5):505–507.
25. Chalam KV, Gandham S, Gupta S, Tripathi BJ, Tripathi RC. Pars plana modified Baerveldt implant versus neodymium:YAG cyclophotocoagulation in the management of neovascular glaucoma. Ophthalmic Surg Lasers 2002;33(5):383–393.
26. Lima FE, Magacho L, Carvalho DM, Susanna R Jr., Avila MP. A prospective, comparative study between endoscopic cyclophotocoagulation and the Ahmed drainage implant in refractory glaucoma. J Glaucoma 2004;13(3):233–237.
27. Kook MS, Yoon J, Kim J, Lee MS. Clinical results of Ahmed glaucoma valve implantation in refractory glaucoma with adjunctive mitomycin C. Ophthalmic Surg Lasers 2000;31(2): 100–106.
28. Costa VP, Azuara-Blanco A, Netland PA, et al. Efficacy and safety of adjunctive mitomycin C during Ahmed Glaucoma Valve implantation: a prospective randomized clinical trial Ophthalmol 2004;111(6):1071–1076.
29. Irak I, Moster MR, Fontanarosa J. Intermediate-term results of Baerveldt tube shunt surgery with mitomycin C use. Ophthalmic Surg Lasers Imaging 2004;35(3):189–196.
30. Lee D, Shin DH, Birt CM, et al. The effect of adjunctive mitomycin C in Molteno implant surgery.Ophthalmology 1997;104(12):2126–2135.
31. Ayyala RS, Zurakowski D, Smith JA, et al. A clinical study of the Ahmed glaucoma valve implant in advanced glaucoma. Ophthalmology 1998;105(10):1968–1976.
32. Susanna R Jr., Nicolela MT, Takaashi WY. Mitomycin C as adjunctive therapy with glaucoma implant surgery. Ophthalmic Surg 1994;25(7):458–462.
33. Perkins TW, Cardakli UF, Eisele JR, Kaufman PL, Heatley GA. Adjunctive mitomycin C in Molteno implant surgery. Ophthalmology 1995;102(1):91–97.
34. Chung AN, Aung T, Wang JC, Chew PT. Surgical outcomes of combined phacoemulsification and glaucoma drainage implant surgery for Asian patients with refractory glaucoma with cataract. Am J Ophthalmol 2004;137(2):294–300.
35. Molteno AC, Whittaker KW, Bevin TH, Herbison P. Otago Glaucoma Surgery Outcome Study: long term results of cataract extraction combined with Molteno implant insertion or trabeculectomy in primary glaucoma. Br J Ophthalmol 2004;88(1):32–35.
36. Bhattacharyya CA, WuDunn D, Lakhani V, Hoop J, Cantor LB. Cataract surgery after tube shunts. J Glaucoma 2000;9(6):453–457.
37. Topouzis F, Coleman AL, Choplin N, et al. Follow-up of the original cohort with the Ahmed glaucoma valve implant. Am J Ophthalmol 1999;128:198–204.
38. Zalloum JN, Ahuja RM, Shin D, Weiss JS. Assessment of corneal decompensation in eyes having undergone molteno shunt procedures compared to eyes having undergone trabeculectomy. CLAO J 1999;25(1):57–60.
39. Al-Torbak A. Graft survival and glaucoma outcome after simultaneous penetrating keratoplasty and Ahmed glaucoma valve implant. Cornea 2003;22(3):194–197.
40. Alvarenga LS, Mannis MJ, Brandt JD, Lee WB, Schwab IR, Lim MC. The long-term results of keratoplasty in eyes with a glaucoma drainage device. Am J Ophthalmol 2004; 138(2):200–205.
41. Kwon YH, Taylor JM, Hong S, et al. Long-term results of eyes with penetrating keratoplasty and glaucoma drainage tube implant. Ophthalmology 2001;108(2):272–278.
42. Joos KM, Lavina AM, Tawansy KA, Agarwal A. Posterior repositioning of glaucoma implants for anterior segment complications. Ophthalmology 2001;108(2):279–284.

43. Luttrull JK, Avery RL, Baerveldt G, Easley KA. Initial experience with pneumatically stented baerveldt implant modified for pars plana insertion for complicated glaucoma. Ophthalmology 2000;107(1):143–149; discussion 149–150.
44. Arroyave CP, Scott IU, Fantes FE, Feuer WJ, Murray TG. Corneal graft survival and intraocular pressure control after penetrating keratoplasty and glaucoma drainage device implantation. Ophthalmology 2001;108(11):1978–1985.
45. Sidoti PA, Mosny AY, Ritterband DC, Seedor JA. Pars plana tube insertion of glaucoma drainage implants and penetrating keratoplasty in patients with coexisting glaucoma and corneal disease. Ophthalmology 2001;108(6):1050–1058.
46. Rothman RF, Sidoti PA, Gentile RC, et al. Glaucoma drainage tube kink after pars plana insertion. Am J Ophthalmol 2001;132(3):413–414.
47. Rumelt S, Rehany U. Implantation of glaucoma drainage implant tube into the ciliary sulcus in patients with corneal transplants. Arch Ophthalmol 1998;116(5):685–687.
48. Al-Torbak AA. Outcome of combined Ahmed glaucoma valve implant and penetrating keratoplasty in refractory congenital glaucoma with corneal opacity. Cornea 23(6):554–559.
49. Luttrull JK, Avery RL. Pars plana implant and vitrectomy for treatment of neovascular glaucoma. Retina 1995;15(5):379–387.
50. Sidoti PA, Dunphy TR, Baerveldt G, et al. Experience with the Baerveldt glaucoma implant in treating neovascular glaucoma. Ophthalmology 1995;102(7):1107–1118.
51. Eid TE, Katz LJ, Spaeth GL, Augsburger JJ. Tube-shunt surgery versus in the management of neovascular glaucoma. Ophthalmology 1997;104(10):1692–1700.
52. Ceballos EM, Parrish RK 2nd, Schiffman JC. Outcome of Baerveldt glaucoma drainage implants for the treatment of uveitic glaucoma. Ophthalmology 2002;109(12):2256–2260.
53. Molteno AC, Sayawat N, Herbison P. Otago glaucoma surgery outcome study : long-term results of uveitis with secondary glaucoma drained by Molteno implants. Ophthalmology 2001;108(3):605–613.
54. Gil-Carrasco F, Salinas-VanOrman E, Recillas-Gispert C, Paczka JA, Gilbert ME, Arellanes-Garcia L. Ahmed valve implant for uncontrolled uveitic glaucoma. Ocul Immunol Inflamm 1998;6(1):27–37.
55. Valimaki J, Airaksinen PJ, Tuulonen A. Molteno implantation for secondary glaucoma in juvenile rheumatoid arthritis. Arch Ophthalmol 1997;115(10):1253–1256.
56. Morad Y, Donaldson CE, Kim YM, Abdolell M, Levin AV. The Ahmed drainage implant in the treatment of pediatric glaucoma. Am J Ophthalmol 2003;135(6):821–829.
57. Coleman AL, Smyth RJ, Wilson MR, Tam M. Initial clinical experience with the Ahmed Glaucoma Valve implant in pediatric patients. Arch Ophthalmol 1997;115(2):186–191.
58. Hill RA, Heuer DK, Baerveldt G, Minckler DS, Martone JF. Molteno implantation for glaucoma in young patients. Ophthalmology 1991;98(7):1042–1046.
59. Englert JA, Freedman SF, Cox TA. The Ahmed valve in refractory pediatric glaucoma. Am J Ophthalmol 1999;127(1):34–42.
60. Budenz DL, Sakamoto D, Eliezer R, et al. Two-staged Baerveldt glaucoma implant for childhood glaucoma associated with Sturge-Weber syndrome. Ophthalmology 2000;107(11):2105–2110.
61. Djodeyre MR, Peralta Calvo J, Abelairas Gomez J. Clinical evaluation and risk factors of time to failure of Ahmed Glaucoma Valve implant in pediatric patients. Ophthalmology 2001;108(3):614–620.
62. Coleman AL, Hill R, Wilson MR, et al. Initial clinical experience with the Ahmed Glaucoma Valve implant. Am J Ophthalmol 1995;120(1):23–31. Erratum in: Am J Ophthalmol 1995;120(5):684.
63. Huang MC, Netland PA, Coleman AL, et al. Intermediate-term clinical experience with the Ahmed Glaucoma Valve implant. Am J Ophthalmol 1999;127:27–33.

64. Francis BA, DiLoreto DA Jr, Chong LP, Rao N. Late-onset bacteria endophthalmitis following glaucoma drainage implantation. Ophthalmic Surg Lasers Imaging 2003;34(2):128–130.
65. Hollander DA, Dodds EM, Rossetti SB, Wood IS, Alvarado JA. *Propionibacterium acnes endophthalmitis* with bacterial sequestration in a Molteno's implant after cataract extraction. Am J Ophthalmol 2004;138(5):878–879.
66. Gutierrez-Diaz E, Montero-Rodriguez M, Mencia-Gutierrez E, Fernandez-Gonzalez MC, Perez-Blazquez E. *Propionibacterium acnes endophthalmitis* in Ahmed glaucoma valve. Eur J Ophthalmol 2001;11(4):383–385.
67. Gedde SJ, Scott IU, Tabandeh H, et al. Late endophthalmitis associated with glaucoma drainage implants. Ophthalmology. 2001;108(7):1323–1327.
68. Eifrig CW, Scott IU, Flynn HW Jr, Miller D. Endophthalmitis caused by *Pseudomonas aeruginosa*. Ophthalmology. 2003;110(9):1714–1717.
69. Fanous MM, Cohn RA. *Propionibacterium endophthalmitis* following Molteno tube repositioning. J Glaucoma. 1997;6(4):201–202.
70. Singh K, Eid TE, Katz LJ, Spaeth GL, Augsburger JJ. Evaluation of Nd:YAG laser membranectomy in blocked tubes after glaucoma tube-shunt surgery. Am J Ophthalmol 1997; 124(6):781–786.
71. Desatnik HR, Foster RE, Rockwood EJ, Baerveldt G, Meyers SM, Lewis H. Management of glaucoma implants occluded by vitreous incarceration. J Glaucoma 2000;9(4):311–316.
72. Nazemi PP, Chong LP, Varma R, Burnstine MA. Migration of intraocular silicone oil into the subconjunctival space and orbit through an Ahmed glaucoma valve. Am J Ophthalmol 2001;132(6):929–931.
73. Han SK, Park KH, Kim DM, Chang BL. Effect of diode laser trans-scleral cyclophotocoagulation in the management of glaucoma after intravitreal silicone oil injection for complicated retinal detachment. Br J Ophthalmol 1999;83:713–717.
74. Al-Jazzaf AM, Netland PA, Charles S. Incidence and management of elevated intraocular pressure after silicone oil injection. J Glaucoma 2005;14(1):40–46.
75. Greenfield DS, Tello C, Budenz DL, Liebmann JM, Ritch R. Aqueous misdirection after glaucoma drainage device implantation. Ophthalmology 1999;106(5):1035–1040.
76. Kee C. Prevention of early postoperative hypotony by partial ligation of silicone tube in Ahmed glaucoma valve implantation. J Glaucoma 2001;10(6):466–469.
77. Syed HM, Law SK, Nam SH, Li G, Caprioli J, Coleman A. Baerveldt-350 implant versus Ahmed valve for refractory glaucoma: a case-controlled comparison. J Glaucoma 2004;13(1):38–45.
78. Molteno AC, Polkinghorne PJ, Bowbyes JA. The vicryl tie technique for inserting a draining implant in the treatment of secondary glaucoma. Aust N Z J Ophthalmol 1986;14(4):343–354.
79. Rose GE, Lavin MJ, Hitchings RA. Silicone tubes in glaucoma surgery: the effect of technical modifications on early postoperative intraocular pressures and complications. Eye 1989;3(Pt 5):553–561.
80. Nguyen QH, Budenz DL, Parrish RK 2nd. Complications of Baerveldt glaucoma drainage implants. Arch Ophthalmol 1998;116(5):571–575.
81. Franks WA, Hitchings RA. Injection of perfluoropropane gas to prevent hypotony in eyes undergoing tube implant surgery. Ophthalmology 1990;97(7):899–903.
82. Garcia-Feijoo J, Cuina-Sardina R, Mendez-Fernandez C, Castillo-Gomez A,Garcia-Sanchez J. Peritubular filtration as cause of severe hypotony after Ahmed valve implantation for glaucoma. Am J Ophthalmol 2001;132(4):571–572.
83. Sidoti PA. Inferonasal placement of aqueous shunts. J Glaucoma 2004;13(6):520–523.
84. The Fluorouracil Filtering Surgery Study Group. Risk factors for suprachoroidal haemorrhage after filtering surgery. Am J Ophthalmol 1998;95:1202–1206.

85. Paysse E, Lee PP, Lloyd MA, et al. Suprachoroidal hemorrhage after Molteno implantation. J Glaucoma 1996;5(3):170–175.
86. Rockwood EJ, Kalenak JW, Plotnik JL, Yoon JS, Sculley L, Medendorp SV. Prospective ultrasonography evauation, of intra-operative and delayed post-operative suprachoroidal haemorrhage from glaucoma filtering surgery. J Glaucoma 1995;4:16–24.
87. Gressel MG, Parrish RK 2nd, Heuer DK. Delayed nonexpulsive suprachoroidal hemorrhage. Arch Ophthalmol 1984;102(12):1757–1760.
88. Wolter JR, Garfinkel RA. Ciliochoroidal effusion as precursor of suprachoroidal hemorrhage: a pathologic study. Ophthalmic Surg 1988;19(5):344–349.
89. Smith MF, Doyle JW, Ticrney JW Jr. A comparison of glaucoma drainage implant tube coverage. J Glaucoma 2002;11(2):143–147.
90. Raviv T, Greenfield DS, Liebmann JM, Sidoti PA, Ishikawa H, Ritch R. Pericardial patch grafts in glaucoma implant surgery. J Glaucoma 1998;7(1):27–32.
91. Tanji TM, Lundy DC, Minckler DS, Heuer DK, Varma R. Fascia lata patch graft in glaucoma tube surgery. Ophthalmology 1996;103(8):1309–1312.
92. Smith MF, Doyle JW, Sherwood MB. Comparison of the Baerveldt glaucoma implant with the double-plate Molteno drainage implant. Arch Ophthalmol 1995;113(4):444–447.
93. Munoz M, Parrish RK 2nd. Strabismus following implantation of Baerveldt drainage devices. Arch Ophthalmol 1993;111(8):1096–1099.
94. Coats DK, Paysse EA, Orenga-Nania S. Acquired Pseudo-Brown's syndrome immediately following Ahmed valve glaucoma implant. Ophthalmic Surg Lasers 1999;30(5): 396–397.
95. Christmann LM, Wilson ME. Motility disturbances after Molteno implants. J Pediatr Ophthalmol Strabismus 1992;29(1):44–48.
96. Rhee DJ, Casuso LA, Rosa RH Jr, Budenz DL. Motility disturbance due to true Tenon cyst in a child with a Baerveldt glaucoma drainage implant. Arch Ophthalmol1 2001;19(3):440–442.
97. Molteno AC, Fucik M, Dempster AG, Bevin TH. Otago Glaucoma Surgery Outcome Study: factors controlling capsule fibrosis around Molteno implants with histopathological correlation. Ophthalmology 2003;110(11):2198–2206.
98. Epstein E. Fibrosing response to aqueous. Its relation to glaucoma. Br J Ophthalmol 1959;43:641–647.
99. Teng CC, Chi HH, Katzin HM. Histology and mechanism of filtering operations. Am J Ophthalmol 1959;47(1 Pt 1):16–33.
100. Molteno AC. New implant for drainage in glaucoma. Clinical trial. Br J Ophthalmol 1969;53(9):606–615.
101. Molteno AC. A new implant for glaucoma. Effect of removing implants. Br J Ophthalmol 1971;55(1):28–37.
102. Hong CH, Arosemena A, Zurakowski D, Ayyala RS. Glaucoma drainage devices: a systematic literature review and current controversies. Surv Ophthalmol 2005;50(1):48–60.
103. Molteno TE, Dempster AG. Methods of controlling bleb fibrosis around drainage implants. In: Mills KB, ed. Fourth International Symposium of the Northern Eye Institute, 1st ed. Manchester, UK: Pergamon Press, 1998;192–211.
104. Ayyala RS, Zurakowski D, Monshizadeh R, et al. Comparison of double-plate Molteno and Ahmed glaucoma valve in patients with advanced uncontrolled glaucoma. Ophthalmic Surg Lasers 2002;33(2):94–101.
105. Tsai JC, Johnson CC, Dietrich MS. The Ahmed shunt versus the Baerveldt shunt for refractory glaucoma: a single-surgeon comparison of outcome. Ophthalmology 2003; 110(9):1814–1821.
106. Nouri-Mahdavi K, Caprioli J. Evaluation of the hypertensive phase after insertion of the Ahmed Glaucoma Valve. Am J Ophthalmol 2003;136(6):1001–1008.

107. Susanna R Jr. Latin American Glaucoma Society Investigators. Partial Tenon's capsule resection with adjunctive mitomycin C in Ahmed glaucoma valve implant surgery. Br J Ophthalmol 2003;87(8):994–998.
108. Hwang JM, Kee C. The effect of surface area expansion with pericardial membrane (preclude) in Ahmed glaucoma valve implant surgery. J Glaucoma 2004;13(4):335–339.
109. Ayyala RS, Harman LE, Michelini-Norris B, et al. Comparison of different biomaterials for glaucoma drainage devices. Arch Ophthalmol 1999;117(2):233–236.
110. Ayyala RS, Michelini-Norris B, Flores A, Haller E, Margo CE. Comparison of different biomaterials for glaucoma drainage devices: part 2. Arch Ophthalmol 2000;118(8): 1081–1084.

14
Pellucid Marginal Corneal Degeneration

Jorge L. Alió, MD, PhD, Mohamed H. Shabayek, MD, MSc, Alberto Artola, MD, PhD, and Hany El Saftawy, MD, PhD

CONTENTS

INTRODUCTION

Pellucid marginal corneal degeneration (PMCD) is a noninflammatory ectatic corneal disorder mostly involving the inferior half of the cornea in a crescentic fashion (Fig. 1). It is a bilateral disease, although one eye may be affected earlier and clinically diagnosed, while the other eye has no clinical features *(1)*. Although classically described as an inferor entity, the site of involvement can be in any quadrant of the cornea, including the superior part, which is termed superior PMCD *(2,3)*. The degeneration is distinguished from other ectatic corneal disorders by its characteristic location and the absence of inflammatory signs. Typically the thinning extends from the 4-o'clock position to the 8-o'clock position, 1 mm from the limbus with intact epithelium and normal corneal thickness superiorly. The area between the limbus and thinning is clear, without scarring, lipid deposition, or vascularization. Usually, present with reduced visual acuity owing to high irregular astigmatism in the fourth to fifth decades of life *(4)*.

DIAGNOSIS

The diagnosis of PMCD is based on the presence of corneal thinning and ectasia, with normal cornea above and below the ectatic (thin) area, with no evidence of scarring, vascularization, or lipid deposition *(4)*. Confirmation of the diagnosis by corneal topography is by the presence of the characteristic feature of PMCD, which shows marked flattening of the cornea along a vertical axis and a steepening of the inferior cornea peripheral to the site of the lesion (Fig. 2) *(4,5)*. PMCD can be associated with keratoconus or keratoglobus

From: *Ophthalmology Research: Visual Prosthesis and Ophthalmic Devices: New Hope in Sight*
Edited by: J. Tombran-Tink, C. Barnstable, and J. F. Rizzo © Humana Press Inc., Totowa, NJ

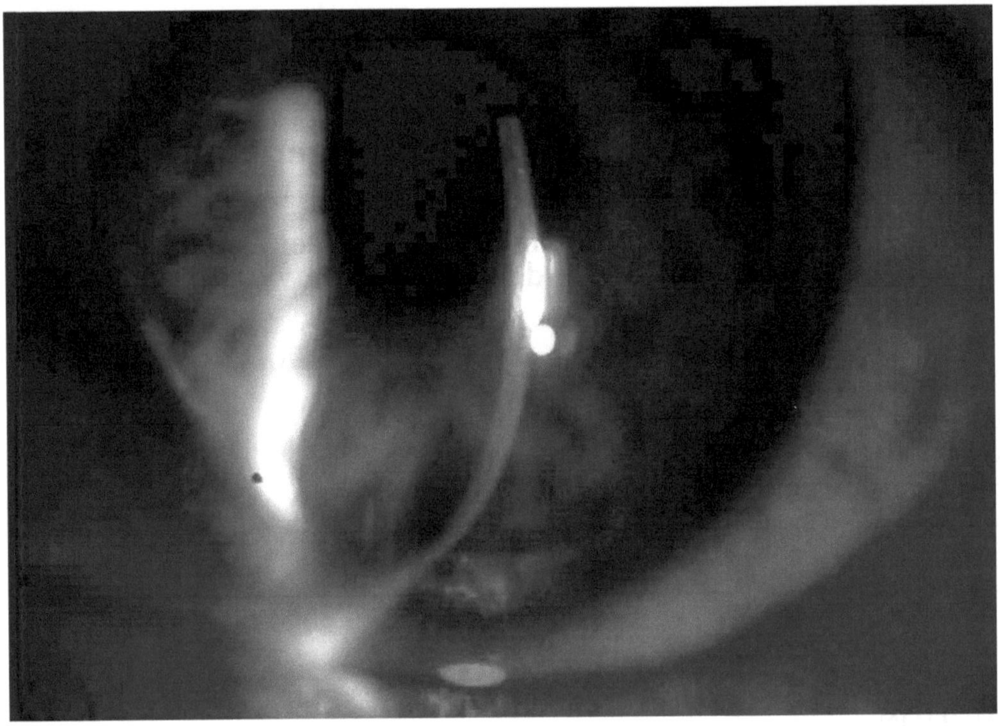

Fig. 1. Clinical picture of PMCD.

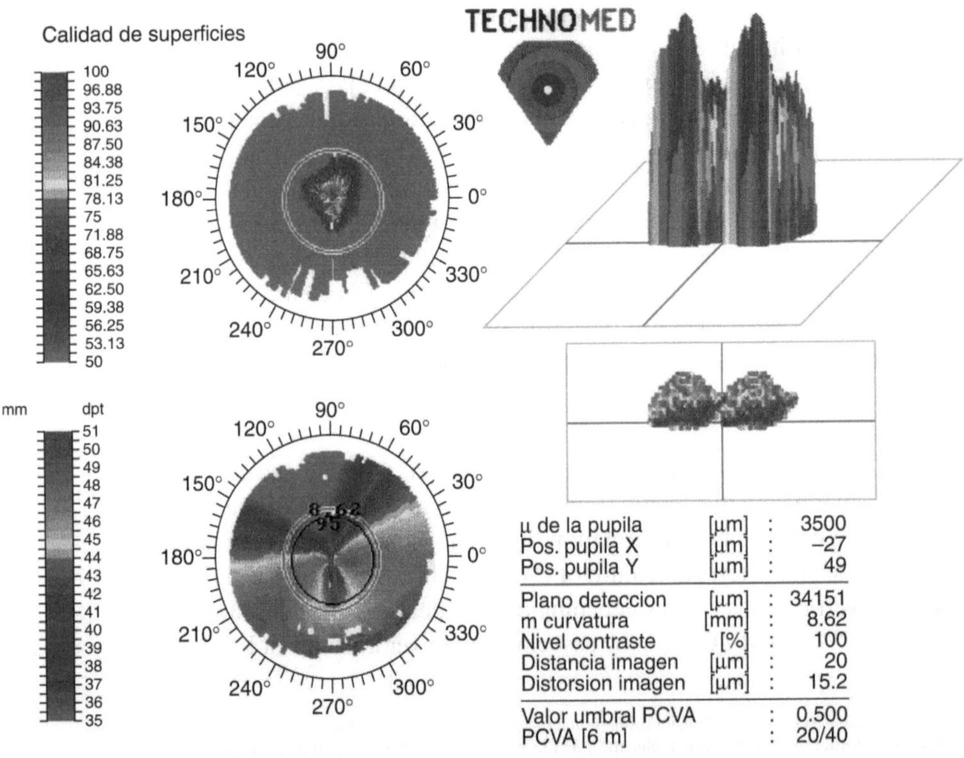

Fig. 2. Corneal topography of PMCD.

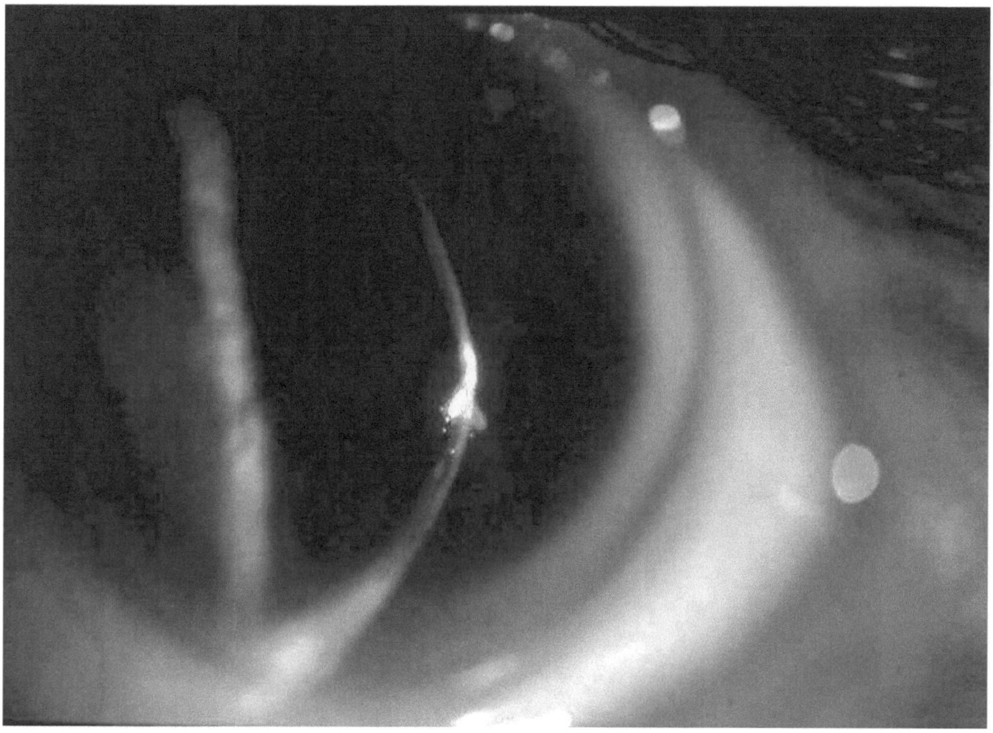

Fig. 3. Clinical picture of keratoconus.

(6,7) in the same or in the contralateral eye, as all three are considered to be a spectrum of noninflammatory ectatic thinning disorders. The incidence of association of PMCD with keratoconus is 10.3% and 12.9% with keratoglobus *(4)*.

Keratoconus

In keratoconus the clinical picture can be distinguished from PMCD by the presence of thinning, protrusion of the cornea reaching the maximum at the site of the cone, where in PMCD the thinning is usually inferior 1 mm of the limbus, with normal cornea above and below the thin area. Also, keratoconus is associated with special clinical signs as Vogt's striae, Fleischer's ring, and Manson sign *(4)* (Fig. 3).

Keratoglobus

In keratoglobus, the ectasia and thinning affects the entire corneal surface as shown from the figure (Fig. 4). Whereas in PMCD the thining is usually 1 mm from the limbus with normal cornea above and below the ectatic (thin) area, with no evidence of scarring, vascularization, or lipid deposition.

INCIDENCE AND COMPLICATIONS

The largest series published on PMCD by Sridhar and collaborators *(4)* included 116 eyes of 58 patients that showed higher incidence in males than females (77.6% were males and 22.4% females). They also stated that the diagnosis was made based on the presence of typical topographic features. The age ranged from 8 to 66 yr. The degree of astigmatism was <5 diopters (D) in 19.2%, 5–10 D in 36.4%, 10–15 D in 23.2%, 15–20 D

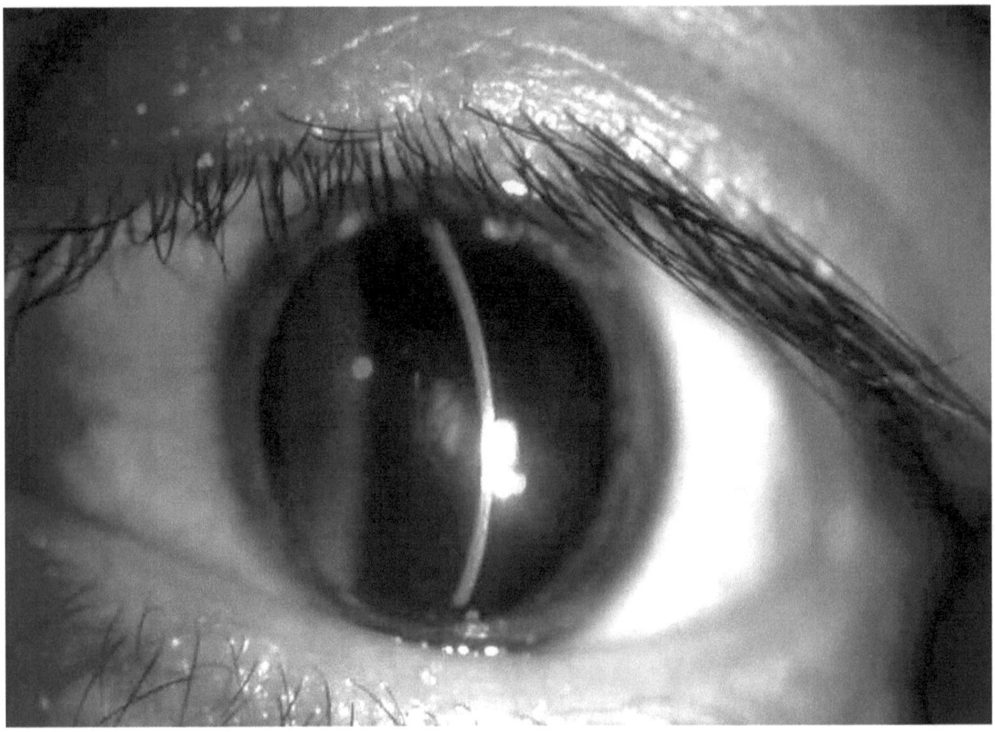

Fig. 4. Clinical picture of keratoglobus.

in 15.2%, and >20 D in 6.1%. Typical inferior PMCD occurs in 85.3%, and superior PMCD occurs in 14.7% *(4)*. The most usual complication of PMCD is decreased visual acuity owing to irregular astigmatism and increase in higher order aberrations (coma-like aberrations), which was even proposed to follow-up the progression of PMCD *(8)*. Although more serious vision threatening complication can occur as perforation and acute hydrops *(4)* as well as ectasia after laser *in situ* keratomileusis *(9)*. Also, keratoconus was reported to progress form PMCD *(10)*.

INTRACORNEAL RING SEGMENTS FOR PMCD CORRECTION

Among many surgical alternatives for PMCD management *(11,12)*, intracorneal implants or more specifically intrastromal rings are one of the new investigated alternatives. Intracorneal ring segments, originally designed to correct low myopia were proposed by Fleming and Reynolds in late 1970 *(13)*. The implant, initially proposed to be almost a complete ring, was inserted through a peripheral single corneal incision. Later on, and owing to technical surgical difficulties, it was refashioned into two halves in a C-shape manner. Finally, it is renamed owing to such modification to be Intrastromal corneal ring segments *(14–17)*.

Two commonly used corneal ring segments are currently available to ophthalmic surgeons. One is known under the trade name INTACS (Addition Technologies® LLC, Fremont, CA) (Fig. 5) and KERARINGS, originally designed by Pablo Ferrara and produced by Mediphacos® Inc. (Belo Horizonte, Brazil) (Fig. 6) the specification for both is shown in Table 1.

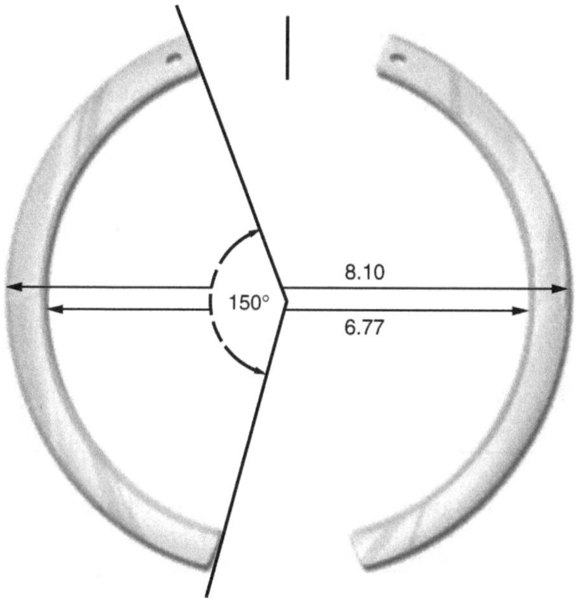

Fig. 5. INTACS design.

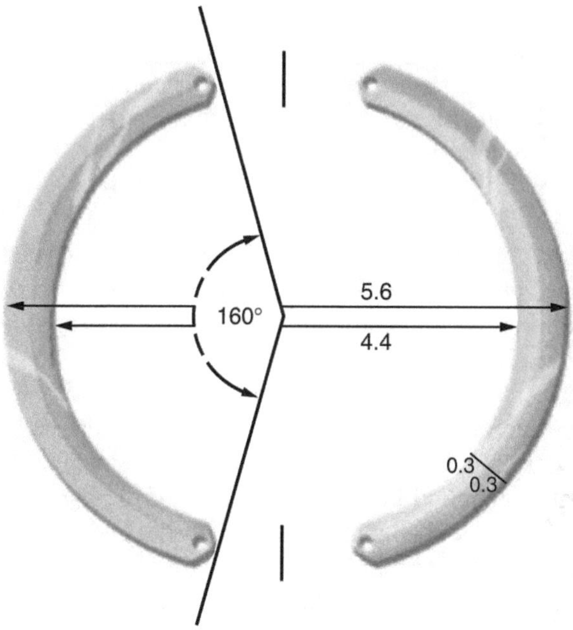

Fig. 6. KERARING design.

Patel and collaborators *(17)* studied different mathematical models to predict the effect of the INTACS on refractive error in relation to corneal asphericity and the spherical aberration of the eye. They found that the large diameter (9 mm) and thin ring (0.1 mm) is less likely to adversely affect corneal asphericity and therefore does not enhance induction of spherical aberration. Also, they concluded that an intracorneal

Table 1
Specification and Differences Between INTACS and KERARING

	INTACS	*KERARINGS*
Design (cross-section)	Hexagonal	Triangular
Inner diameter (mm)	6.77	5.40
Outer diameter (mm)	8.10	6.60
Implantation in respect to	Center of the cornea	Center of the pupil
Implantation depth	70% of the corneal thickness	80% of the corneal thickness
Arc length	150°	120 and 160°
Available segment thickness (mm)	0.25, 0.30, 0.35, 0.40, and 0.45	0.15,0.20. 0.25, 0.30, and 0.35
Material	Polymethyl methacrylate	Polymethyl methacrylate or Acrylic Perspex CQ

ring could not correct more than –4 D of myopia without significantly increasing the spherical aberration, which in turn, will compromise the final visual outcome *(17)*.

Mechanism of Action

Intracorneal ring segments act as a spacer element between arching bundles of corneal lamellae leading to shortening of the central arc length (arc shortening effect). Furthermore, there is a near-linear relationship between the thickness of the spacer elements and the degree of the corneal flattening *(18)*. In this way they can reduce the corneal steepening and astigmatism associated with PMCD. The aim of implantation is not to treat or eliminate the existing disease but to decrease the corneal abnormality and thus increase the visual acuity to acceptable limits.

Surgical Technique

The procedure is performed in the majority of cases under topical anaesthesia. Preoperative medication includes proparacaine 0.5%, ciprofloxacin 0.3%, and oxibuprocaine CIH 0.2% *(19,20)*. Marking the geometrical center of the cornea is a must in implanting INTACS, as they are implanted in respect to the corneal center wherein, KERARINGS are implanted in respect to the pupil center. Intraoperative ultrasonic pachymetry is performed at the site of the incision. Seven readings should be taken and the highest and lowest readings are discarded and the average of the remaining five readings is taken *(19,20)*. Calibrated diamond knife is set at 70% of the mean measured corneal thickness (Fig. 6), a redial incision 1.8 mm in length is made. The incision is sited at 7 mm from the optical zone for INTACS implantation and 5 mm for KERARINGS.

The stromal pocket is dissected on both sides of the incision using a modified Suarez spatula. For KERARINGS implantation, widening the tunnel is made manually with 270° dissecting spatula followed by wound suturing after segments implantation. As for INTACS a semiautomated vacuum device is needed, this device contains a suction ring that can be placed around the limbus guided by the previously marked geometrical center of the cornea. Following careful checking on the suction force, two semicircular lamellar dissectors are placed sequentially into the lamellar pocket to be steadily advanced by a rotational movement. As a result, two 180° semicircular dissections into

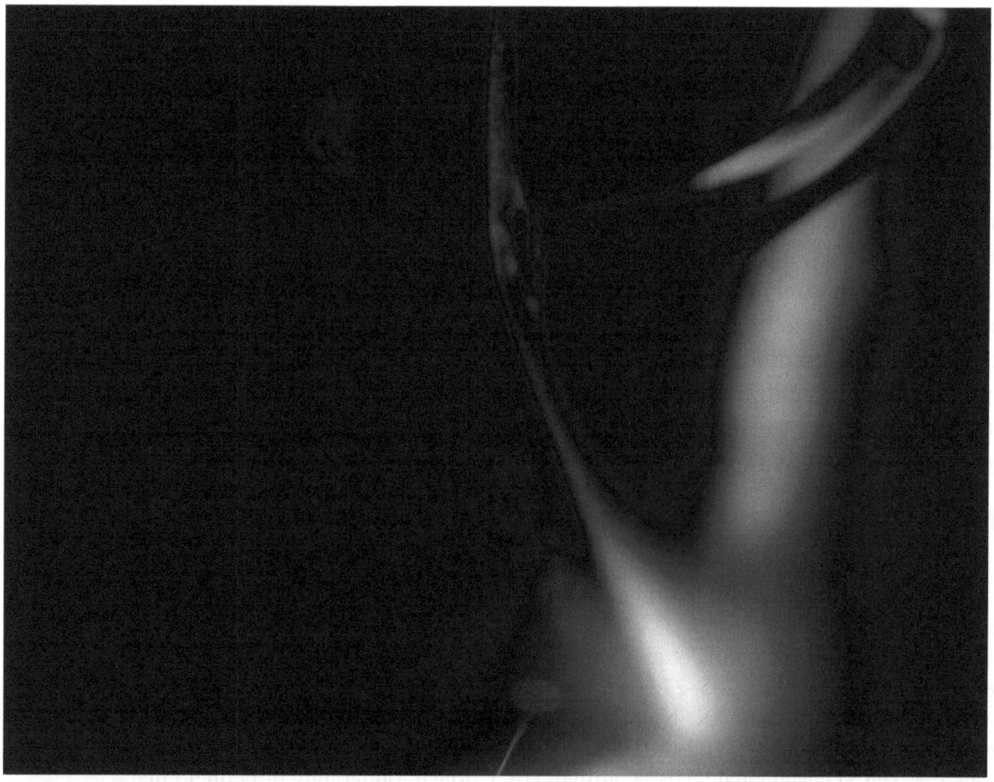

Fig. 7. INTACS segments showing excellent corneal tolerance. Note the corneal thinning lower in the segment.

the stroma are achieved with an approximate diameter of 7.5 mm. After removing the suction device, the two segments of the INTACS are inserted into each of the semicircular channels, the placement of both segments of the INTACS will leave a gap of approx 15° nasally and 35–40° temporally. The radial incision "wound" is then gently hydrated or closed with one or two carefully embedded 10 zero-nylon sutures. The edges of the stroma are then approximated to prevent epithelial ingrowth. Topical antibiotic and corticosteroidal eyedrops were applied four times daily for 2 wk. Removing the corneal suture 2–3 wk following surgery to minimize the potential occurrence of induced *(19,20)*.

Results

The intracorneal ring segments whether INTACS or KERARING are implanted more centrally to the ectasia (thin area) as shown in (Fig. 7). The purpose of implanting intracorneal ring segments is not to eliminate the existing pathology of PMCD but to correct the resulting irregular astigmatism, which is the most common complication of the disease in a way to improve visual acuity. All available reports *(21–23)* reported that both types reduce the spherical equivalent, keratometric values, and improve corneal topographic pattern, resulting in more regular corneal shape with astigmatism reduction as well as improving the visual acuity. However, the residual refractive error can be corrected with contact lenses with better tolerance than before intracorneal ring segments implantation *(21)*. Although the results are encouraging, concern still exists regarding the long-term effect of this approach for the management of patients with PMCD *(21–23)*.

Complications

Inspite of being nonvision threatening, complications have been reported after intracorneal ring segments implantation such as: keratitis, superficial corneal neovascularization, channel deposits, and migration of the segment, which might lead to their explantation *(24,25)*. In conclusion, intracorneal ring segments is a different surgical approach not treating the existing pathology of PMCD as lamellar keratoplasty *(11,12)*, but correcting the most common consequence of PMCD, which is irregular astigmatism without affecting the central visual axis or causing permanent change in the corneal tissue. In the coming years we should expect less need of other more invasive surgical alternatives as keratoplasty with the use of intracorneal ring segments in PMCD.

REFERENCES

1. Krachmer JH. Pellucid marginal corneal degeneration. Arch Ophthalmol 1978;96:1217–1221.
2. Taglia DP, Sugar J. Superior pellucid marginal corneal degeneration with hydrops. Arch Ophthalmol 1997;115:274–275.
3. Rao SK, Fogla R, Padmanabhan P, et al. Corneal topography in atypical pellucid marginal degeneration. Cornea 1999;18:265–272.
4. Sridhar MS, Mahesh S, Bansal AK, Nutheti R, Rao GN. Pellucid marginal corneal degeneration. Ophthalmol 2004;111:1102–1107.
5. Maguire LJ, Klyce SD, McDonald MB, Kaufman HE. Corneal topography of pellucid marginal degeneration. Ophthalmol 1987;94:519–524.
6. Kayazawa F, Nishimura K, Kodama Y, et al. Keratoconus with pellucid marginal corneal degeneration. Arch Ophthalmol 1984;102:895–896
7. Varley GA, Macsai MS, Krachmer JH. The results of penetrating keratoplasty for pellucid marginal corneal degeneration. Am J Ophthalmol 1990;110:149–152
8. Kamiya K, Hirohara Y, Mihashi T, Hiraoka T, Kaji Y, Oshika T. Progression of pellucid marginal degeneration and higher-order wavefront aberration of the cornea. Japan J Ophthalmol 2003;47:523–525.
9. Fogla R, Rao SK, Padmanabhan P. Keratectasia in 2 cases with pellucid marginal corneal degeneration after laser in situ keratomileusis. J Cataract Refract Surg 2003;29:788–791.
10. Karabatsas CH, Cook SD. Topographic analysis in pellucid marginal corneal degeneration and keratoglobus. Eye 1996;10:451–455.
11. Cheng CL, Theng JT, Tan DT. Compressive C-shaped lamellar keratoplasty: a surgical alternative for the management of severe astigmatism from peripheral corneal degeneration. Ophthalmology 2005;112:425–430.
12. Javadi MA, Karimian F, Hosseinzadeh A, Noroozizadeh HM, Sa'eedifar MR, Rabie HM. Lamellar crescentic resection for pellucid marginal corneal degeneration. J Refract Surg 2004;20:162–165.
13. Fleming JR, Reynolds AI, Kilmer L. The intrastromal corneal ring-two cases in rabbits. J Refractive Surg 1987;3:227–232.
14. Burris TE, Baker PC, Ayer et al. Flattening of the curvature with intrastromal corneal rings of increasing thickness-an eye bank eye study. J Refractive Surg 1993;19:182–187.
15. Nosé W, Neves RA, Schanzlin DJ, et al. Intrastromal corneal ring-one year results of first implant in humans: a preliminary non-functional eye study. Refract Corneal Surg 1993; 9:452–458.

16. Nosé W, Neves RA, Burris TE, et al. Intrastromal corneal ring-12months sighted myopic eyes. J Refract Surg 1996;12:20–28.
17. Patel S, Marshall J, Fitzke FW. Model for deriving the optical performance of the myopic eye corrected with an intracorneal ring. J Refract Surg 1995;11:248–252.
18. Pinsky PM, Datye DV, Silvestrini TA. Numerical simulation of topographical alterations in the cornea after intrastromal corneal ring (ICR) placement. Invest Ophthalmol Vis Sci 1995;36 suppl:308.
19. Alió JL, Shabayek MH, Belda JI, Correas P, Feijoo ED. Analysis of results related to good and bad outcome of INTACS implantation for correction of Keratoconus. J Cataract Refract Surg 2006;32:756–761.
20. Alió AJ, Artola A, Hassanein A, Haroun H, Galal A. One or two INTACS segments for the correction of keratoconus. J Cataract Refract Surg 2005;31:943–953.
21. Rodriguez-Prats J, Galal A, Garcia-Lledo M, De La Hoz F, Alió JL. Intracorneal rings for the correction of pellucid marginal degeneration. J Cataract Refract Surg 2003;29:1421–1424.
22. Kymionis GD, Aslanides IM, Siganos CS, Pallikaris IG. INTACS for early pellucid marginal degeneration. J Cataract Refract Surg 2004;30:230–233.
23. Akaishi L, Tzelikis PF, Raber IM. Ferrara intracorneal ring implantation and cataract surgery for the correction of pellucid marginal corneal degeneration. J Cataract Refract Surg 2004;30:2427–2430.
24. Shehadeh-Masha'our R, Modi N, Barbra A Grazozi HJ. Keratitis after implantation of intrastromal ring segments. J cataract Refract Surg 2004;30:1802–1804.
25. Hofling-Lima AL, Branco BC, Romano AC, et al. Corneal infections after implantation of intracorneal ring segments. Cornea 2004;23:547–549.
26. Bourcier T, Borderie V, Laroche L. Late bacterial keratitis after implantation of intrastromal corneal ring segments. J Cataract Refract Surg 2003;29:407–409.

15

Artisan Toric Lens Implantation for Correction of Postkeratoplasty Astigmatism

Rudy M. M. A. Nuijts, MD, PhD **and Nayyirih G. Tahzib,** MD

CONTENTS

INTRODUCTION

Although visual function may show substantial improvement after keratoplasty *(1)*, many series after corneal grafting report significant astigmatism of four to five diopters (D) *(2–5)*. It is well known that patients do not tolerate spectacle correction for more than three to four D of astigmatism or anisometropia. Therefore, contact lenses are fitted in 10–30% of patients, and in keratoconus this value may increase up to 50% *(6,7)*. Contact lenses may be effective in 80% of cases but contact lens intolerance in the postkeratoplasty population may be high owing to ocular problems such as topographical abnormalities, blepharitis and dry eye, poor manual dexterity, occupational problems, and lack of motivation. Surgical correction of postkeratoplasty astigmatism has been performed by corneal relaxing incisions *(8,9)*, wedge resections *(10)*, and intraocular lens (IOL) exchange or piggy-back implantation *(11,12)*. Recently, excimer laser photorefractive keratectomy and laser *in situ* keratomileusis (LASIK) have been used for corneal tissue ablation. However, photorefractive keratectomy causes significant haze in corneal grafts and induces irregular astigmatism and regression, although recent adjunctive use of mitomycin C may prevent these side-effects *(13–16)*. LASIK surgery is able to treat a greater range of postkeratoplasty refractive error but the corneal graft thickness and the amount of ametropia and astigmatism suitable for correction, limit the efficacy of the procedure. In addition, wound dehiscence may occur owing to the high vacuum pressure, flap complications in steep corneas, and a high rate of retreatments *(17–25)*. Recently, the short-term preliminary results of Artisan toric lens implantation for the correction of postkeratoplasty refractive error was published *(26)*. The

From: *Ophthalmology Research: Visual Prosthesis and Ophthalmic Devices: New Hope in Sight*
Edited by: J. Tombran-Tink, C. Barnstable, and J. F. Rizzo © Humana Press Inc., Totowa, NJ

present series provides the results with respect to safety and endothelial cell loss in a larger series with a minimal follow-up of 1 yr.

PATIENTS AND METHODS

Patient Population

The 37 eyes of 36 patients included in this study could not be corrected by wearing spectacle because of anisometropia, the magnitude of refractive cylinder, and/or contact lens intolerance. Exclusion criteria were as follows: a preoperative spectacle corrected visual acuity worse than 20/50, an anterior chamber depth less than 3 mm, glaucoma, retinal pathology, or an endothelial cell count lower than 500 cells/mm^2. Investigational review board approval was obtained from the Academic Hospital Maastricht.

Patient Examination

Patients were examined preoperatively and at day 1, week 1, month 1, month 3, month 6, and from then at 6-mo intervals. Preoperatively, uncorrected visual acuity (UCVA) and spectacle best-corrected visual acuity (BCVA) with subjective refraction, cycloplegic refraction, slit-lamp microscopy, applanation tonometry, corneal topography (Alcon Eyemap EH-290, Alcon, Fort Worth, TX), dilated fundus examination, and pupil size measurement at dim illumination with the Colvard pupillometer (Oasis, Glendora, CA) were performed. On postoperative day 1, uncorrected and BCVA and biomicroscopic examination with registration of intraocular pressure were performed. Thereafter, UCVA and BCVA with subjective refraction, slit-lamp microscopy, applanation tonometry, and corneal topography were assessed. The simulated keratometry values of the steep and flat meridians were used for calculation of the topographically induced surgical astigmatism. Preoperatively and at 6 mo, 1, 2, and 3 yr postoperatively, specular microscopy of the corneal endothelium using a noncontact specular microscope (Konan Noncon Robo, SP 8000, Konan, Hyogo, Japan) was performed *(27,28)*. Preoperatively and at 6 mo postoperatively the patients were asked to rate the quality of vision of the eye with their present correction on a visual analog scale of 1–10 (1 is very poor, 10 is excellent), based on a previously validated questionnaire *(29)*.

Surgical Technique

The Artisan toric IOL has a convex–concave toric optic with a spherical anterior surface and a sphero-cylindrical posterior surface (Ophtec BV, Groningen, the Netherlands). This single-piece lens is made up of polymethyl methacrylate and is manufactured using compression molding technology. The toric lens is iris claw-fixated and has a 5-mm optical zone (Fig. 1). Refractive error, refractive cylinder power, anterior chamber depth, and topographically derived keratometric dioptric values were inserted into the Van der Heijde *(30)* formula to calculate the dioptric power of the lens for two meridians. The axis of the cylinder identified by the subjective refraction was used to determine the axis of surgical enclavation. The power of the lens was chosen to obtain emmetropia ($n = 32$) or a postoperative spherical equivalent of –1 D ($n = 2$) or +1 D ($n = 1$) to match the ametropia in the untreated eyes. The IOL is available in dioptric powers of –3 to –20.5 D, +2 to +12 D and in cylindrical powers of 2–7.5 D. The cylinder is in line with the haptics or at an angle of 90° with the haptics. In 9 of 37 eyes the cylinder dioptric power of the toric lens was less

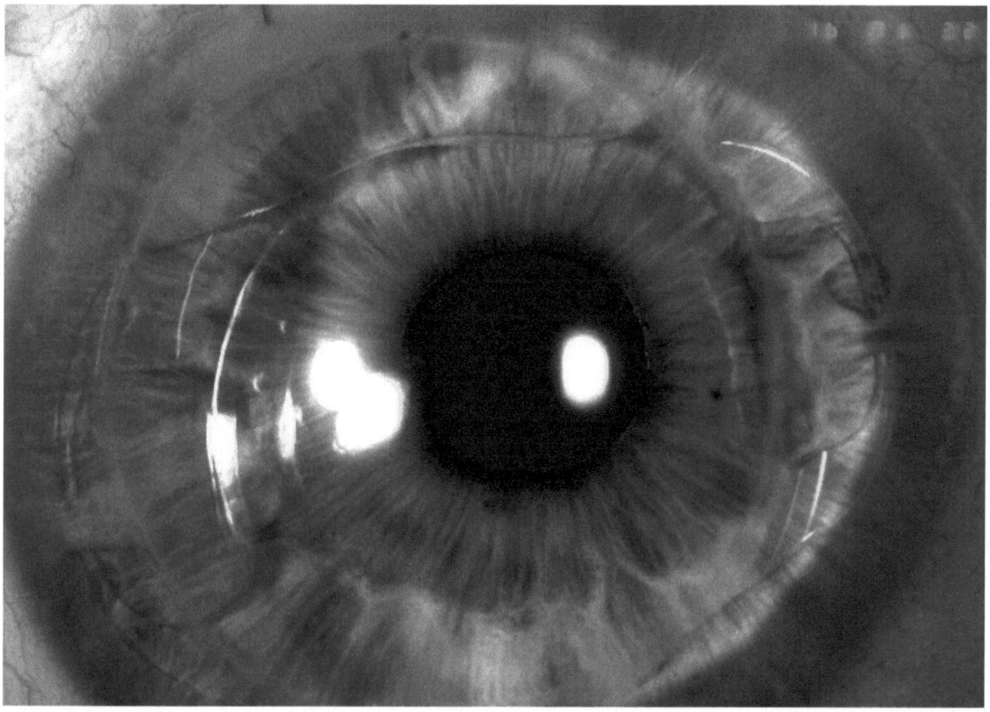

Fig. 1. Artisan toric lens implantation for correction of postkeratoplasty astigmatism (patient no. 9; Table 1). Preoperatively, best corrected visual acuity was 20/50 with –9 –6 × 165°. Six months after implantation of an Artisan toric lens with a power of –8.5 –5.5 × 0° and enclavated in the axis 165°, best-corrected visual acuity increased to 20/32 with pl –2 × 50° and UCVA to 20/40.

than 75% of the calculated power required for full correction of the cylinder. All surgery was performed under general anaesthesia by one surgeon (RN). Preoperatively, the horizontal and vertical axis of the eye was marked at the limbus and postoperatively the axis for lens enclavation was marked. A two-plane 5.3 mm corneoscleral incision was centered at 12-o'clock. Two stab incisions were performed at 2- and 10 o'clock and directed toward the enclavation sites. After an intracameral injection of acetylcholine and under an ophthalmic viscoelastic (Healon GV, Pharmacia, Uppsala, Sweden) device the lens was introduced by forceps with a Budo forceps (Duckworth and Kent, Ltd, Baldock Herts, England). After gentle rotation the lens was fixated with a disposable enclavation needle in the marked axis (Ophtec BV, Groningen, Netherlands). A slit iridotomy was performed at 12 o'clock to avoid pupillary block glaucoma. The viscoelastic material was exchanged with BSS Plus (Alcon, Fort Worth, TX). The wound was sutured with three to four interrupted 10 - 0 nylon sutures (Alcon, Fort Worth, TX). Postoperatively, topical tobramycin 0.3% combined with dexamethasone 0.1% (Tobradex, Alcon, Couvreur, Belgium) and ketorolactrometamol 0.5% (Acular, Westport Co., Mayo, Ireland) were used four times daily for 3 wk in a tapered schedule and three times daily for 1 wk, respectively. In cases with an initial diagnosis of herpes simplex virus keratitis acyclovir 400 mg three times daily (GlaxoSmithKline, Zeist, the Netherlands) was used for 6 mo. Selective suture removal was performed depending on the subjective refraction.

Outcome Measures and Statistics

Main outcome measures were refractive and visual outcome as reflected by UCVA and efficacy (reduction in refractive and topographic astigmatism, reduction in anisometropia of spherical and defocus equivalent and number of eyes losing/gaining lines of UCVA), BCVA, and short-term safety (number of eyes losing more than two lines of BCVA). A patient satisfaction questionnaire, specular microscopy, and incidence of complications were assessed. Snellen uncorrected or BCVA was converted to logMar values to facilitate statistical analysis. Comparison between preoperative data and postoperative data was performed by paired *t*-test (SPSS for Windows [SPSS Inc., Chicago, IL]). All averages in the text are mean ± standard deviation (SD).

Analysis of Astigmatism

Both vector analysis and nonvector analysis of the cylinder was performed. The efficacy of the procedure (i.e., the proportion of astigmatism correction achieved) was quantified using the correction index, expressed as a percent of the surgically induced astigmatism (SIA), divided by the target-induced astigmatism (TIA) for each individual eye and aggregated *(31)*. The Holladay method to convert polar values (cylinder and axis) to a Cartesian (X and Y) coordinate system was used to determine the mean ± SD value of the refractive and topographical keratometric astigmatism *(32)*. The coordinates were plotted in a doubled-angle plot and the centroid was determined.

RESULTS

Patient Population

Twenty-nine patients were female and eight were male. The mean age was 63.9 ± 15.8 (range, 23–82 yr). The mean time interval between penetrating keratoplasty and toric lens implantation was 58.4 ± 30.2 mo (range, 26–168 mo) and the interval between suture removal and lens implantation was 31 ± 27.9 mo (range, 3–144 mo). Twenty-five eyes were pseudophakic after previous implantation of a posterior chamber IOL. The initial diagnosis requiring corneal transplantation was Fuchs endothelial dystrophy (40%), (pseudophakic) bullous keratopathy (14.3%), herpes simplex keratitis (17.1%), keratoconus (14.3%), corneal scarring (11.4%), and high astigmatism (2.9%). The baseline parameters were a mean sphere of 0.18 ± 4.40 D (range, +9 to –10 D), a mean spherical equivalent refraction of –3.31 ± 4.33 D (range, +5.5 to –14.25 D), a mean baseline refractive cylinder power of –6.99 ± 2.02 (range, –3 to –11 D), a mean defocus equivalent of 7.01 ± 2.56 D (range, 3.25–14.25 D), and a mean baseline topographically derived simulated keratometric cylinder of 6.92 ± 2.29 D (range, +3.51 to +11.65 D). The mean follow-up was 24.5 ± 11.4 mo (range, 12–48 mo).

Visual Acuity Outcome

The mean logMar UCVA preoperatively was 1.40 ± 0.44 and increased to 0.56 ± 0.34 at the last follow-up ($p < 0.001$, paired *t*-test). Postoperatively, 25.7% of eyes had a UCVA better than 20/40 as compared with 0% preoperatively (Fig. 2). The mean number of gained lines of UCVA was 8.51 ± 5.50 (range, –2 to 18 lines). The mean logMar BCVA preoperatively was 0.25 ± 0.16 and postoperatively was 0.26 ± 0.23

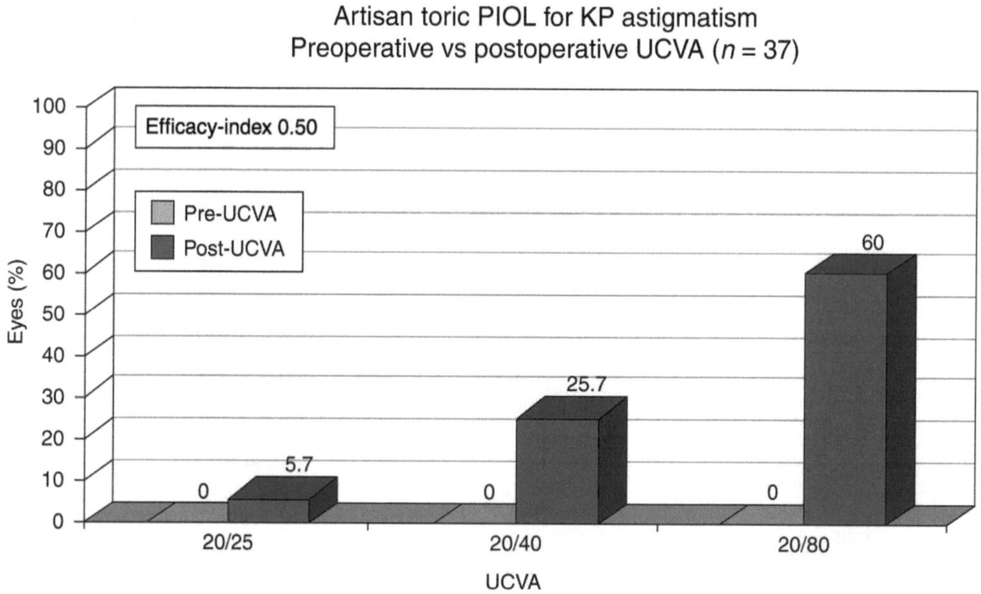

Fig. 2. Percent of eyes within given range of UCVA preoperatively and at last follow-up after toric lens implantation. The efficacy-index (mean postoperative UCVA divided by mean preoperative BCVA) was 0.50.

($p = 0.802$, paired *t*-test). Postoperatively, 77.1% of eyes had a BCVA better than 20/40 as compared with 69.4% preoperatively, and 20% had a BCVA better than 20/25 (Fig. 3). The mean number of gained lines of BCVA was –0.14 ± 2.15 (range, –6 to 4 lines). There was a loss of BCVA of greater than two lines in 8.1% of eyes and a gain of at least two lines in 8.1% of eyes (Fig. 4). The predictability of intended against achieved cylinder correction showed 20 of 37 eyes (54.1%) within 2 D and 9 of 37 eyes (24.3%) within 1 D of the intended correction. The predictability of intended against achieved defocus equivalent showed 27 of 37 eyes (73%) within 2 D and 18 of 37 eyes (48.6%) within 1 D of the intended correction.

Refractive Outcome

The mean sphere at the last follow-up was 0.12 ± 1.23 D (range, +2.50 to –3.75 D), the mean spherical equivalent refraction was –0.92 ± 1.23 D (range, +1 to –5.25 D), and the mean defocus equivalent was +1.65 ± 1.65 D (range, +7.13 to +0 D). The refractive cylinder was reduced to –1.74 ± 1.23 D, –1.68 ± 1.14 D, –1.75 ± 1.12 D, –2.01 ± 2.07 D, and –2.08 ± 1.46 D at 6 mo ($n = 37$), 1 yr ($n = 37$), 2 yr ($n = 20$), 3 yr ($n = 13$), and the final follow-up examination (24.5 ± 11.4 mo), respectively ($p < 0.001$ for all time-points, paired *t*-test). Concerning stability, there was no significant change in refractive cylinder values from 6 mo postoperatively to 3 yr postoperatively (Fig. 5) ($p = 0.065$, paired *t*-test). At the last follow-up 9 of 37 eyes (24.3%) had a refractive cylinder less than 1 D, 20 of 37 eyes (54.2%) had a refractive cylinder less than 2 D, and 34 of 37 eyes (91.9%) had a cylinder less than 4 D. The mean topographically derived simulated keratometric cylinder did not change significantly from 6.50 ± 2.49 D at 3 mo postoperatively to 6.56 ± 3.09 D at 12 mo ($p = 0.994$, paired *t*-test), and to 6.82 ± 3.37 D ($p = 0.596$, paired *t*-test)

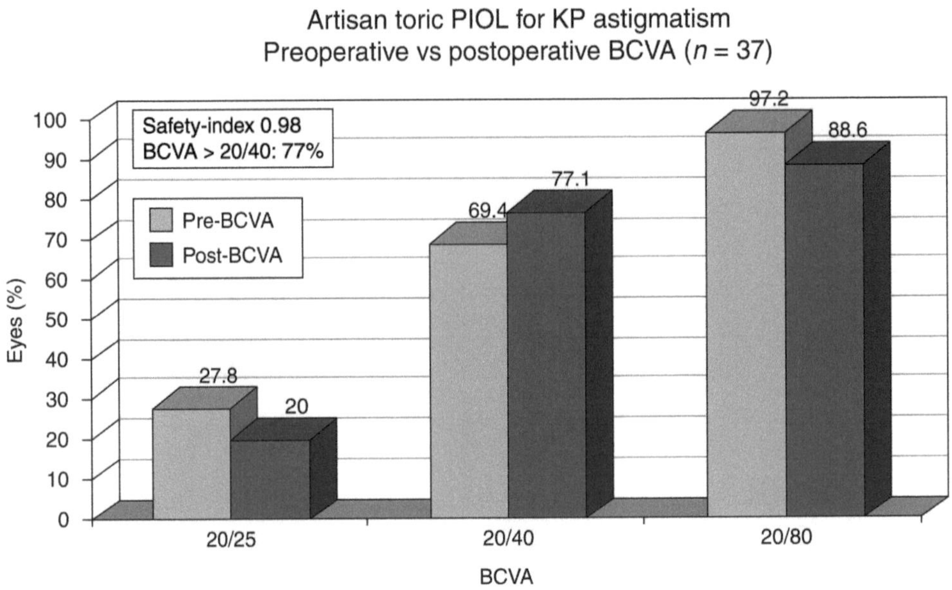

Fig. 3. Percent of eyes within given range of BCVA preoperatively and at last follow-up after toric lens implantation. The safety-index (mean postoperative BCVA divided by mean preoperative BCVA) was 0.98.

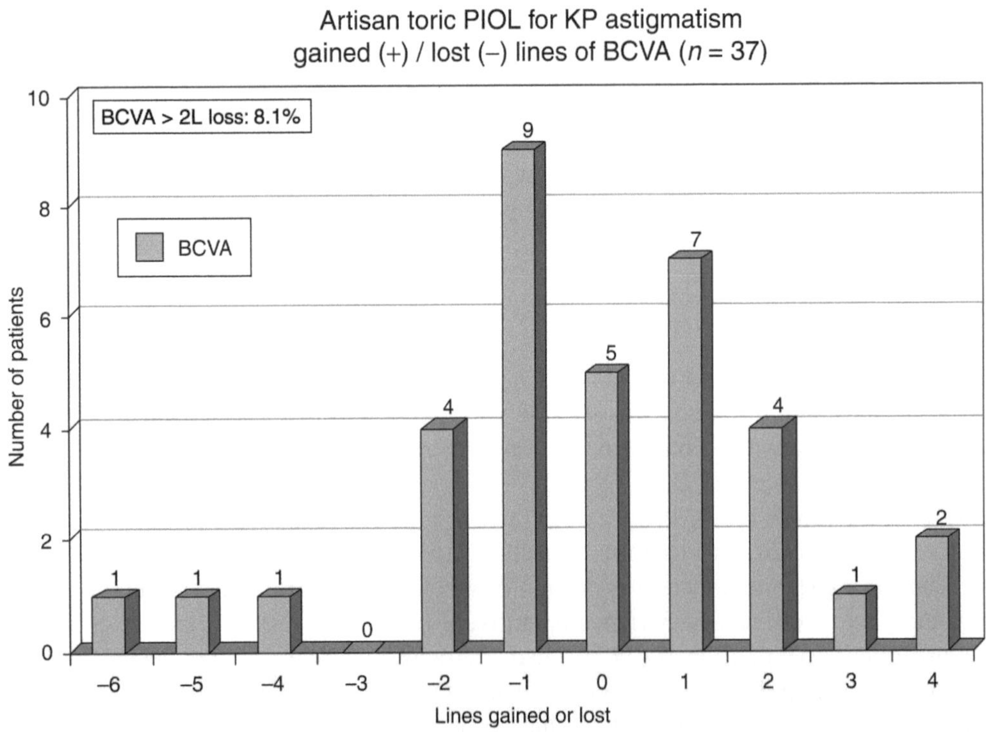

Fig. 4. The number of gained or lost lines of BCVA at last follow-up after toric lens implantation. BCVA = best-corrected visual acuity.

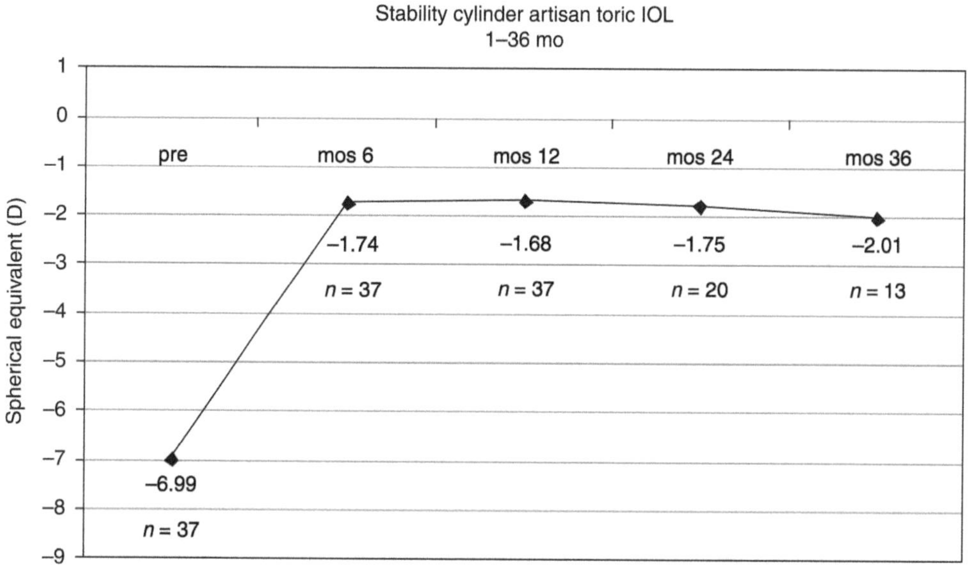

Fig. 5. Mean refractive cylinder preoperatively and at postoperative time intervals. Number of patients at each time-point are shown. D = diopters.

at the last follow-up. The percent reduction in refractive and topographical astigmatism was 85.6 ± 30.9% and 1.73 ± 48.8%, respectively (Table 1). There was a reduction of 107.5 ± 31.6% and 77.5 ± 19.6% in sphere and defocus equivalent (for eyes with preoperative defocus values >3 D, $n = 25$), respectively. Based on the limited correction that could be achieved in the nine eyes that required a dioptric power that exceeded the available cylindrical power of the toric lens the reduction in refractive astigmatism was 90.5 ± 21.9%. The correction index (SIA/TIA) was 94.2 ± 24% at the last follow-up. The centroid (±SD) in the double-angled plot changed from –2.83 D at 141.3° (±6.43 D) preoperatively to –0.51 D at 91.9° (±2.15 D) postoperatively (Figs. 6A and 6B, $p < 0.001$, paired t-test). The mean SIA of the topographical cylinder by the placement of the corneoscleral incision centered at 90° was 2.58 ± 1.78 D (range, 0.02–6.25 D) at 6 mo postoperatively. Patient satisfaction increased from 3.6 preoperatively to eight postoperatively ($p < 0.001$, paired t-test). The intraocular pressure (IOP) was 14.5 ± 2.9 mmHg preoperatively, 15.2 ± 4.2 mmHg at 1 mo postoperatively, 13.5 ± 3.2 mmHg at 6 mo postoperatively, and 13 ± 2.8 mmHg at the last follow-up (p = ns, paired t-test for all time-points). The endothelial cell loss was 13.4 ± 18.6% ($n = 36$), 20.2 ± 22% ($n = 33$), 34.5 ± 23.7% ($n = 19$), and 26.6 ± 34.1% ($n = 13$) at 6 mo ($p = 0.002$), 1 yr ($p = 0.001$), 2 yr ($p < 0.001$), and 3 yr postoperatively ($p = 0.011$), respectively.

Complications

In two patients irreversible graft rejections occurred and in one patient gradual endothelial decompensation occurred. A 77-yr-old male underwent penetrating keratoplasty of the right eye in April 1997 for HSV stromal keratitis. Seven months after implantation of an Artisan toric lens with a power of +5–7 × 0°, BCVA was increased to 20/25 with +0.75–1.50 × 155°. One month later metastasized lung cancer was diagnosed

Table 1
Results of Comparative Studies for Correction of Postkeratoplasty Astigmatism

Reference	No. of eyes	Technique	Follow-up months; mean (range)	Refractive astigmatism (mean ± SD) Preop	Refractive astigmatism (mean ± SD) Postop	Refractive astigmatism Reduction (%)	Spherical equivalent or sphere[a] (mean ± SD) Preop	Spherical equivalent or sphere[a] (mean ± SD) Postop	Spherical equivalent or sphere[a] Reduction (%)	UCVA ≥ 20/40 (%)	BCVA ≥ 20/40 (%)	BCVA ≥ 2 lines gain (%)	BCVA ≥ 2 lines loss (%)	Additional procedures or complications
Donnenfeld *(33)*	23	LASIK	7.6 (1–14)	3.64 ± 1.7	1.64 ± 1.14	64.6	6.88 ± 4.4	1.42 ± 1.05	79.3	36	74	26	4.3	9.1% enhancements
Webber *(2)*	25	LASIK	(1–12)	8.67 ± 3.22	2.92 ± 1.71	66.3	–5.2 ± 2.31	–1.31 ± 1.63	74.8	28	83	12	0	53.8% arcuate keratotomy one perforation
Forseto *(22)*	22	LASIK	10.1 (6–18)	4.44 ± 2.1	1.75 ± 1.1	60.6	–4.55 ± 3.66	–0.67 ± 1.24	85.3	54.5	90.9	18	9.1	One flap perforation one flap dislocation
Nassarella *(21)*	8	LASIK	6	3.5 ± 1.22	1.25 ± 0.74	64.3	–4.50 ± 1.52	–0.75 ± 0.75	83.3	37.5	87.5	25	0	none
Rashad *(20)*	19	LASIK	7 (6–10.5)	9.21 ± 1.95	1.09 ± 0.33	87.9	–2.14 ± 2.11[a]	+0.43 ± 0.82	79.9	73.7	100	42	0	53% enhancements
Koay *(19)*	8	LASIK	8.6	6.79 ± 3.3	1.93 ± 1.2	71.6	–6.79 ± 4.17	–0.64 ± 1.92	90.6	12.5	NA	37.5	0	one buttonhole

Kwitko *(18)*	14	LASIK	26.9 (12–42)	5.37 ± 2.1	2.82 ± 2.4	47.5	–7.51 ± 3.87	–1.25 ± 2.30	83.3	28.6	85.7	21	7.1	42.9% enhancements one buttonhole two epithelial ingrowths
Busin *(17)*	9	Flap cut	3	5.03 ± 1.35	3.42 ± 1.29	32.0	–5.40 ± 1.69	–4.37 ± 1.72	19.1	0	77.8	33	0	none
Malecha *(34)*	19	LASIK	5 (1–14)	4.05 ± 1.71	1.22 ± 1.14	69.9	4.24 ± 2.81	0.85 ± 0.84	80	73.7	89.5	0	5.2	Three DLK three enhancements
Gayton *(12)*	7	Piggyback lens	7.5	NA	NA	NA	3.41 ± 1.15	0.98 ± 0.81[b]	71.3	42.9	71	NA	0	none
Current study	37	Toric iris-fixated lens	8.4 (3–18)	–6.99 ± 2.02	–2.08 ± 1.46	85.6	0.18 ± 4.40[a]	0.12 ± 1.23[a]	107.5[a]	25.7	77.1	50	0	none

UCVA, uncorrected visual acuity; BCVA, best-corrected visual acuity; preop, preoperatively; postop, postoperatively; NA, not available.
[a]Sphere
[b]Myopic sphere

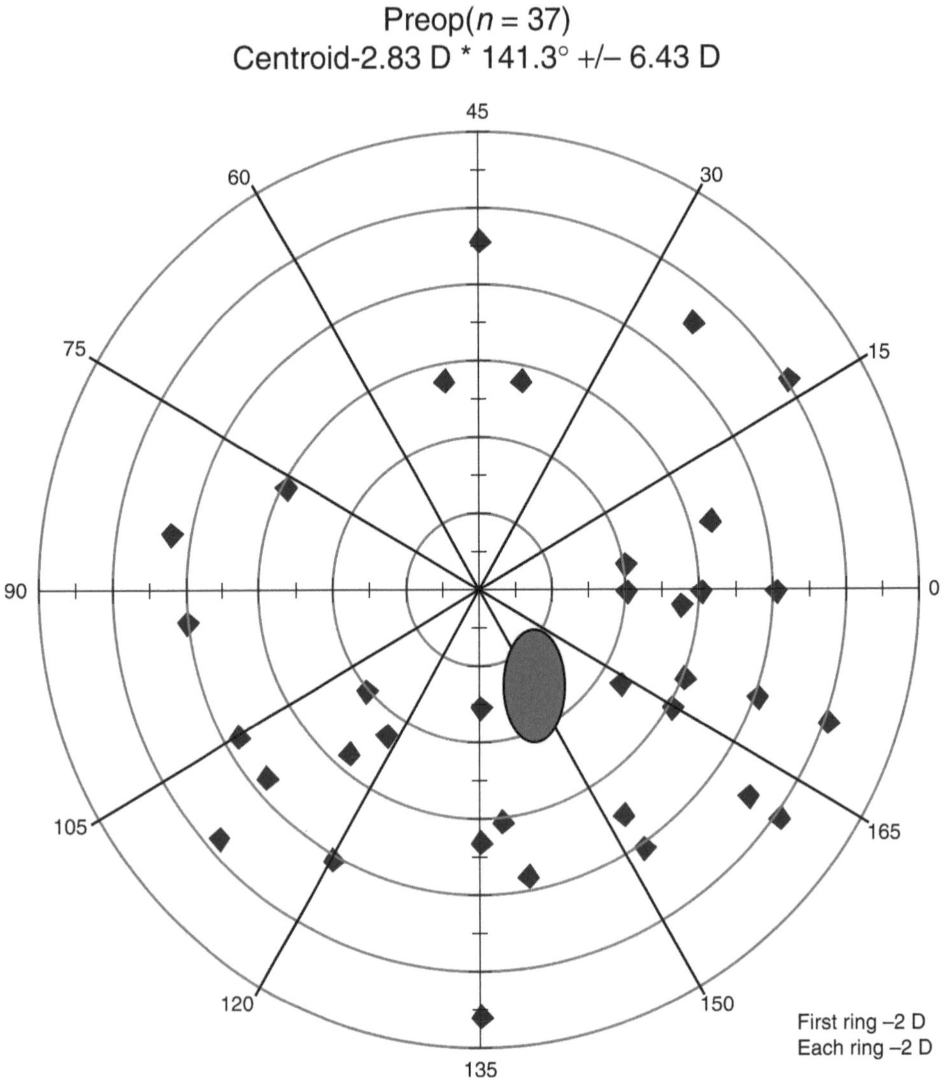

Fig. 6. *(Continued)*

and a recurrence of HSV keratitis followed by irreversible graft rejection developed. Twenty-five months after Artisan implantation BCVA was 20/100 with +0.5–1.50 × 160°. No further surgical treatment followed.

A 81-yr-old female underwent penetrating keratoplasty of the left eye in September 1999 for pseudophakic bullous keratoplasty. Before Artisan implantation the endothelial cell density was 1384 cells/mm^2. Five months after Artisan implantation (lens power of –1.50–7 × 0°, enclavation axis 162°) the BCVA was 20/30 with –0.75–2.25 × 63°. At 12 mo after Artisan toric lens implantation the endothelial cell density had decreased to 385 cells/mm^2. Twenty months after implantation gradual endothelial decompensation occurred and BCVA decreased to 20/100 after metastasized colon cancer was diagnosed. Twenty-eight months after Artisan implantation a rekeratoplasty with

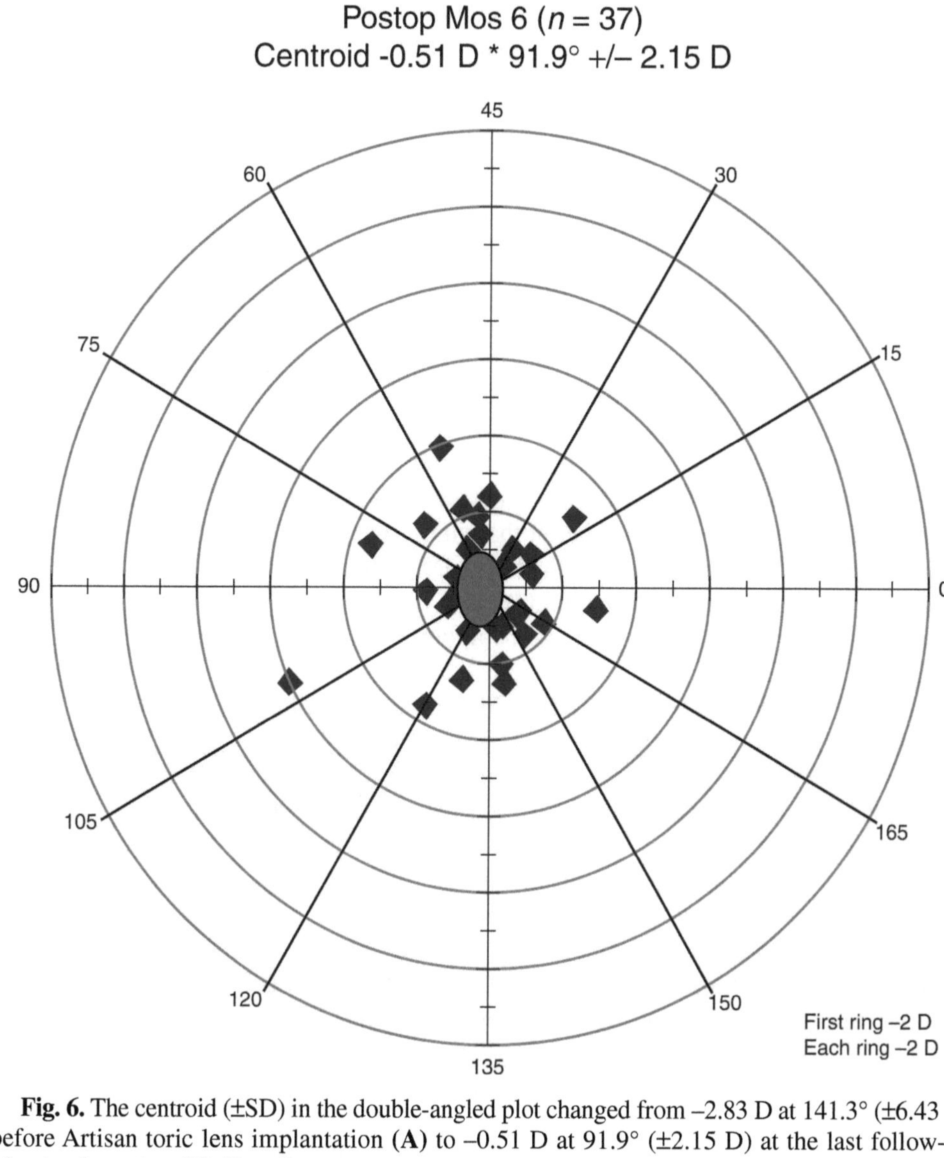

Fig. 6. The centroid (±SD) in the double-angled plot changed from –2.83 D at 141.3° (±6.43 D) before Artisan toric lens implantation (**A**) to –0.51 D at 91.9° (±2.15 D) at the last follow-up after implantation (**B**). D = diopters.

explantation of the Artisan lens was performed. One year after rekeratoplasty the corneal graft was clear and BCVA was 20/30 with –6.5 0 × 0°.

A 62-yr-old female underwent a repenetrating keratoplasty of the right eye for graft failure in July 1999 after an intial keratoplasty for keratoconus in 1984. In May 2001, phacoemulsification with IOL implantation was performed followed by a reversible graft rejection in September 2001. In May 2003, 7 mo after Artisan implantation (lens power of +8 –7.5 × 0°, enclavation axis 138°) the BCVA was 20/40 with –0.50 –1.50 × 50°. One month later an irreversible immunological graft failure occurred. In March

2004, a rekeratoplasty was performed with explantation of the toric Artisan lens. At the last follow-up in May 2005, the UCVA was 20/40 with a clear graft. No other complications like cystoid macular edema, chronic inflammation of the anterior chamber, or retinal detachment in any of the patients were noted.

DISCUSSION

This prospective study of 37 eyes demonstrates the efficacy and stability of the Artisan toric IOL for correction of postkeratoplasty astigmatism. Until now, LASIK appears to be the preferred technique for correction of anisometropia and astigmatism after keratoplasty (Table 1) *(17–25)*. The use of the Artisan toric IOL with a power range of 7.5 D of cylinder and –20.5 D of myopia to +12 D of hyperopia, provides a wide field for correction of postkeratoplasty astigmatism, and ametropia. In the series this is reflected by the magnitude of baseline spherical error (range +9 to –10 D) and cylindrical error (range –3 to –11 D), which is much higher than in most postkeratoplasty LASIK series. As regards the knowledge, the reduction of the refractive cylinder by 86 ± 31% (without any enhancements) is better than in most reported LASIK series. The reduction of refractive astigmatism after LASIK varies from 48 to 88% (Table 1). However, enhancements were reported in 9.1% *(33)*, 15% *(34)*, 42.9% *(18)*, 45% *(35)*, and 53% *(20)* of cases and in one study LASIK was combined with arcuate incisions in the stromal bed in 56% of eyes *(2)*.

Improving the UCVA of 20/40 or better from 0 to 26% of the cases illustrates the efficacy of the Artisan toric IOL procedure in this patient group with highly ametropic eyes. In most LASIK series with lower preoperative ametropia, UCVA better than 20/40 varied from 28 to 74% *(2,18,20,22,33,34)*. With respect to safety, there was a loss of BCVA of greater than two lines in 8.1% of eyes and a gain of at least two lines in 8.1% of eyes. This is in accordance with two recent randomized studies in routine refractive surgery for the correction of high myopia that showed a greater gain of BCVA with Artisan phakic IOL implantation as compared with a greater loss of BCVA with LASIK and a better quality of vision with the Artisan lens in moderate to high myopia *(36,37)*. The loss of greater than two lines of BCVA in 8.1% in the series is comparable with series of LASIK for postkeratoplasty astigmatism that show a greater than two lines loss of BCVA in 4.3% *(38)*, 7.1% *(18)*, and 9.1% *(22)*. However, the pattern of complications induced by the two techniques is very different. LASIK surgery may be complicated by flap complications in steep corneas and has limitations owing to corneal graft thickness and amount of ametropia and astigmatism suitable for correction *(17–25)*. LASIK related complications like diffuse lamellar keratitis *(34)*, buttonhole flaps *(18,22)*, wound dehiscence *(39)*, and epithelial ingrowth *(18)* have been reported. Because the majority of eyes in the reported LASIK series were grafted in young patients for keratoconus with a rapid wound healing, wound dehiscence problems were less likely to occur than in a group of older patients grafted for Fuchs endothelial dystrophy or bullous keratopathy *(40)*. As the effect of the flap cut alone may induce a significant reduction of refractive astigmatism in up to 50% in some patients and because of the high enhancement rate a two-stage LASIK procedure has been proposed *(17,41–43)*. However, it is unclear whether a two-stage procedure bears a higher risk for complications like epithelial ingrowth, wound healing problems, and flap dislocation. In the present Artisan series, irreversible corneal decompensation occurred in two patients after metastasized cancer was diagnosed. Before the diagnosis of malignant disease BCVA was 20/25 in both patients and no signs of immunological rejection had been noted. It is believe that changes in the immune system owing to the concomitant development of malignant systemic disease might have initiated the graft failures. In a third patient with a rekeratoplasty for a graft failure after an initial diagnosis of keratconus, an immunological irreversible graft failure occurred. This was the second rejection period after the rekeratoplasty following a previous reversible rejection period of 4 mo after cataract surgery. It can not be excluded that this rejection, although 8 mo after surgery, may have been related to the Artisan toric lens implantation. Two of the three cases with corneal decompensation underwent successful regrafting with explantation of the toric Artisan IOL. No other complications like cystoid macular edema, chronic inflammation of the anterior chamber, or retinal detachment in any of the patients were noted.

The stability of the postoperative refractive cylinder after Artisan toric lens implantation up to 36 mo was excellent. After LASIK however, progressive changes were seen in refraction and topography in 35.7% of cases after a mean follow-up time of 26.9 mo *(18)*. A potential limitation of the Artisan toric IOL for the correction of postkeratoplasty astigmatism is SIA, by implantation of the rigid polymethylmethacrylate IOL through a 5.3 mm incision. In a recent series of implantation of the Artisan toric IOL for correction of myopia or hyperopia with astigmatism the SIA was 0.53 D *(44)*. However, after keratoplasty the biomechanical response of the corneoscleral tissue to the incision might be somewhat unpredictable and a greater variability in SIA may be seen. Indeed, in the series the mean SIA was 2.58 D 6 mo postoperatively and varied from 0.02 to 6.25 D. Because of this variability it is believed that the SIA cannot be incorporated into the power calculation of the lens. As the goal of correcting postkeratoplasty astigmatism is mainly to reduce the refractive astigmatism and ametropia to enable patients to wear spectacles, it is felt that a lesser predictability of astigmatism reduction may be acceptable.

Concerns have been raised, especially with respect to the development of complications like endothelial cell loss, chronic inflammation, and cystoid macular edema after Artisan toric lens implantation. A study using fluorometry showed inflammation comparable with cataract surgery at 6 mo postoperatively *(45)*, whereas a study using a flare-cell meter found chronic inflammation 1–2 yr after implantation of the older Worst–Fechner IOL *(46)*. No chronic inflammation was found by slit-lamp examination in the present study and cystoid macular edema is not to be expected, as none of the eyes lost best-corrected visual acuity in the immediate postoperative phase. The mean endothelial cell loss was 13.4 ± 18.6%, 20.2 ± 22%, and 34.5 ± 23.7% at 6, 12, and 24 mo. There was a significant continuing progressive endothelial cell loss at each time-point as compared with preoperative cell density levels. The cell loss is much higher than the reported cell loss in other studies of Artisan lens implantation for correction of high myopia that show values under 8% at 1 yr postoperatively *(36,37,47,48)*. Since the time period between penetrating keratoplasty and Artisan toric lens implantation ranges from 26 to 168 mo, it is unclear how the cell loss in the present series compares with the natural endothelial cell loss after penetrating keratoplasty that has an annual rate of 7.8% from 3 to 5 yr after transplantation and of 4.2% from 5 to 10 yr after transplantation *(49–51)*. Therefore, it cannot be excluded that the Artisan iris-fixated IOL in the presence of a corneal graft with low cell densities may cause a higher rate of endothelial cell loss owing to the compromised endothelium. In addition, the accuracy of noncontact specular microscopy for determining endothelial cell density, which is usually around 5% is not known in grafts with low cell counts and might hamper the interpretation of the results *(52,53)*. Risk factors for endothelial decompensation in corneal grafts with low cell densities have not been clearly defined. Cell counts as low as 370 cells/mm^2 and 515 cells/mm^2 have been measured before decompensation *(54–57)*. Nevertheless, it is felt that an endothelial cell density of at least 500 cells/mm^2 as exclusion criterium is permitted, as no other treatment modalities exist but corneal regrafting, and the Artisan lens is perfectly removable at future regrafting procedures, as has been shown in two patients in the series. Of course, a larger number of patients followed for longer periods of time are needed to assess the effect of Artisan toric lens implantation on the corneal graft endothelium.

Based on the objective medical outcomes, the subjective patient satisfaction that increased from 3.6 preoperatively to 8 postoperatively (scale 1–10), and the suitability of all patients for spectacle correction Artisan toric lens implantation appears to be a valuable option for correction of postkeratoplasty astigmatism and anisometropia. However, more patients with a longer follow-up up to 5 yr are needed to identify the risk factors for progressive endothelial cell loss and a randomized study of Artisan toric lens implantation vs LASIK with larger numbers of patients could clarify the advantages and disadvantages of both techniques with respect to efficacy, safety, and complications.

REFERENCES

1. Brahma A, Ennis F, Harper R, Ridgway A, Tullo A. Visual function after penetrating keratoplasty for keratoconus: a prospective longitudinal evaluation. Br J Ophthalmol 2000; 84:60–66.
2. Webber SK, Lawless MA, Sutton GL, Rogers CM. LASIK for post penetrating keratoplasty astigmatism and myopia. Br J Ophthalmol 1999;83:1013–1018.
3. Troutman RC, Lawless MA. Penetrating keratoplasty for keratoconus. Cornea 1987;6: 298–305.
4. Binder PS. The effect of suture removal on postkeratoplasty astigmatism. Am J Ophthalmol 1988;106:507.
5. Seitz B, Langenbucher A, Kus MM, Kuchle M, Naumann GO. Nonmechanical corneal trephination with the excimer laser improves outcome after penetrating keratoplasty. Ophthalmology 1999;106:1156–1164; discussion 1165.
6. Lopatynsky M, Cohen EJ, Leavitt KG, Laibson PR. Corneal topography for rigid gas permeable lens fitting after penetrating keratoplasty. Clao J 1993;19:41–44.
7. Eggink FA, Nuijts RM. A new technique for rigid gas permeable contact lens fitting following penetrating keratoplasty. Acta Ophthalmol Scand 2001;79:245–250.
8. Solomon A, Siganos CS, Frucht-Pery J. Relaxing incision guided by videokeratography for astigmatism after keratoplasty for keratoconus. J Refract Surg 1999;15:343–348.
9. Koay PY, McGhee CN, Crawford GJ. Effect of a standard paired arcuate incision and augmentation sutures on postkeratoplasty astigmatism. J Cataract Refract Surg 2000;26:553–561.
10. Lugo M, Donnenfeld ED, Arentsen JJ. Corneal wedge resection for high astigmatism following penetrating keratoplasty. Ophthalmic Surg 1987;18:650–653.

11. Fenzl RE, Gills JP, 3rd, Gills JP. Piggyback intraocular lens implantation. Curr Opin Ophthalmol 2000;11:73–76.
12. Gayton JL, Sanders V, Van der Karr M, Raanan MG. Piggybacking intraocular implants to correct pseudophakic refractive error. Ophthalmology 1999;106:56–59.
13. Bilgihan K, Ozdek SC, Akata F, Hasanreisoglu B. Photorefractive keratectomy for post-penetrating keratoplasty myopia and astigmatism. J Cataract Refract Surg 2000;26: 1590–1595.
14. Tuunanen TH, Ruusuvaara PJ, Uusitalo RJ, Tervo TM. Photoastigmatic keratectomy for correction of astigmatism in corneal grafts. Cornea 1997;16:48–53.
15. Amm M, Duncker GI, Schroder E. Excimer laser correction of high astigmatism after keratoplasty. J Cataract Refract Surg 1996;22:313–317.
16. Chang DH, Hardten DR. Refractive surgery after corneal transplantation. Curr Opin Ophthalmol 2005;16:251–255.
17. Busin M, Arffa RC, Zambianchi L, Lamberti G, Sebastiani A. Effect of hinged lamellar keratotomy on postkeratoplasty eyes. Ophthalmology 2001;108:1845–1851; discussion 1851–1852.
18. Kwitko S, Marinho DR, Rymer S, Ramos Filho S. Laser in situ keratomileusis after penetrating keratoplasty. J Cataract Refract Surg 2001;27:374–379.
19. Koay PY, McGhee CN, Weed KH, Craig JP. Laser in situ keratomileusis for ametropia after penetrating keratoplasty. J Refract Surg 2000;16:140–147.
20. Rashad KM. Laser in situ keratomileusis for correction of high astigmatism after penetrating keratoplasty. J Refract Surg 2000;16:701–710.
21. Nassaralla BR, Nassaralla JJ. Laser in situ keratomileusis after penetrating keratoplasty. J Refract Surg 2000;16:431–437.
22. Forseto AS, Francesconi CM, Nose RA, Nose W. Laser in situ keratomileusis to correct refractive errors after keratoplasty. J Cataract Refract Surg 1999;25:479–485.
23. Hardten DR, Chittcharus A, Lindstrom RL. Long term analysis of LASIK for the correction of refractive errors after penetrating keratoplasty. Cornea 2004;23:479–489.
24. Buzard K, Febbraro JL, Fundingsland BR. Laser in situ keratomileusis for the correction of residual ametropia after penetrating keratoplasty. J Cataract Refract Surg 2004;30: 1006–1013.
25. Barraquer CC, Rodriguez-Barraquer T. Five-year results of laser in-situ keratomileusis (LASIK) after penetrating keratoplasty. Cornea 2004;23:243–248.
26. Nuijts RM, Abhilakh Missier KA, Nabar VA, Japing WJ. Artisan toric lens implantation for correction of postkeratoplasty astigmatism. Ophthalmology 2004;111:1086–1094.
27. Waring GO 3rd. Standard graphs for reporting refractive surgery. J Refract Surg 2000; 16:459–466.
28. Koch DD, Kohnen T, Obstbaum SA, Rosen ES. Format for reporting refractive surgical data. J Cataract Refract Surg 1998;24:285–287.
29. Nijkamp MD, Dolders MG, de Brabander J, van den Borne B, Hendrikse F, Nuijts RM. Effectiveness of multifocal intraocular lenses to correct presbyopia after cataract surgery: a randomized controlled trial. Ophthalmology 2004;111:1832–1839.
30. van der Heijde GL, Fechner PU, Worst JG. Optical consequences of implantation of a negative intraocular lens in myopic patients. Klin Monatsbl Augenheilkd 1988;193:99–102.
31. Alpins N. Astigmatism analysis by the Alpins method. J Cataract Refract Surg 2001;27: 31–49.
32. Holladay JT, Cravy TV, Koch DD. Calculating the surgically induced refractive change following ocular surgery. J Cataract Refract Surg 1992;18:429–443.
33. Donnenfeld ED, Solomon R, Biser S. Laser in situ keratomileusis after penetrating keratoplasty. Int Ophthalmol Clin 2002;42:67–87.

34. Malecha MA, Holland EJ. Correction of myopia and astigmatism after penetrating keratoplasty with laser in situ keratomileusis. Cornea 2002;21:564–569.
35. Guell JL, Gris O, de Muller A, Corcostegui B. LASIK for the correction of residual refractive errors from previous surgical procedures. Ophthalmic Surg Lasers 1999;30:341–349.
36. Malecaze FJ, Hulin H, Bierer P, et al. A randomized paired eye comparison of two techniques for treating moderately high myopia: LASIK and artisan phakic lens. Ophthalmology 2002;109:1622–1630.
37. El Danasoury MA, El Maghraby A, Gamali TO. Comparison of iris-fixed Artisan lens implantation with excimer laser in situ keratomileusis in correcting myopia between -9.00 and -19.50 diopters: a randomized study. Ophthalmology 2002;109:955–964.
38. Donnenfeld ED, Kornstein HS, Amin A, et al. Laser in situ keratomileusis for correction of myopia and astigmatism after penetrating keratoplasty. Ophthalmology 1999;106:1966–1974; discussion 1974–1975.
39. Ranchod TM, McLeod SD. Wound dehiscence in a patient with keratoconus after penetrating keratoplasty and LASIK. Arch Ophthalmol 2004;122:920–921.
40. Abou-Jaoude ES, Brooks M, Katz DG, Van Meter WS. Spontaneous wound dehiscence after removal of single continuous penetrating keratoplasty suture. Ophthalmology 2002;109:1291–1296; discussion 1297.
41. Busin M, Zambianchi L, Garzione F, Maucione V, Rossi S. Two-stage laser in situ keratomileusis to correct refractive errors after penetrating keratoplasty. J Refract Surg 2003;19:301–308.
42. Alio JL, Javaloy J, Osman AA, Galvis V, Tello A, Haroun HE. Laser in situ keratomileusis to correct post-keratoplasty astigmatism; 1-step versus 2-step procedure. J Cataract Refract Surg 2004;30:2303–2310.
43. Dada T, Vajpayee RB, Gupta V, Sharma N, Dada VK. Microkeratome-induced reduction of astigmatism after penetrating keratoplasty. Am J Ophthalmol 2001;131:507–508.
44. Dick HB, Alio J, Bianchetti M, et al. Toric phakic intraocular lens: European multicenter study. Ophthalmology 2003;110:150–162.
45. Perez-Santonja JJ, Bueno JL, Zato MA. Surgical correction of high myopia in phakic eyes with Worst-Fechner myopia intraocular lenses. J Refract Surg 1997;13:268–281; discussion 281–284.
46. Perez-Santonja JJ, Iradier MT, Benitez del Castillo JM, Serrano JM, Zato MA. Chronic subclinical inflammation in phakic eyes with intraocular lenses to correct myopia. J Cataract Refract Surg 1996;22:183–187.
47. Landesz M, Worst JG, van Rij G. Long-term results of correction of high myopia with an iris claw phakic intraocular lens. J Refract Surg 2000;16:310–316.
48. Pop M, Payette Y. Initial results of endothelial cell counts after Artisan lens for phakic eyes: an evaluation of the United States Food and Drug Administration Ophtec Study. Ophthalmology 2004;111:309–317.
49. Bourne WM, Nelson LR, Hodge DO. Continued endothelial cell loss ten years after lens implantation. Ophthalmology 1994;101:1014–1022; discussion 1022–1023.
50. Bourne WM, Hodge DO, Nelson LR. Corneal endothelium five years after transplantation. Am J Ophthalmol 1994;118:185–196.
51. Patel SV, Hodge DO, Bourne WM. Corneal endothelium and postoperative outcomes 15 years after penetrating keratoplasty. Am J Ophthalmol 2005;139:311–319.
52. Binder PS, Akers P, Zavala EY. Endothelial cell density determined by specular microscopy and scanning electron microscopy. Ophthalmology 1979;86:1831–1847.
53. Laing RA. Specular microscopy of the cornea. Curr Top Eye Res 1980;3:157–218.
54. Hoffer KJ. Corneal decomposition after corneal endothelium cell count. Am J Ophthalmol 1979;87:252–253.

55. Nuyts RM, Boot N, van Best JA, Edelhauser HF, Breebaart AC. Long term changes in human corneal endothelium following toxic endothelial cell destruction: a specular microscopic and fluorophotometric study. Br J Ophthalmol 1996;80:15–20.
56. Breebaart AC, Nuyts RM, Pels E, Edelhauser HF, Verbraak FD. Toxic endothelial cell destruction of the cornea after routine extracapsular cataract surgery. Arch Ophthalmol 1990;108:1121–1125.
57. Bates AK, Hiorns RW, Cheng H. Modelling of changes in the corneal endothelium after cataract surgery and penetrating keratoplasty. Br J Ophthalmol 1992;76:32–35.

16
Phakic Intraocular Lenses for the Treatment of High Myopia

Maria I. Kalyvianaki, MD, George D. Kymionis, MD, PhD, and Ioannis G. Pallikaris, MD, PhD

CONTENTS

INTRODUCTION

High myopia remains a challenge for refractive surgeons. The use of laser *in situ* keratomileusis (LASIK), which has become increasingly popular for the treatment of myopia *(1,2)*, is limited in high myopic eyes, because a residual corneal stromal thickness of 250 μm is essential to minimize the risk of corneal ectasia *(3)*. Apart from that, large excimer laser corrections may result in poor optical quality of the reshaped cornea *(4)*. Clear lens extraction, as another method for treating high refractive errors, results in accommodation loss, at the same time it carries all the risks of cataract surgery *(5–7)*. An evolving technique for the treatment of myopia in young phakic patients is the implantation of phakic intraocular lenses (PIOLs). In this chapter the main types of these lenses, the criteria for their implantation, their advantages, and their possible complications will be discussed.

HISTORY OF PIOLS—PIOLS CATEGORIES

Since their inception, PIOLs have undergone many improvements of their design and surgical technique. In the 1950s, Strampelli and Barraquer were the first to implant

From: *Ophthalmology Research: Visual Prosthesis and Ophthalmic Devices: New Hope in Sight*
Edited by: J. Tombran-Tink, C. Barnstable, and J. F. Rizzo © Humana Press Inc., Totowa, NJ

intraocular lenses in the anterior chamber of high myopic eyes. Owing to the poor quality of these lenses and the lack of advanced surgical instruments, most of these lenses had to be explanted. In the early 1990s, Baikoff *(8)* reintroduced angle-supported PIOLs for treating high myopia. Nowadays, after modifications of their design, there are several new types of angle-supported PIOLs, such as Nu Vita lens (Bausch and Lomb Surgical, St. Louis, MO), Phakic-6 (Ophthalmic Innovations International Inc., Ontario, Canada), and Morcher model (Morcher GMBH, Stuttgart, Germany).

In 1986, Worst and Fechner *(9)* designed an anterior chamber lens, which was fixated on the iris with claws. This constitutes the second category of iris-fixated PIOLs (Artisan™, Ophtek BV, Groningen, Netherlands and Verisyse™, AMO, Santa Ana, CA). In 1986, Fyodorov and his colleagues *(10)* designed a posterior chamber lens made of silicone. This lens passed through three generations before the development of the silicone phakic refractive lens (PRL™, Medennium Inc., Irvine, CA). To the same category of posterior chamber lenses belongs the implantable contact lens (ICL™, Staar Surgical Co, Monrovia, CA), which is made of collamer, a copolymer of (hydroxyethyl)methacrylate (HEMA) and porcine collagen.

PREOPERATIVE EVALUATION—INCLUSION CRITERIA

High myopes over the age of 18, with a stable refraction who wish refractive surgery are candidates for the implantation of a PIOL for the treatment of their myopia. Exclusion criteria include age <18 yr, previous intraocular surgery, anterior chamber depth (ACD) less than 2.8–3 mm, glaucoma or intraocular pressure at initial measurement greater than 20 mmHg, any sign of cataract and any intraocular or systemic disease. A thorough preoperative examination is mandatory in all patients. This includes manifest and cycloplegic refraction, corneal topography, pachymetry, A-scan ultrasonography, slit-lamp microscopy, pupil size measurement under scotopic conditions, white-to-white corneal diameter measurement with the use of a caliper, applanation tonometry, and dilated fundus examination.

ANTERIOR CHAMBER ANGLE-SUPPORTED PIOLS

Anterior chamber lenses have the advantage of a comparatively simple surgical technique. Two types of angle-supported PIOLs are the ZB5MF (Chiron Domilens, Lyon, France) and the ZSAL-4 model (Morcher, Stuttgart, Germany) that are made of single-piece poly-methyl-methacrylate (PMMA) with Z-shaped haptics, to increase haptic flexibility and decrease compression forces against angle structures *(11)*. The power of the lens is calculated by the company using ACD, dioptric power of the cornea, patient's refraction, and desired postoperative refraction. The lens length is selected by measuring the horizontal white-to-white distance and adding 1 mm to this distance.

Surgery is performed after constriction of the pupil under periocular anesthesia. The lens is inserted through a 6-mm corneal incision in the steepest meridian attempting to correct some of the preoperative astigmatism. The lens is introduced into the anterior chamber with the use of forceps on a slide sheet. A peripheral iridectomy is always performed. The lens must be well centered, the pupil must be round, and no traction forces have to be present from a haptic footplate. New models are developed

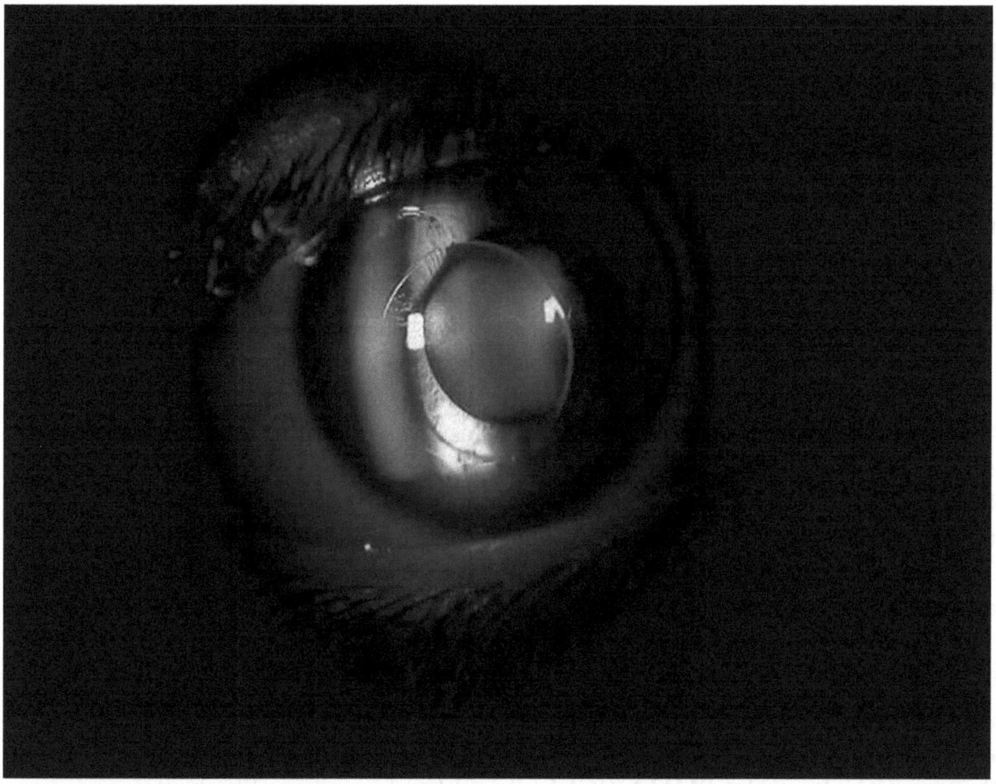

Fig. 1. Slit-lamp photograph of a myopic eye 4 yr post an anterior chamber angle-supported PIOL (Phakic-6) implantation. The decent ration of the implant in relation to the pupil is apparent.

attempting to decrease the size of the incision (Duet-Kelman lens, TEKIA, Irvine, CA *(12)* or a foldable angle-supported Baikoff lens).

Refractive results are shown to be very satisfactory after the implantation of an anterior chamber angle-supported PIOL *(8,11,13)*. Uncorrected visual acuity (UCVA) is significantly better than preoperative values, whereas BCVA is also improved. This improvement in postoperative best-corrected visual acuity (BCVA) is attributed to the magnification of the retinal image by the intraocular position of the PIOL *(14)*. The refractive results are stable and 82% of the eyes are reported to be within 1 Diopter of target refraction *(11)*.

ANTERIOR CHAMBER ANGLE-SUPPORTED PIOLS COMPLICATIONS

Giant papillary conjunctivitis can occur in case of exposed nylon sutures *(11)* and can be resolved with the removal of the sutures. Halos and glare are reported in about 20% *(15)* to 26% *(11)* of the treated eyes. These problems could be caused by the decentration of the PIOL in relation to the pupil (Fig. 1), or by the small optical zone (4 or 4.5 mm) of some PIOL models. Acute postoperative anterior uveitis can be observed immediately after surgery or at three and six postoperative intervals (4.5% *[15]* to 8.7% *[11]*). Another complication is the elevation of intraocular pressure (IOP) (7.2%), which can be attributed to the use of steroids postoperatively *(11)* and, therefore, is transient. Pupillary block by the implant can also occur in case of nonpatent iridectomies *(13)*.

Another complication of angle-supported PIOLs is pupil ovalization (6% *[14]* to 22.6% *[13]*), in some cases owing to sectorial iris atrophy. It might also be caused by oversized angle-supported PIOLs as a result of compression forces by the haptics against the iris root.

PIOL rotation is noticed in many cases during the follow-up *(11,16)* (43.5%). Cataract has been reported post anterior chamber PIOL implantation *(15,17)* (3.4%), with age over 40 and long axial myopia being related to its development. The authors speculated that pre-existing early nuclear sclerosis that could not be recognized before surgery might have led to nuclear cataract after surgery *(17)*. Therefore, careful patient selection is important to avoid this complication.

An important risk of anterior chamber lenses is the damage to the corneal endothelium. With previous models the high endothelial cell loss was related to intermittent contact between the optic edge and the endothelium, because of their short distance *(18,19)*. Recent models have a different design that allows more space between the implant and the endothelium. Endothelial cell density has been found significantly decreased one and two years after surgery *(20)*, but this decrease seems to be caused by the surgical trauma.

IRIS-FIXATED PIOLS

The main representative in this category is the Artisan lens (Ophtec BV, Groningen, Netherlands), which was formerly known as the Worst-Fechner Claw Lens *(21,22)*. It is made of Perspex CQ ultraviolet PMMA (ICI Ltd) and has two pincer-like haptics (claws) that are enclavated into the midperipheral iris stroma, thus requiring a more sophisticated surgical technique. Two different types of the implant have an optical zone of 5 and 6 mm, respectively. The lens power needed for emmetropia is calculated by inserting spherical equivalent refraction, mean corneal curvature, and adjusted ACD-0.8 into the Van der Heijde formula *(23)*.

Ocular anesthesia can be retrobulbar or peribulbar before surgery. After constriction of the pupil, the lens is inserted through a scleral tunnel, a limbal incision, or a corneal incision. The main incision must be equal to the lens optic diameter. Two paracenteses are made at 10 and 2 o'clock in instruments assessed to fixate the lens. Then the lens is rotated 90° to be perpendicular to the direction of incision. With the use of a disposable enclavation needle a small knuckle of iris is drawn through the pincer of each haptic. A peripheral iridectomy is performed and the wound is closed with sutures.

In order to decrease the incision size, a foldable version of iris-fixated PIOL, the Artiflex lens (Ophtec BV, Groningen, Netherlands), has been recently developed, which can be inserted through an incision of 3.2 mm. Iris-fixated lenses are reported to have rather encouraging clinical results *(22,24–26)*. The technique appears to be effective, safe, stable, and predictable. UCVA improves in all cases, and a gain in BCVA is characteristic of the implantation of this type of PIOL as well *(22,26)*. The majority (71 *[25]* to 83% *[26]*) of the treated eyes are within 1 Diopter of emmetropia.

An advantage of iris-fixated lenses is their postoperative positional stability inside the eye, which minimizes the risk of contact between them and ocular structures (endothelium, crystalline lens). With the use of Scheimpflug photography it has been demonstrated that they show the highest overall stability in comparison to other PIOLs (angle-supported and posterior chamber) *(16)*. In this study rotation was typical for anterior chamber angle-supported lenses, whereas posterior chamber PIOLs had the tendency to decrease in distance to the crystalline lens.

IRIS-FIXATED LENSES COMPLICATIONS

During the enclavation of the implant haptics into the iris, an iris trauma can occur. Apart from that, intraoperative bleeding may be a consequence of surgical iridectomy or of excessive handling of the iris while fixating the haptics *(26)*. A complication that might follow the PIOL implantation is iritis, usually in the early postoperative period *(25,26)*. An important issue concerning the iris-fixation is constant iris irritation by the implant. Fechner supported that no signs of such irritation were observed *(21)*. In another study *(27)*, however, evidences of chronic subclinical inflammation were found 1 and 2 yr following implantation of both Worst-Fechner and Baikoff PIOLs.

An uncommon complication of iris-fixated PIOLs (one case described) is the Urrets-Zavalia syndrome, which is a permanent wide dilation of the pupil *(26)*. IOP elevation can also occur during the postoperative steroid treatment, as after implantation of other PIOLs types. Decentration of the PIOL can occur especially in cases of inexperienced surgeons *(22,26)*. After the implantation a significant trauma may result in dislocation of the PIOL *(28)*. When the iris is enclavated asymmetrically, pupil irregularity is provoked *(25)*. In case of poor centration of the implant or in cases of large pupils, patients may complain of glare and halo *(24)*. The model with 6-mm optical zone reduces this possibility, but requires a larger incision and has to be thicker than the 5-mm optical zone model with the same dioptric power.

A threatening complication of anterior chamber lenses is the damage to the corneal endothelium. After iris-fixated PIOLs implantation, endothelial cell loss has been found more evident at the sixth postoperative month *(26)*. A significant negative correlation was observed between endothelial cell loss and ACD, and a positive correlation between cell loss and the dioptric power of the IOL. At 2 yr the hexagonality and coefficient variation in cell size were close to preoperative levels. It is suggested that endothelial cell loss is attributed to the surgical manipulations. Although iris-fixated lenses have the advantage of a stable position and maintain a safe distance from the endothelium, the surgical manipulations for the attachment of the haptics to the iris might contribute to greater endothelial damage *(20)*.

POSTERIOR CHAMBER PIOLS

Two kinds of posterior chamber lenses, the ICL and the PRL, are currently in clinical use. ICL is made of Collamer, a copolymer of hydroxyethyl methacrylic acid and 0.1% porcine collagen, with a high refractive index. It is designed to vault anteriorly to the crystalline lens having no contact with the latter. PRL is made of silicone and because of its hydrophobic material it is designed to float in the posterior chamber having no contact with the crystalline lens. The selection of the model of the lens to be inserted is based on the white-to-white diameter. The power of the lens is calculated by the company using the preoperative cycloplegic refraction, the desired postoperative refraction, the ACD, and the keratometry readings.

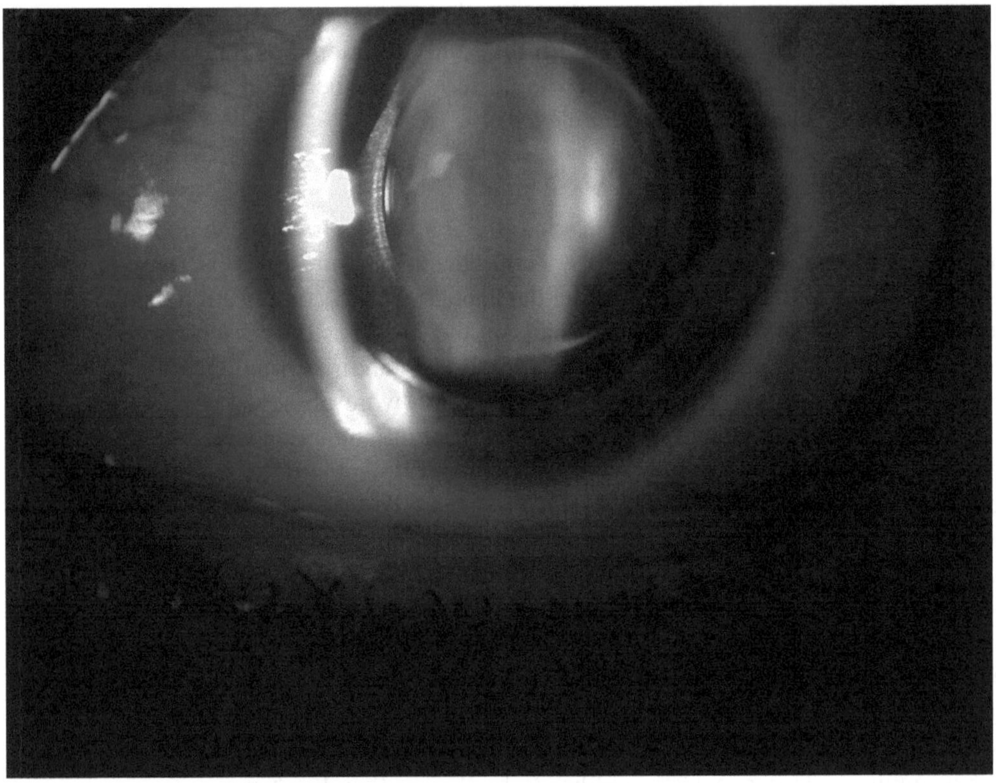

Fig. 2. Slit-lamp photograph of a PRL implanted eye 6 mo postoperatively. The PIOL is rotated in the posterior chamber.

Both posterior chamber lenses are foldable and can be inserted through a 3–3.5 mm incision. The procedure requires gentle manipulations in order not to touch the crystalline lens, the zonular fibers, or any other ocular structures. The lens is implanted with the use of an injector or special forceps and with the use of a manipulator the haptics of the lens are placed in the posterior chamber. Two neodymium-doped yttrium aluminum garnet (Nd:YAG) laser iridotomies in the upper peripheral iris a week before surgery or a surgical iridectomy at 12 o'clock are performed in all cases, in order to avoid postoperative pupillary block.

Published clinical results of posterior chamber PIOLs have shown the effectiveness, the predictability, the stability, and the safety of this technique in treating high myopia *(29–38)*. All previous studies reported an improvement of UCVA after posterior chamber PIOL implantation. The refractive outcome demonstrated stability from the first postoperative month *(32–34)* in which 57% *(35)* to 85% *(36)* of the treated eyes were within ±1.00 D of target refraction. An important outcome in the previous studies is that the majority of eyes presented an improvement in BCVA postoperatively *(29,34–38)*. The line-gain ranged between one and five Snellen lines, probably owing to the image magnification resulting from the existence of the PIOL inside the eye.

To estimate the exact intraocular position of posterior chamber PIOL ultrasound biomicroscopy can be performed postoperatively. In a recent study of PRL-implanted

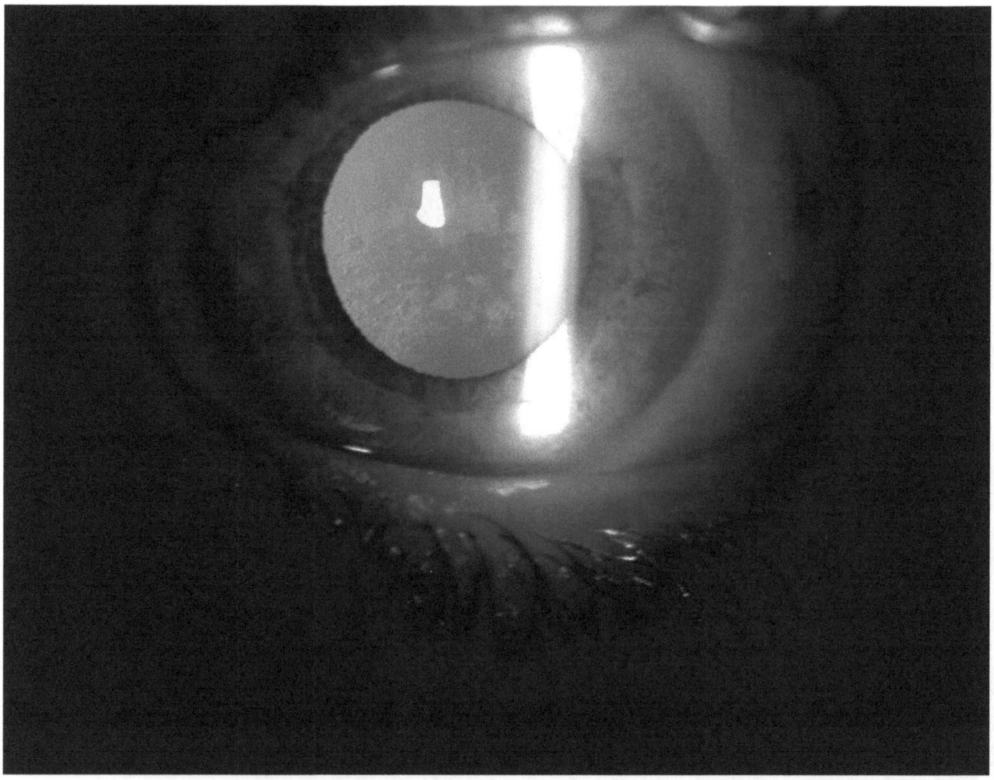

Fig. 3. Slit-lamp photograph of a myopic eye 4 yr post ICL implantation. The poor vaulting of the posterior chamber PIOL resulted in anterior subcapsular opacities of the crystalline lens.

eyes ultrasound microscopy demonstrated that haptics of the PRL were in the zonules, in the ciliary sulcus, or in the ciliary body *(39)*. Rotation of the lens can occur when its haptics rest on the zonules (Fig. 2). The different positions of the lens in several eyes can be explained by the difference between the PRL diameter and the sulcus diameter. Relatively small implants rest on the zonules, whereas relatively larger PRLs fit in the sulcus. A positioning in the sulcus can lead to pigment dispersion and inflammation.

POSTERIOR CHAMBER PIOLS COMPLICATIONS

The main short or long-term risk of the implantation of a posterior chamber lens is cataract formation in case of touch of the crystalline lens during surgery or because of postoperative contact between the phakic and the crystalline lens *(35,40,41)*. The presence of poor PIOL vaulting (clearance between PIOL and crystalline lens) is associated with the development of late anterior subcapsular opacities *(41,42)* (Fig. 3). In case of focally-stable (not progressing) anterior capsule opacification that can occur during surgical iridectomy or because of touch of the crystalline lens during surgery, no further treatment is needed *(29,32,38)*, wherein progressive cataract formation that results in BCVA loss the lens has to be explanted and cataract extraction has to be performed *(40,41)*.

Another uncommon complication is iris atrophy (Fig. 4) owing to surgical manipulations, and bleeding during surgical iridectomy that usually ceases in few minutes

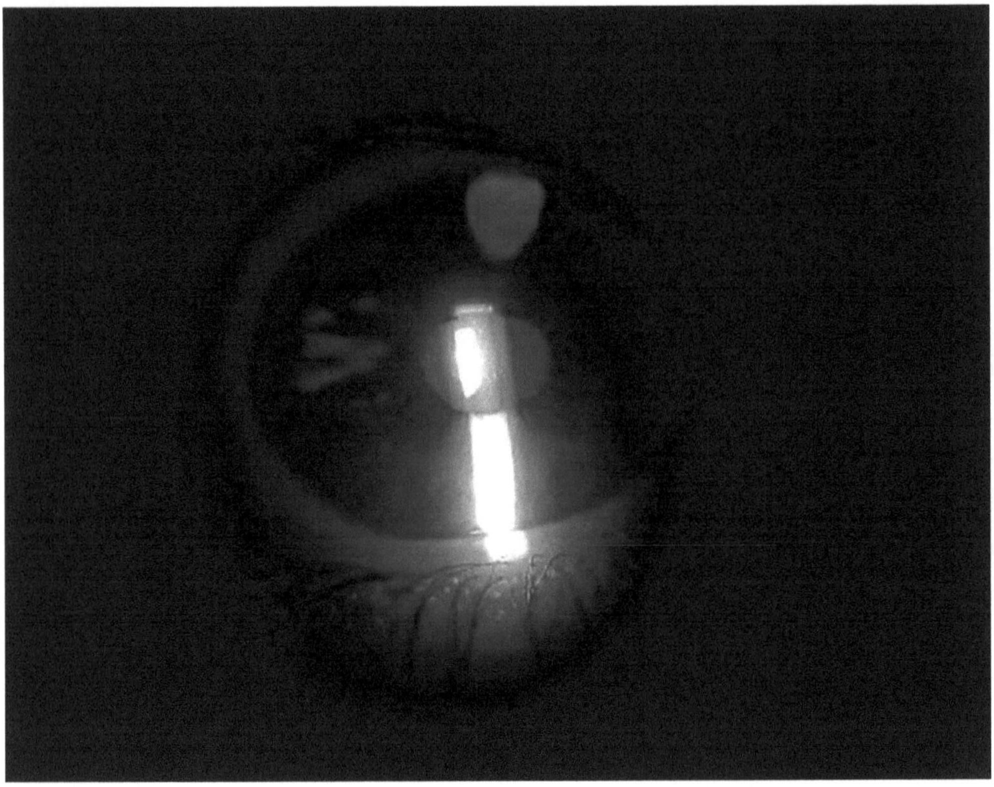

Fig. 4. Slit-lamp photograph of a PRL implanted eye 1 yr postoperatively. A large patent iridectomy, along with temporal iris damage has been observed since the first postoperative day.

with no further consequences. Of great importance is the evaluation of intraocular pressure at all postoperative intervals, as well as the correlation of the findings to the preoperative values, in order to early estimate any IOP changes owing to the presence of the phakic lens. An IOP increase is noticed during the first postoperative month *(34,38)*, probably owing to residual viscoelastic during the first postoperative days and to corticosteroid-response, which is frequent in high myopic patients, because intraocular pressure returns to normal levels after discontinuation of steroid drops *(32,34,38)*.

Posterior chamber lenses have also been reported to induce pigment dispersion *(37,43)*, owing to contact of the PIOL with the posterior surface of the iris. Although more common in hyperopic eyes, closed iridectomies can lead to pupillary block glaucoma *(32,33,43)*, which can be resolved with further iridotomies. According to each study and the used questionnaire, a percent of patients between 20 *(38)* and 53% *(35)* complained of glare and halo during night postoperatively. These night phenomena could be explained by the patients' large pupils, 6 and 7 mm at scotopic conditions, while the optic zone of the implant is 5–5.5 mm. Because not all patients with large pupils experience these phenomena, perhaps other factors besides the size of the pupil contribute to their existence. However, the quick visual recovery during the first postoperative week, the stable refractive outcome, the satisfactory UCVA, and the gain in

BCVA Snellen lines compensate the patients for these problems at night. Moreover, another study reports the improvement of these symptoms after ICL implantation compared with before surgery *(43)*.

Decentration of the PIOL can occur if its size is not appropriate (too short) for the specific eye *(32,37)*. Two recent studies *(44,45)* reported three cases of dislocation of PRL into the vitreous cavity 2, 4, and 22 mo after surgery. A pars plana vitrectomy had to be performed in all cases to remove the lens. This dislocation could be attributed to preoperative zonular weakness, common in high axial myopic eyes. Previous ocular trauma that might have caused zonular defects or Marfan's syndrome or other systemic diseases with zonular weakness should be considered as contraindications for a posterior chamber PIOL implantation. Other causes for the dislocation could be a surgical damage of the zonular fibers during the positioning of PRL haptics into the posterior chamber and/or postoperative rotation of the PRL in cases of zonule-haptic contact.

POSTOPERATIVE TREATMENT—EVALUATION

Postoperatively each patient is given one tablet of Acetazolamide 250 mg. Antibiotic-steroid combination drops are prescribed in a tapered dose for 1 mo. In case of elevated intraocular pressure during the first postoperative month IOP-suppressants are used until the normalization of IOP. Patients are typically examined on the first postoperative day, at 1 wk and at 1, 3, 6, 9, and 12 mo postoperatively. After the first postoperative day, the examination includes uncorrected visual acuity, best-corrected visual acuity, manifest refraction, corneal topography, slit lamp microscopy, and tonometry. At 6 and 12 mo the examination also includes gonioscopy and dilated fundoscopy.

CONCLUSIONS

PIOLs offer a very promising alternative in the correction of high myopia in which excimer laser cannot be applied and clear lens extraction is more aggressive and risky, especially in young myopic patients. PIOLs implantation has very satisfactory refractive results, quick visual recovery, and stable outcomes. The most impressive outcome of all kinds of PIOLs implantation is the improvement of BSCVA in the majority of high myopic eyes. Despite all these encouraging results and the patients' need for a surgical solution to their high myopia, surgeons have to take into consideration all the inclusion criteria and carefully select the candidates for this technique, because of the intraocular nature of the procedure and its possible complications. More studies with long follow-up are needed to evaluate the long-term safety of the technique and to ensure that it is the best procedure for the surgical treatment of high myopic eyes.

REFERENCES

1. Perez-Santonja JJ, Bellot JJ, Claramonte P, et al. Laser in situ keratomileusis to correct high myopia. J Cataract Refract Surg 1997;23:372–385.
2. Duffey RG, Leaming D. US trends in refractive Surgery:2002 ISRS survey. J Refract Surg 2003;19:357–363.
3. Pallikaris IG, Kymionis GD, Astyrakakis NL. Corneal ectasia induced by LASIK. J Cataract Refract Surg 2001;27:1796–1802.
4. Applegate RA, Howland HC. Refractive surgery, optical aberrations, and visual performance. J Refract Surg 1997;13:295–299.

5. Goldbeg MF. Clear lens extraction for axial myopia. Ophthalmology 1987;94:571–582.
6. Lyle WA, Jin GJC. Clear lens extraction for the correction of high refractive error. J Cataract Refract Surg 1994;20:273–276.
7. Lee KH, Lee JH. Long-term results of clear lens extraction for severe myopia. J Cataract Refract Surg 1996;22:1411–1415.
8. Baikoff G, Joly P. Comparison of minus power anterior chamber lenses. Refract Corneal Surg 1990;6:245–251.
9. Landesz MA, Worst JGF, Siertsema JV, Van Rij G. Negative implant, a retrospective study. Doc Ophthalmol 1993;83:261–270.
10. Fyodorov SN, Zuyev VK, Aznabayev BM. Intraocular correction of high myopia with negative posterior chamber lens. Ophthalmosurgery 1991;31:57–58.
11. Perez-Santonja JJ, Alio JL, Jimerez-Alfaro I, Zato MA. Surgical correction of myopia with an angle-supported phakic intraocular lens. J Cataract Refract Surg 2000;26:1288–1302.
12. Alio JL, Kelman C. The Duet-Kelman lens: A new exchangeable angle-supported phakic intraocular lens. J Refract Surg 2003;19:488–495.
13. Baikoff G, Arne JL, Boko bza Y, et al. Angle-fixated anterior chamber phakic intraocular lenses for myopia -7 to -19 Diopters. J Refract Surg 1998;14:282–293.
14. Garcia M, Gonzalez C, Pascual I, Fimia A. Magnification and visual acuity in highly myopic phakic eyes corrected with an anterior chamber intraocular lens versus other methods. J Cataract Refract Surg 1996;22:1416–1422.
15. Alio JL, de la Hoz F, Perez-Santonja JJ, Ruiz-Moreno JM, Quesada JA. Phakic anterior chamber lenses for the correction of myopia. A 7-year cumulative analysis of complications in 263 cases. Ophthalmology 1999;106:458–466.
16. Baumeister M, Bühren J, Kohnen T. Position of angle-supported, iris-fixated and ciliary sulcus-implanted myopic phakic intraocular lenses evaluated by Scheimpflug photography. Am J Ophthalmol 2004;138:723–731.
17. Alio JL, de la Hoz F, Ruiz-Moreno JM, Salem TF. Cataract surgery in highly myopic eyes corrected by phakic anterior chamber angle-supported lenses. J Cataract Refract Surg 2000; 26:1303–1311.
18. Mimouni F, Colin J, Koffi V, Bonnet P. Damage to the corneal endothelium from anterior chamber intraocular lenses in phakic myopic eyes. Refract Corneal Surg 1991;7:277–281.
19. Saragoussi JJ, Cotinat J, Renard G, et al. Damage to the corneal endothelium by minus power anterior chamber intraocular lenses. Refract Corneal Surg 1991;7:282–285.
20. Perez-Santonja JJ, Iradier MT, Sanz-Iglesias L, et al. Endothelial changes in phakic eyes with anterior chamber lenses to correct high myopia. J Cataract Refract Surg 1996; 22:1017–1022.
21. Fechner PU, Strobel J, Wichmann W. Correction of myopia by implantation of a concave Worst-iris claw lens into phakic eyes. Refract and corneal Surg 1991;7:286–298
22. Menezo JL, Cisneros A, Hueso JR, Harto M. Long term results of surgical treatment of high myopia with Worst-Fechner intraocular lenses. J Cataract Refract Surg 1995;21:93–98.
23. Van der Heijde GL. Some optical aspects of implantation of an IOL in a myopic eye. Eur J Implant Ref Surg 1989;1:245–248.
24. Budo C, Hessloehl J, Izak M, et al. Multicenter study of the Artisan phakic intraocular lens. J Cataract Refract Surg 2000;26:1163–1171.
25. Maloney RK, Nguyen LH, John ME. Artisan Phakic Intraocular Lens for myopia. Short-term results of a prospective, multicenter study. Ophthalmology 2002;109:1631–1641.
26. Menezo JL, Cisneros AL, Rotriquez-Salvador V. Endothelial study of iris-claw phakic lens: four year follow up. J Cataract Refract Surg 1998;24:1039–1049.
27. Perez-Santonja JJ, Iradier MT, Benitez del Castillo JM, Serrano JM, Zato MA. Chronic subclinical inflammation in phakic eyes with intraocular lenses to correct myopia. J Cataract Refract Surg 1996;22:183–187.

28. Yoon H, Macaluso DC, Moshirfar M, Lundergan M. Traumatic dislocation of an Ophtec Artisan phakic intraocular lens. J Refract Surg 2002;18:481–483.
29. Hoyos JE, Dementiev DD, Cigales M, Hoyos-Chacon J, Hoffer KJ. Phakic refractive lens experience in Spain. J Cataract Refract Surg 2002;28:1939–1946.
30. Koirula A, Petrelius A, Zetterstrom C. Cinical outcomes of practice refractive lens in myopic and hyperopic eyes: 1-year results. J Cataract Refract Surg 2005;31:1145–1152.
31. Pesando PM, Ghiringhello MP, Tagliavacche P. Posterior chamber Collamer phakic intraocular lens for myopia and hyperopia. J Refract Surg 1999;15(4):415–423.
32. Zaldivar R, Davidorf JM, Oscherow S. Posterior chamber phakic intraocular lenses for myopia –8 to –19 diopters. J Refract Surg 1998;14:294–305.
33. Rosen E, Gore C. Staar collamer posterior chamber phakic intraocular lens to correct myopia and hyperopia. J Cataract Refract Surg 1998;24:596–606.
34. Jimerez-Alfaro I, Benitez del Castilo JM, Garcia-Feijoo J, Gil de Bernabe JG, Serrano de la Iglesia JM. Safety of posterior chamber phakic intraocular lenses for the correction of high myopia-anterior segment changes after posterior chamber phakic intraocular lens implantation. Ophthalmology 2001;108:90–99.
35. Arne LA, Leseur LC. Phakic posterior chamber lenses for high myopia: Functional and anatomical outcomes. J Cataract Refract Surg 2000;26:369–374.
36. The Implantable contact lens in treatment of myopia study group. US Food and Drug administration clinical trial of the Implantable contact lens for moderate to high myopia. Ophthalmology 2003;110:255–266.
37. Asseto V, Benedetti S, Pesando P. Collamer intraocular contact lens to correct high myopia. J Cataract Refract Surg 1996;22:551–556.
38. Pallikaris IG, Kalyvianaki MI, Kymionis GD, Panagopoulou SI. Phakic refractive lens implantation in high myopic patients: One-year results. J Cataract Refract Surg 2004; 30:1190–1197.
39. Garcia-Feijoo J, Hernandez-Matamoros JL, Castillo-Gomez A, et al. Ultrasound biomicroscopy of silicone posterior chamber phakic intraocular lens for myopia. J Cataract Refract Surg 2003;29:1932–1939.
40. Fink AM, Gore C, Rosen E. Cataract development after implantation of the Staar Collamer posterior chamber phakic lens. J Cataract Refract Surg 1999;25(2):1278–1282.
41. Trindade F, Pereira F. Cataract formation after posterior chamber phakic intraocular lens implantation. J Cataract Refract Surg 1998;24:1661–1663.
42. Sanders DR, Vukich JA. Incidence of lens opacities and clinically significant cataracts with the Implantable Contact Lens: Comparison of two lens designs. J Refract Surg 2002; 18:673–682.
43. Brandt JD, Moskovac ME, Chayet A. Pigmentary dispersion syndrome induced by a posterior chamber phakic refractive lens. Am J Ophthalmol 2001;131:260–263.
44. Martinez-Castillo V, Elies D, Boixadera A, et al. Silicone posterior chamber intraocular lens dislocated into the vitreous cavity. J Refract Surg 2004;20:773–777.
45. Eleytheriadis H, Amoros S, Bilbao R, Teijeiro MA. Spontaneous dislocation of a phakic refractive lens into the vitreous cavity. J Cataract Refract Surg 2004;30:2013–2016.

17
A Telescope Prosthetic Device for Patients With End-Stage AMD

Henry L. Hudson, MD, FACS

CONTENTS

INTRODUCTION

Age-related macular degeneration (AMD) is the leading cause of blindness in the United States and has a significant impact on public health *(1)*. Almost 1.8 million people have advanced AMD, which is defined as choroidal neovascularization (CNV) or geographic atrophy that results in scarring or visual loss in the central visual field *(2,3)*. The advanced stages of AMD can be considered "end-stage" when the result is bilateral, untreatable macular scars causing central scotomata and associated moderate to profound visual impairment that limits a patient's ability to engage in daily activities requiring good central vision. Advanced forms of AMD are associated with legal blindness, elevated risk of depression, reduced independence, increased risk of accidents, and a significant decrease in quality of life *(4–6)*.

Although, rapid development has been seen in recent years of a number of therapies for AMD, including antiangiogenic drugs, gene therapies, and combinations of various laser and drug treatments, end-stage AMD is still medically untreatable. The only option for these patients has been the use of magnifying glasses or cumbersome, cosmetically unappealing, handheld or spectacle-mounted telescopes. These devices provide a very restricted field of view, which limits the usefulness of magnification. Furthermore, external devices are commonly used for static spotting tasks and head movements can be intolerable because of the vestibular ocular reflex conflict.

From: *Ophthalmology Research: Visual Prosthesis and Ophthalmic Devices: New Hope in Sight*
Edited by: J. Tombran-Tink, C. Barnstable, and J. F. Rizzo © Humana Press Inc., Totowa, NJ

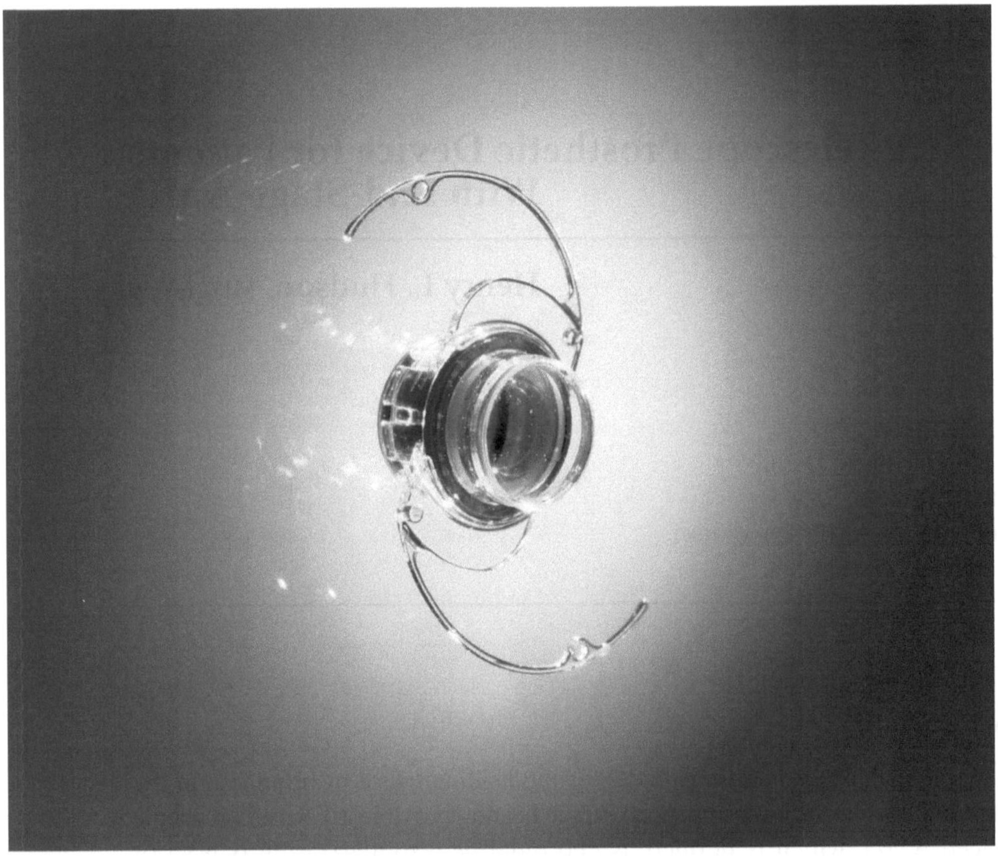

Fig. 1. Telescope visual prosthetic device. The quartz glass telescope is embedded in a carrier apparatus. The light restrictor (blue) is designed to keep light from passing in between the telescope cylinder and the iris.

VISUAL PROSTHETIC DEVICE

Implantable Telescope

Recently, a visual prosthetic device has been developed that provides the eye with an enlarged retinal image of the central visual field. This device was designed to improve central vision in patients with moderate to profound visual impairment from bilateral, untreatable, advanced AMD. The implantable miniature telescope (IMT™ by Dr. Isaac Lipshitz), (Vision Care Ophthalmic Technologies, Saratoga, CA), is designed for unilateral implantation in the posterior chamber of the anterior segment during an outpatient surgical procedure (Fig. 1). By implanting the device inside the eye, it can provide a relatively large field of view (20–24°) whereas reducing the effective size of the scotoma.

The ultraprecision wide-angle micro-optics of this visual prosthetic device work with the cornea to function as a fixed-focus telephoto system (Fig. 2). The micro-optics are housed by a quartz cylinder that is 3.6 mm in diameter and 4.4 mm in length (Fig. 3). The quartz cylinder is embedded in a rigid polymethyl methacrylate carrier apparatus with haptic extensions that are 13.5 mm in diameter. Optimal uncorrected focusing

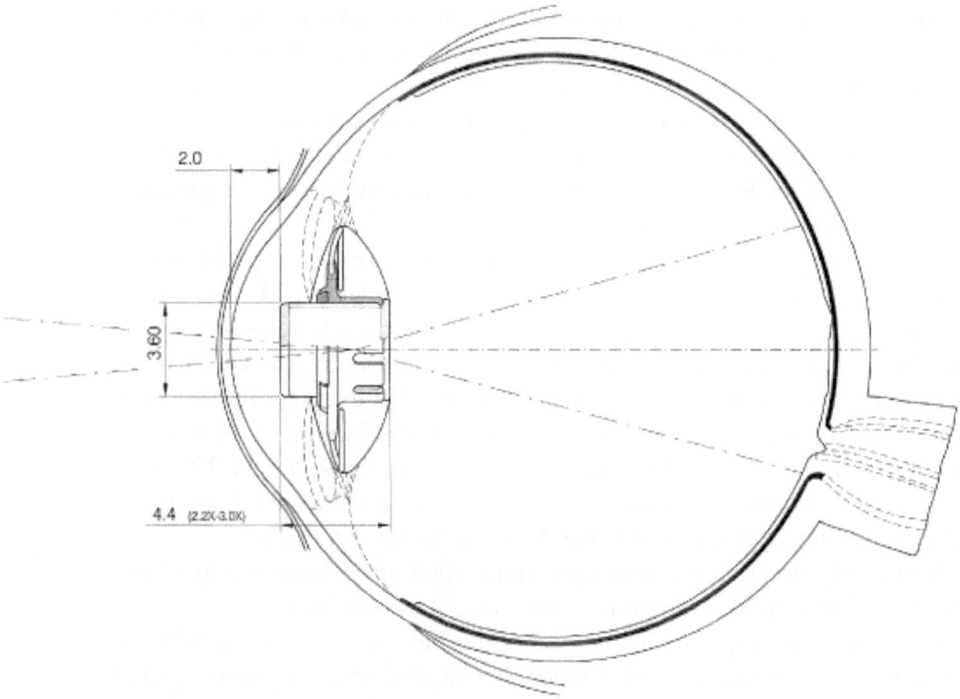

Fig. 2. The telescope prosthetic device is secured in the capsular bag to render an enlarged retinal image onto more than 50° of the retina.

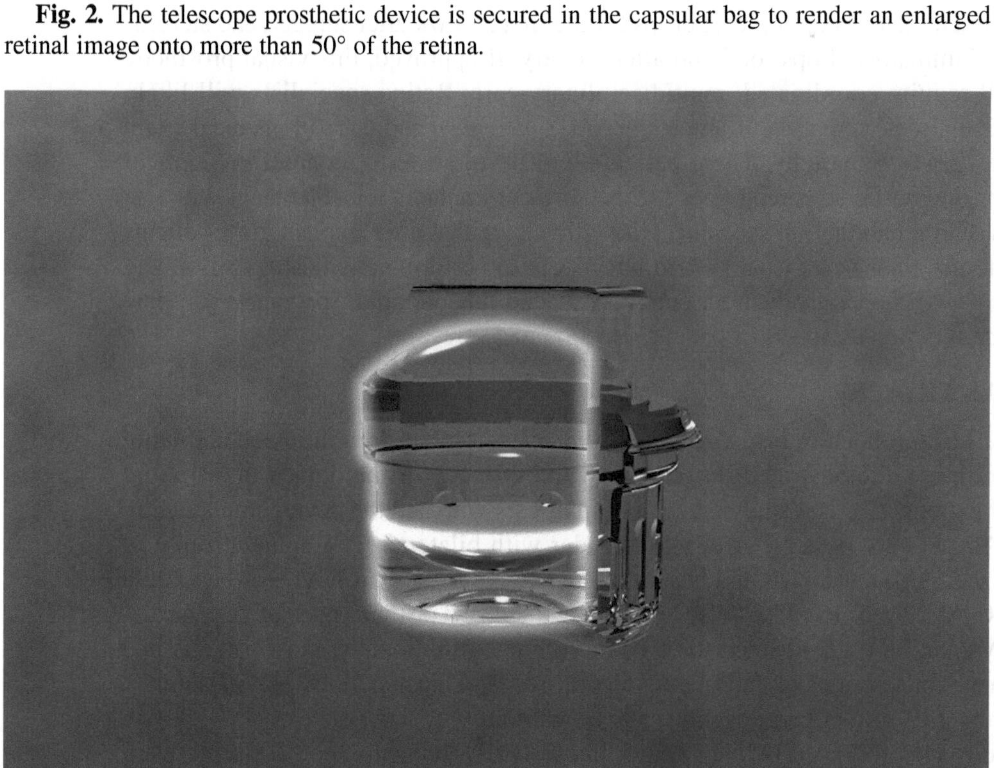

Fig. 3. Cross-sectional view of the telescope implant. The anterior and posterior micro-optical elements provide retinal image enlargement of × 2.2 or × 3 (two model options).

distance is set at 1.5–10 m. Distance and near focus are attained with standard spectacle prescription. With this device, light is transmitted from the object in the central visual field through the cornea, anterior window, refractive air space, anterior high-plus microlens, refractive air space, posterior high-minus microlens, refractive air space, posterior window, and retina in that order. This creates a telephoto effect on the retina, and the resultant image is rendered over 55° of the retina. In comparison, the dysfunctional macula covers approx 5–10° of the retina.

Because the image projected onto the retina is enlarged, the size of the patient's scotoma is reduced relative to the objects in the patient's central visual field. This allows patients to see more visual information in the central field. When implanted in one eye of a patient with bilateral macular disease, the prosthesis improves central vision. The unimplanted fellow eye provides peripheral vision for orientation and mobility. Placement in the posterior chamber of the anterior segment provides stable fixation in the eye and proper alignment to achieve the desired retinal image of the central visual field. The prosthesis cannot be used in conjunction with the natural lens or an intraocular lens (IOL). The implantation procedure is uniquely different than cataract surgery and IOL implantation. The 4.4-mm profile necessitates special techniques and viscoelastic use to ensure proper implantation with avoidance of trauma to the corneal endothelium.

Device implantation is accompanied by removal of the lens, by any method, and must be performed by a skilled anterior segment surgeon experienced with large incisions and wound management. Six to eight sutures are typically required. The postoperative medication regimen includes cycloplegic drops for the first 3–4 wk after surgery, and steroids and anti-inflammatory drops for 3 mo after surgery. If approved, this visual prosthetic device will require coordinated, multidisciplinary care. Retina specialists will likely make the initial determination of late-stage AMD patients' candidacy by medical exam and fluorescein angiography. Then, patients will be referred to an anterior segment surgeon for diagnostic screening tests for the surgical implantation. The team will also require a visual rehabilitation specialist who will ensure that after implantation, patients can learn to use their new visual system effectively to perform activities of daily living in a better way. Preoperatively, both optometrists and rehabilitation specialists can play a role in the functional aspects of patient selection and screening.

CLINICAL TRIALS

A phase I trial (under an Investigational Device Exemption from the Food and Drug Administration) demonstrated initial safety and efficacy of the device in a limited number of patients *(7)*. The study included 14 patients who had the device implanted in one eye. All patients were 60 yr of age or older with bilateral geographic atrophy or disciform scar AMD, cataract, and best-corrected visual acuity (BCVA) between 20/80 and 20/400. At 12 mo, 10 (77%) of 13 patients gained two or more lines of near or distance BCVA, and 8 (62%) gained three or more lines. Additionally, most patients had improvements in activities of daily living test scores, and mean endothelial cell density decreased by 13%. All adverse events resolved without sequelae.

This study was followed by a 28 center phase II/III study in which 206 eyes were successfully implanted with the device. Twelve-month data are currently available.

To be enrolled in the study, patients had to have bilateral, stable macular disease, and cataract. Preoperative best-corrected distance vision had to be between 20/80 and 20/800, with a minimum five-letter improvement on the visual acuity chart using an external telescope. Exclusion criteria included evidence of active CNV or treatment of active CNV within the preceding 6 mo, retinal vascular disease, diabetic retinopathy, history of retinal detachment, pathology compromising peripheral vision, previous intraocular or corneal surgery, and endothelial cell density of <1600 cells/mm^2. The primary efficacy end point was the proportion of patients improving two lines or more in either near or distance BCVA at 1 yr.

A total of 217 patients were enrolled. Eleven procedures were aborted because of intraoperative posterior capsule rupture (eight cases), intraoperative choroidal hemorrhage (two cases), or suspected choroidal hemorrhage (one case). This left a study population of 206 eyes. Study participants' mean age was 76 yr, and 52% were men. At 1 yr, patients' mean distance BCVA in the study eye improved from 20/316 to 20/141, a 3.5-line improvement. Similarly, near BCVA improved by 3.2 lines from baseline. At 1 yr, 96.3% (185/192) of patients had improved distance BCVA or no change. Additionally, 89.6% (172/192) of patients achieved a two-line or greater improvement in near or distance BCVA—exceeding the primary efficacy end point of the pivotal trial (Fig. 4). Visual acuity results were not related to lesion type.

A secondary end point was improvement in quality-of-life outcomes, which was demonstrated by the (NEI VFQ-25) questionnaire. Baseline scores on the National Eye Institute Visual Function Questionnaire (NEI-VFQ) quality-of-life questionnaire were very low, at 44/100. At 1 yr, the mean composite VFQ score had improved to 50. Of the 12 VFQ subscales, eight were considered relevant to the intervention, and seven of these showed statistically significant improvement (a 7–14-point increase): general vision, near activities, distance activities, social functioning, mental health, role difficulties, and dependency.

Additionally, there was a trend toward improvement in color vision. Of the four other subscales, ocular pain, and driving were unchanged, whereas general health and peripheral vision declined. At 1 yr, corneal endothelial cell loss was 25.3%. There were two incidents of corneal decompensation, in cases in which wound leak and a shallow anterior chamber were reported. In terms of intraoperative retinal complications, one device was dropped into the vitreous, and two patients had choroidal hemorrhages. Postoperative anterior segment complications included transient elevated intraocular pressure and corneal edema in the first month postoperatively, both of which were manageable. More common adverse anterior segment events included inflammatory deposits (21%), guttata (8%), and posterior synechiae (6%). There were no retinal detachments or other retinal adverse events or complications >1%. One recurrence of CNV in an implanted eye was seen 6 mo postoperatively and was successfully treated with argon laser photocoagulation through the optical portion of the device. Results of this study suggest that, in patients with advanced, bilateral end-stage macular degeneration, this AMD telescope prosthesis is safe and effective in improving both distance and near visual acuity as well as quality of life.

Patient selection and vision rehabilitation are important, and surgical success is dependent on surgical technique. Patients must be educated on the benefits of potential

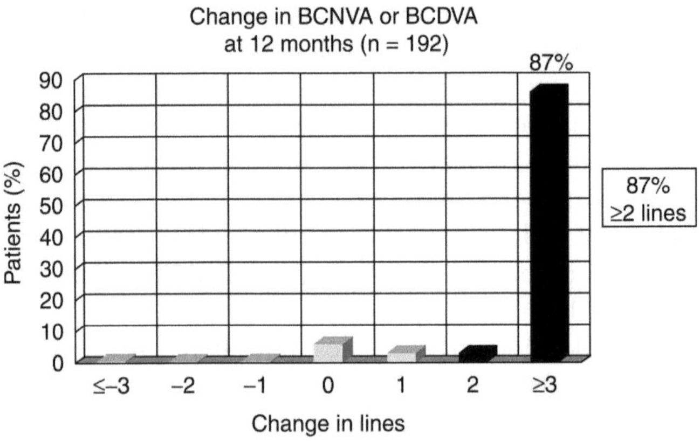

Fig. 4. Distribution of patients by lines gained in BCVA at near or distance (BCNVA and BCDVA, respectively). The visual acuity efficacy end point was achieved by 90% of patients in the phase II/III study (dark gray bars).

permanent central vision improvement that allows natural eye movements vs the tradeoff of restricted peripheral vision in the implanted eye. They must also understand that to integrate their new vision into daily activities, they will need to work with a vision rehabilitation professional postoperatively. Because of its relatively large three-dimensional volume, implanting this telescope prosthetic device requires considerably more skill and technique than cataract extraction with standard IOL implantation. For example, surgeons must avoid touching the cornea during implantation to minimize endothelial cell loss. In this study, although cell loss was more than targeted, overall corneal health remained good. Improved vision and improved quality of life justify the surgical risks in appropriately selected patients. Complications early in the postoperative period are manageable, and safe explantation is possible.

THE FUTURE OF AMD TREATMENT

If this device is approved by the Food and Drug Administration, many AMD patients will benefit. In fact, it is possible that with better early interventions for wet AMD, scotomata resulting from disciform scars might be smaller in size, giving future candidates for this device an even better chance of a higher level of visual functioning. Other visual prostheses, such as subretinal and epiretinal implants are on the horizon, but this implantable telescope may soon be available to us within the year. Thus, during the next decade, this device may play an important role in the treatment of visual disability associated with end-stage AMD.

REFERENCES

1. Gohdes DM, Balamurugan A, Larsen BA, Maylahn C. Age-related eye diseases: An emerging challenge for public health professionals. Prev Chronic Dis 2005;2:A17.
2. Friedman DS, O'Colmain BJ, Munoz B, et al. For the Eye Diseases Prevalence Research Group. Prevalence of age-related macular degeneration in the United States. Arch Ophthalmol 2004;122:564–572.

3. Age-Related Eye Disease Study Research Group. Potential public health impact of age-related eye disease study results. AREDS report no. 11. Arch Ophthalmol 2003;121: 1621–1624.
4. Williams RA, Brody BL, Thomas RG, Kaplan RM, Brown SI. The psychosocial impact of macular degeneration. Arch Ophthalmol 1998;116:514–520.
5. Casten RJ, Rovner BW, Tasman W. Age-related macular degeneration and depression: a review of recent research. Curr Opin Ophthalmol 2004;15:181–183.
6. Brown MM, Brown GC, Sharma S, Kistler J, Brown H. Utility values associated with blindness in an adult population. Br J Ophthalmol 2001;85:327–331.
7. Lane SS, Kuppermann BD, Fine IH, et al. A prospective multicenter clinical trial to evaluate the safety and effectiveness of the Implantable Miniature Telescope. Am J Ophthalmol 2004;137:993–1001.

18
The Use of Intracorneal Ring Segments for Keratoconus

Sérgio Kwitko, MD, PhD

CONTENTS

INTRODUCTION

Keratoconus is one of the leading causes of corneal blindness and a frequent indication for penetrating keratoplasty (PK) *(1)*. One of the most effective methods for visual rehabilitation of keratoconus patients is rigid contact lenses use, but a high percent of these patients are contact lens intolerant. Various types of surgical procedures have been suggested for keratoconus as an alternative for PK, including apex cauterization, epikeratophakia, sectorial keratotomy, photorefractive keratotomy (PRK), Lasik, and lamellar keratoplasty, but with worse results when compared to PK *(2)*.

Intracorneal implantation of a synthetic material to correct spherical and cylindrical refractive errors is not a new idea. Barraquer *(3)* in the mid-1950, had suggested the possibility of correcting myopia and astigmatism using intrastromal implants in the corneal midperiphery. Polymethylmethacrylate (PMMA) biocompatibility, low rejection rate to PMMA implants, and reasonable predictability of these implants were stressed by many published series *(4–8)*.

Intracorneal ring segments (INTACS®) for correction of low myopias (up to –3.00 D) were recently approved by Food and Drug Administration because of its effectiveness, precision, stability, safety, and reversibility *(9–13)*. Furthermore, it tends to preserve corneal asphericity, with obvious contribution to better visual acuity and contrast sensitivity *(14,15)*.

In 1986, Ferrara started to implant modified PMMA rings in rabbit corneas, and in 1994 developed a better technique of corneal stromal tunnel construction for implantation

From: *Ophthalmology Research: Visual Prosthesis and Ophthalmic Devices: New Hope in Sight*
Edited by: J. Tombran-Tink, C. Barnstable, and J. F. Rizzo © Humana Press Inc., Totowa, NJ

of these rings *(16)*. In 1996, this author has substituted the ring for 2 segments of 160° of arc, improving his results for high myopias, and has also started to implant them in corneas with keratoconus and after PK *(16)*. The effect of intracorneal rings or segments in correcting myopia, reducing keratoconus, and irregular astigmatism is explained by Barraquer thickness law, which states that a central corneal flattening is achieved by adding tissue to corneal periphery *(3)*.

Ferrara's ring segments (FICRS) differ from INTACS mainly in two aspects: (1) fixed radius of curvature of 2.5 mm, instead of variable from 2.5 to 3.5 mm of INTACTS®, and (2) triangular anterior shape instead of a flat anterior surface of INTACS *(20)*. More recently, other authors have suggested INTACS ring segments in keratoconus *(17)*, with safe and good results, avoiding or delaying PK in many of these patients *(18)*.

The triangular anterior shape and flat posterior shape of FICRS are a key for a stronger effect on central corneal flattening in keratoconus patients *(8,14,16)*. Besides adding tissue to the corneal midperiphery, the end portion of FICRS lifts anteriorly inside the corneal stroma after implantation, pushing down the segment body, adding an extra flattening effect in the opposite meridian of implantation.

Several studies have shown that intracorneal PMMA ring implantation is a safe and predictable procedure, with preservation of best-corrected spectacle visual acuity (BCSVA) in most cases, with a stable result over time *(11,13)*. This surgery has the advantages of preserving the central cornea and of being a reversible procedure, recovering preoperative corneal and refractive parameters after ring removal most of the times *(9,10,12,18)*.

There is also some evidence of less blood-aqueous barrier disturbance with this surgery than PRK and Lasik, as well as preservation of endothelial cells integrity. In addition, there is an excellent corneal tolerance to PMMA rings (Fig. 1), with a short-term, low-grade inflammatory stromal reaction, with a discrete concentration of inflammatory cells adjacent to the rings (Fig. 2) *(21,22)*.

Improvement of implantation techniques, ring quality manufacture, and precise methods of corneal surface evaluation, has allowed a safer and a more precise implantation of both FICRS and INTACS in keratoconus *(17,18,24)*. The possibility of implanting two FICRS through one incision had significantly reduced the extrusion and infection rate *(25)*. Furthermore, the implantation of asymmetric segments in regard to its thickness, according to preoperative topography, had also improved the results *(26)*.

Ring segments are made of acrylic Perspex CQ® (Mediphacos, Brazil), with an inner radius of curvature of 2.5 mm thickness varying from 150 to 350 μm, and arc length between 120° and 160°. These ring segments have a prism format with the flat posterior surface implanted facing corneal endothelium. Optical correction is achieved with central corneal flattening, which is directly proportional to ring thickness *(15)*. Ring segments' thickness and arc length were selected according to a previous Ferrara's nomogram *(16)*. This nomogram suggests changing the ring thickness depending on the keratoconus intensity, i.e., 200 μm segment for stage I, 250 μm segment for stage II, 300 μm segment for stage III, and 350 μm segment for stage IV *(16)*. In addition, asymmetric rings are selected based on preoperative corneal topography.

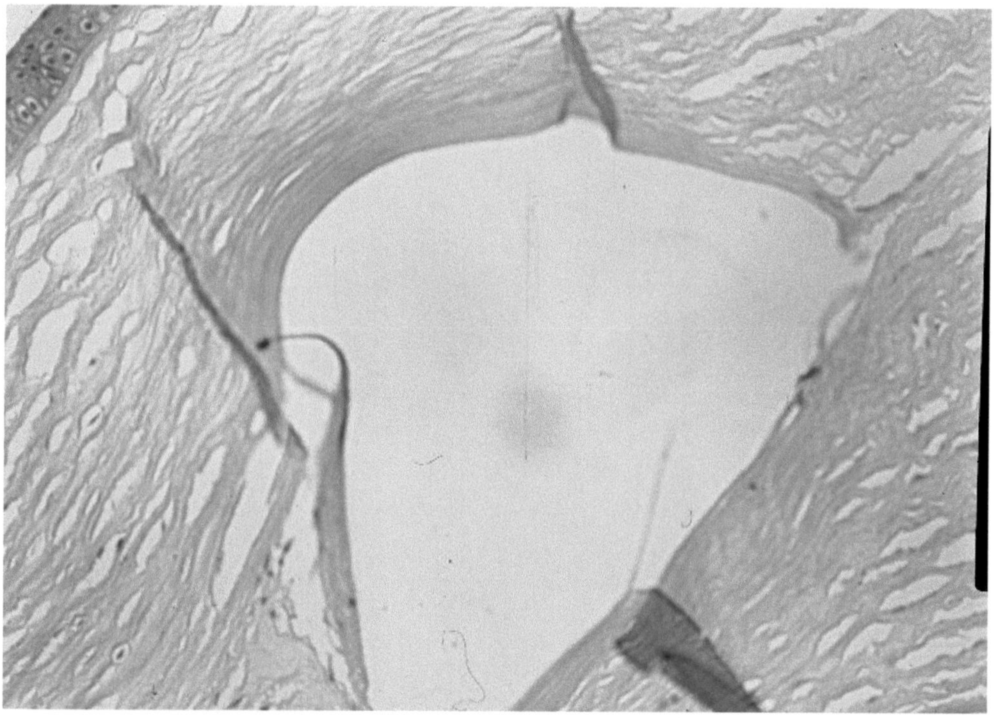

Fig. 1. Histological detail of a Ferrara ring segment inside a human corneal stroma, revealing a good biocompatibility with virtually no inflammatory reaction.

Fig. 2. Slit-lamp detail of inflammatory cells adjacent to a Ferrara intracorneal ring segment.

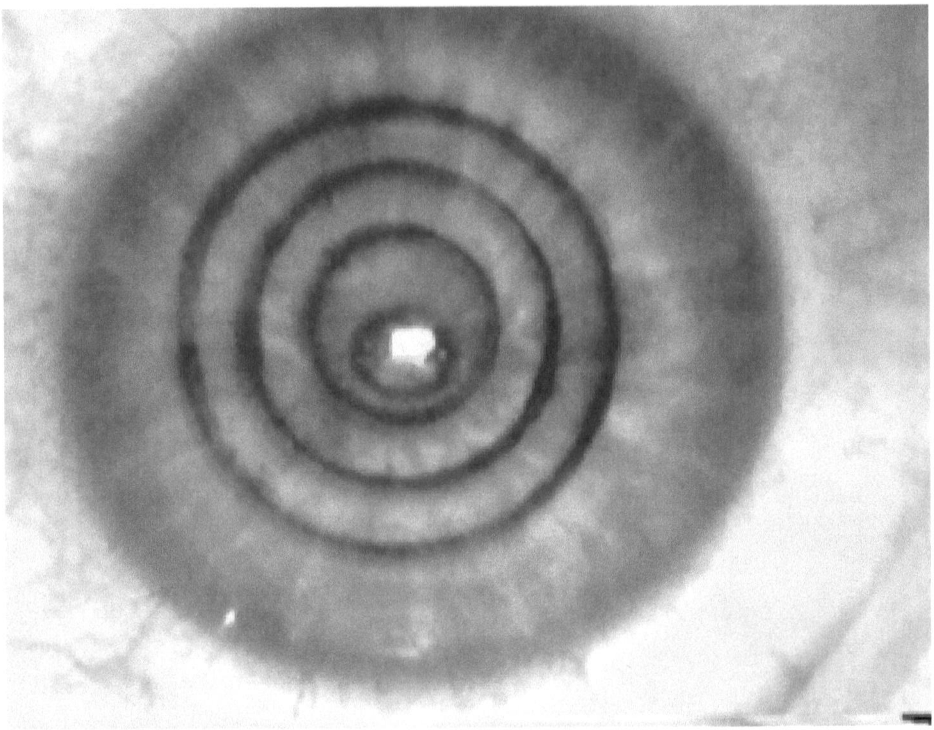

Fig. 3. Optical zones of 3, 5, and 7 mm marked with an appropriate marker tinted with methylen blue.

SURGICAL TECHNIQUE

Preoperative topical medication includes 0.3% gatifloxacin and ketorolac drops every 30 min, 2 h before surgery. Surgery is performed under topical anesthesia (proximetacaine 0.5%), and under an operating microscope (Topcon OMS-610®, Japan). Central corneal reflex is marked by asking the patient to look at the center of the coaxial microscope bulb filament, having the surgeon keep patient's both eyes opened. Optical zones of 3.0, 5.0, and 7.0 mm are marked with an appropriate marker tinted with methylene blue (Fig. 3).

One radial corneal incision of 1 mm-length is performed at the steep corneal meridian based on preoperative corneal topography, avoiding the inferior quadrant between optical zones of 5 and 6 mm, with a double-faced guided radial keratotomy diamond knife set at a depth of 80% of local corneal pachymetry. This is done in order to try to perform a stromal tunnel at approx 50% of corneal thickness. The radial incision is performed in the steep corneal meridian, so that the ring segments are implanted in the flat corneal meridian, to achieve a corneal flattening at the opposite steep meridian.

Two concentric stromal corneal tunnels, with an internal radius of curvature of 2.5 mm and an extension of 170°, are then constructed with either a double (Fig. 4) or single (Fig. 5) Ferrara curved spatula® (Ferrara Ophthalmics, Brazil), and then ring segments are implanted in these tunnels (Fig. 6). Ketorolac drops are used every 15 min for 3 h

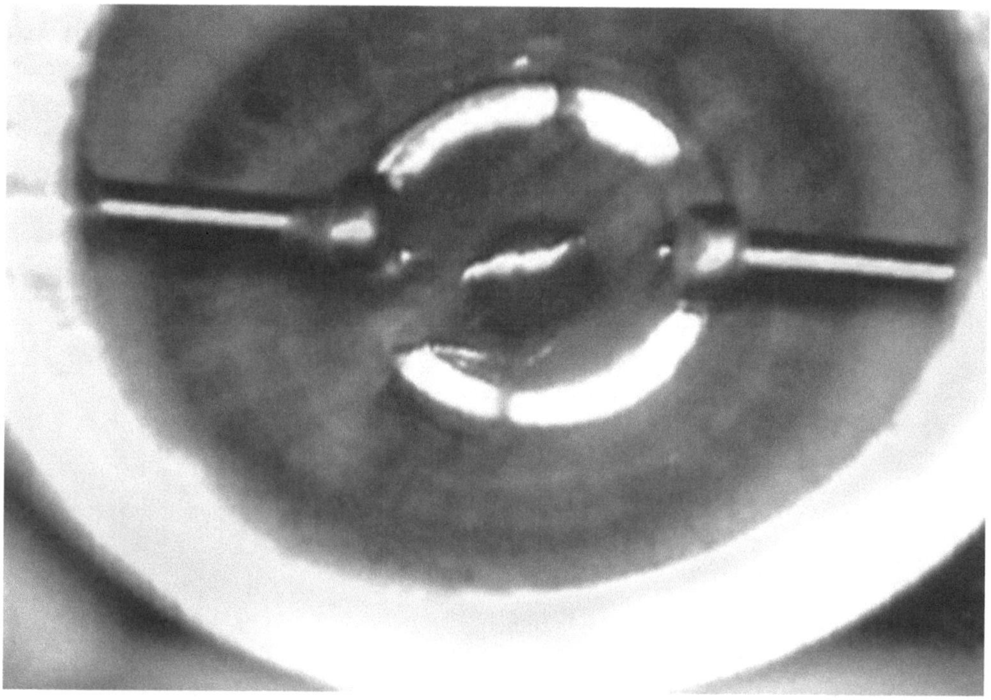

Fig. 4. Stromal tunnels being constructed with a Ferrara double curved spatula.

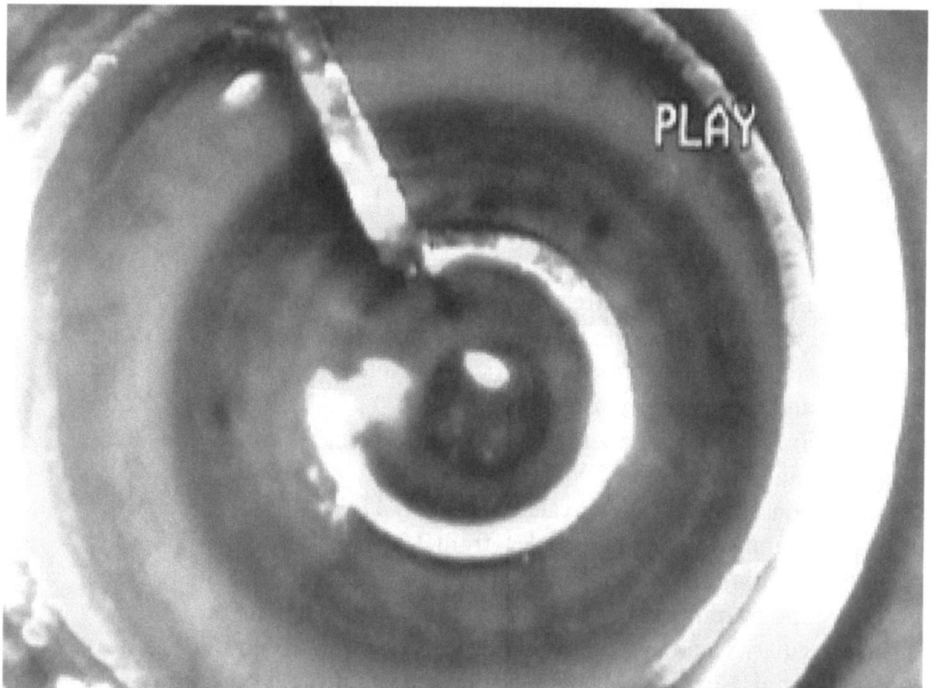

Fig. 5. Stromal tunnels being constructed with a Ferrara single curved spatula.

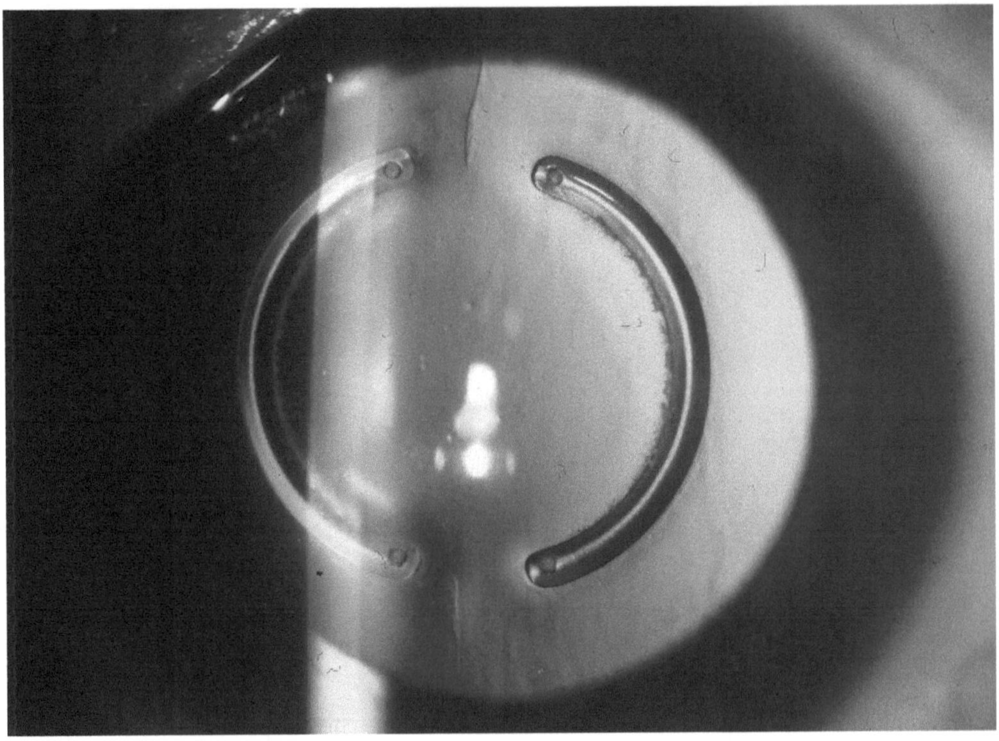

Fig. 6. Ring segments implanted inside stromal tunnels.

after surgery, and a combination of 0.1% dexametasone and gatifloxacin drops are used every 4 h for 7 d, as well as 0.5% methylcellulose drops every 6 h for 30 d.

RESULTS

Ferrara was one of the first authors to suggest implantation of PMMA ring segments to correct keratoconus and irregular astigmatism *(16)*. In Ferrara's initial study, this author reported using FICRS to correct high myopia (up to –20 diopters), myopic and compound regular astigmatism, as well as irregular astigmatism owing to keratoconus and PK. He reported a significant reduction in spherical equivalent (from –10.20 ± 5.98 D to –2.02D ± 2.02 D) and in cylinder (from –4.09 ± 2.42 D to –1.89 ± 1.31 D) after surgery, with preservation of corneal asphericity, and improvement in contrast sensitivity, best-corrected spectacle visual acuity (BCSVA), and topography pattern *(16)*.

A significant reduction in spherical equivalent (from –5.47 ± 4.36 D preoperatively to –2.08 ± 3.52 D postoperatively) and in cylinder (from 3.90 ± 2.30 D to 2.25 ± 1.83 D) in a series of 154 eyes, during the mean follow-up period of 29.1 ± 19.4 mo has been obtained. However, topographic corneal regularity and symmetry, as measured by topography surface regularity and symmetry indexes, did not show significant difference. Holmes-Higgin et al. *(23)* have also not found a correlation between pre- and postoperative BSCVA and predicted corneal visual acuity, a topographically derived index, in patients having undergone intrastromal corneal ring implantation.

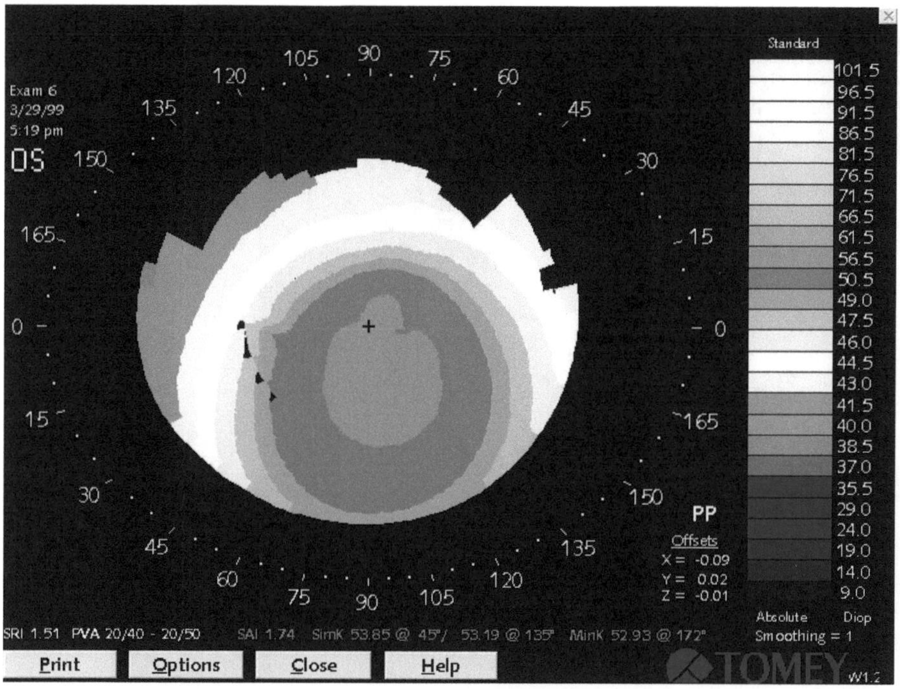

Fig. 7. Example of a preoperative topography of a keratoconus eye.

Our series of 154 keratoconus eyes had BCSVA improvement in 137 eyes (88.9%), having 120 eyes (77.9%) achieved a BCSVA 20/60 or better, during a follow-up period of 29.0 ± 20.37 mo. During this period, 76.5% of these patients did not need a PK for visual recovery *(24)*. Colin et al also found an improvement in mean uncorrected visual acuity (UCVA) and BCSVA in 10 keratoconic patients with INTACS, with a follow-up of 12 mo *(17)*.

Our series of patients revealed that central keratoconus eyes had better results than inferior keratoconus eyes regarding topographic astigmatism, spherical equivalent, and refraction cylinder. A thinner central cornea (far away from the peripheral ring implant location), and a more normal peripheral corneal thickness in central keratoconus eyes might explain the better result in theses eyes, because the thinner central cornea might be more able to relax than a thicker cornea (when the keratoconus apex is inferiorly located). Another fact that might contribute to this finding is that a less resistant thinned inferior cornea may not very well support the implanted ring in that area (in inferior keratoconus). Keratoconus apex preoperatively located in the center of the implanted segment rings may also contribute to a better postoperative central corneal flattening.

A better effect of FICRS in low to moderate keratoconus has been observed (Figs. 7 and 8), with the advanced cases of keratoconus showing the worst results. With a modification of technique suggested by Fabri *(27) (28)*, i.e., performing a 160 μm corneal flap with a Lasik microkeratome before implanting the intrastromal corneal ring segments (Figs. 9–11), we have improved the results in selected cases, with a possibility of further excimer laser photoablation. This modified technique has been performed in keratoconus patients with a minimum corneal thickness of

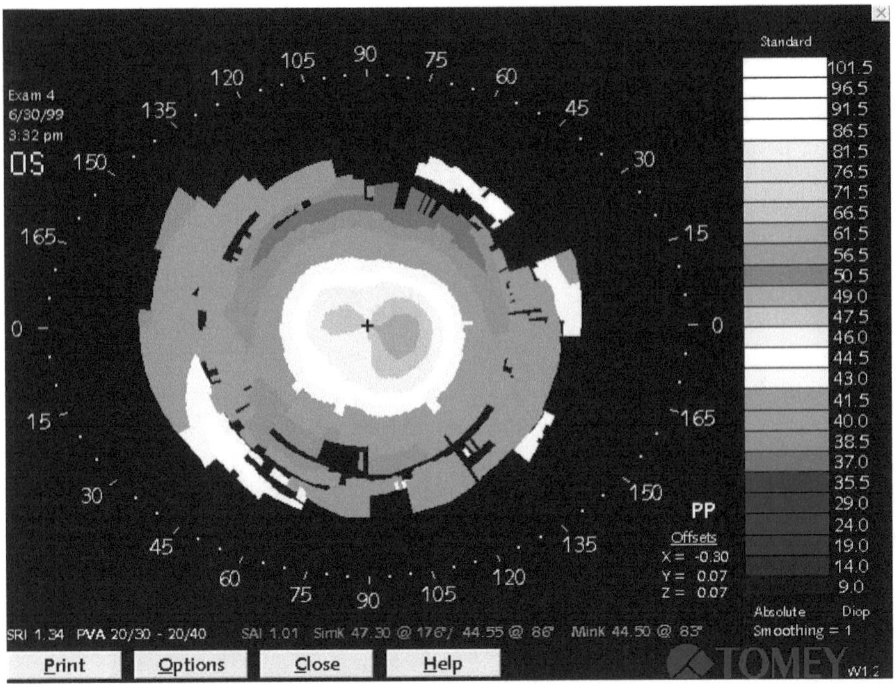

Fig. 8. Same eye of Fig. 5, 6 mo after FICRS implantation.

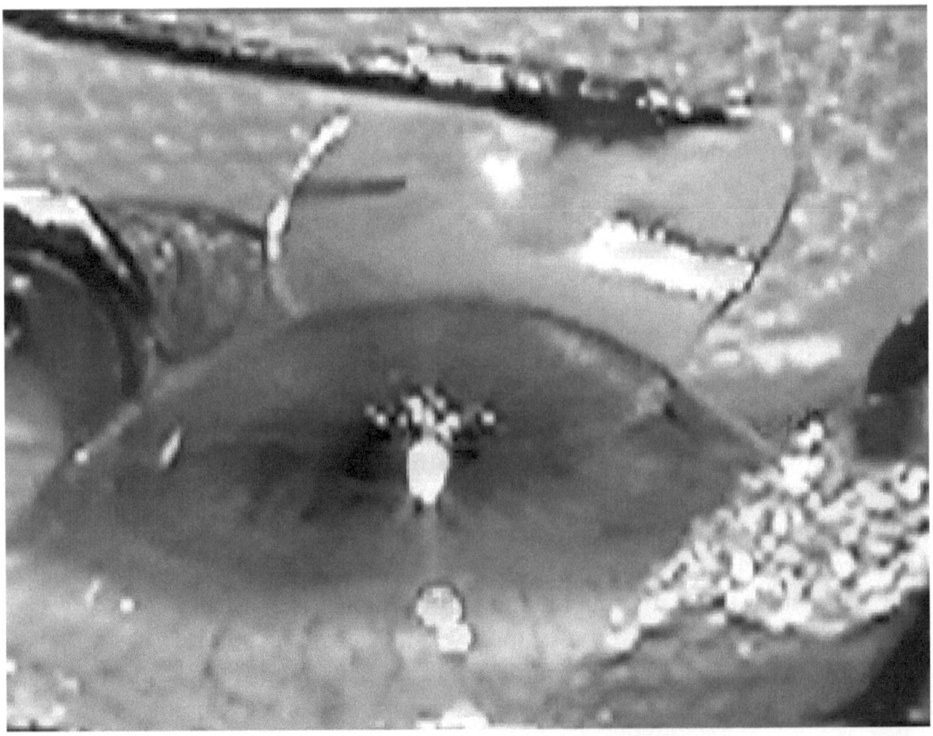

Fig. 9. A 160 μm corneal flap being lifted.

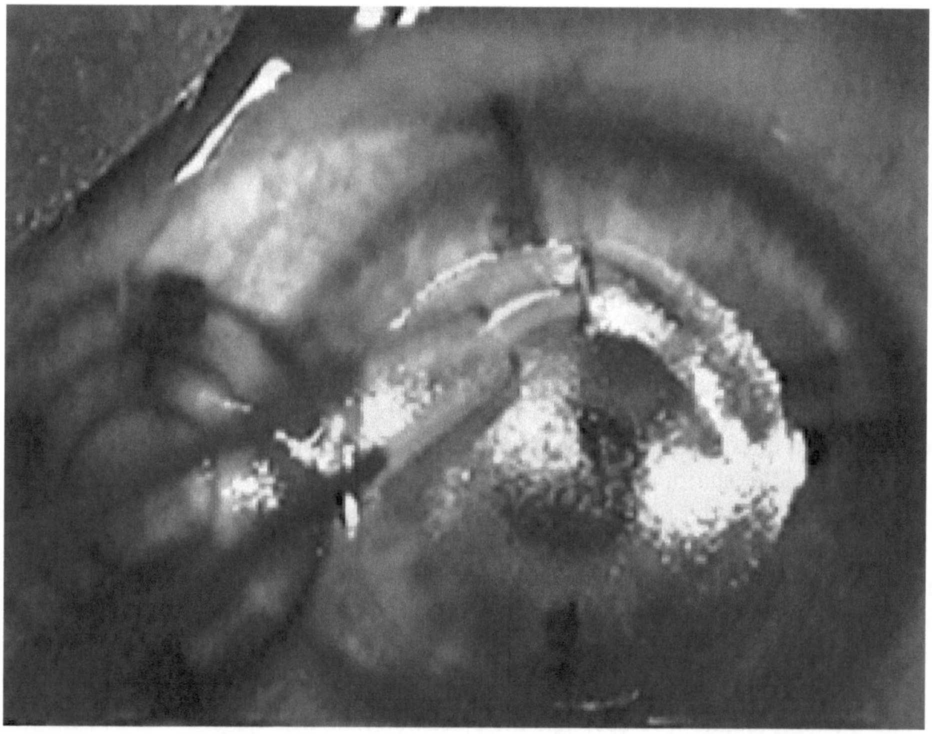

Fig. 10. Ring segment being implanted inside stromal tunnel under the corneal flap.

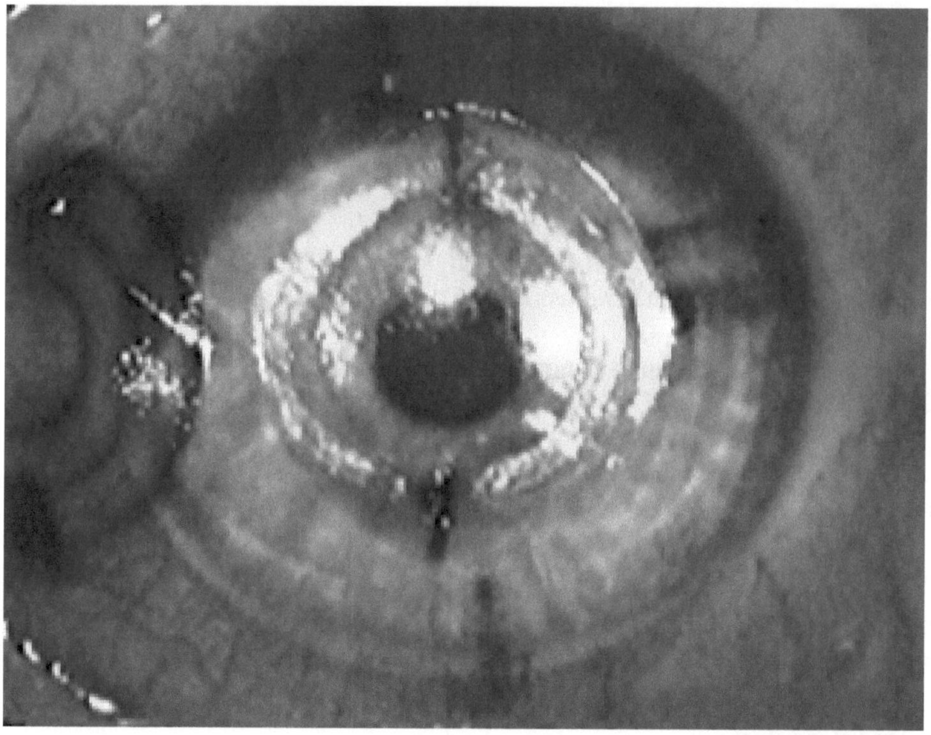

Fig. 11. Ring segments implanted inside stromal tunnels under the corneal flap.

500 μm in the thinnest point, and a faster and more uniform topographic flattening has been observed (Figs. 12 and 13), probably because Bowman's layer was cut by the flap and became less resistant to the ring effect. There was no case of infection or ring extrusion in 28 eyes in which modified technique was performed, and an improvement in BCSVA in 81.5% of eyes was obtained *(28)*. Patients eligible to this technique had a less severe keratoconus, in order to have a minimum corneal thickness of 500 μm, which might explain a better result in these cases.

COMPLICATIONS

A possible intraoperative complication of intracorneal ring implantation is perforation during tunnel construction, as reported by Schanzlin et al. with INTACS *(20)*. With the Ferrara technique of tunnel construction *(16)*, fortunately this complication in non-grafted corneas was not encountered. The curved spatula that constructs the stromal tunnel must be blunt and not sharp, to avoid this complication. In the series, postoperative complications were ring descentration in 1.9%, and ring extrusion (Fig. 14) in 10.2%. INTACS technique implantation revealed a rate of extrusion of 2–10% *(7,20)*. Colin et al had to explant two segments of INTACS because of superficial implantation (10% of their series) in keratoconus patients *(18)*. A significant higher rate of ring segment extrusion was observed in patients with two incisions performed instead of one incision (7.8% and 2.4%, respectively). Therefore, it is suggested to construct the stromal tunnels from one incision location, to decrease the extrusion risk.

It is not possible to be certain about the exact tunnel depth. Different depth of segments implantation might contribute to a better or worse response of this procedure. Furthermore, if the two ring segments are implanted in different depths, or if the same segment is not exactly parallel to the anterior surface of the cornea, induction of astigmatism and lack of keratoconus correction may occur.

One advantage of this surgery is that, as it is reversible when removing the ring, if needed, one can re-implant the segments in a deeper tunnel in the majority of cases 30 d after ring extrusion. Furthermore, if the achieved result is not as expected, one can exchange for a thicker ring to obtain a stronger effect *(12)*.

In the series of 154 keratoconus eyes, there was only one eye that had a ring-related complication, which was a disciform keratitis (Fig. 15), probably as an immunological reaction to the foreign body.

Bacterial keratitis after INTACS implantation is reported from 1% to 20% in the literature *(7,20,29)*. The incidence of FICRS postoperative infection is 1.9% (Fig. 16), all occurring in patients with ring extrusion. All theses cases were cured after ring explantation and vigorous topical and intracorneal antibiotic treatment, with BSCVA returning to the preoperative levels.

CONCLUSIONS

FICRS implantation technique is relatively easy for a corneal surgeon, but one should pay attention to some important details for the intended success, such as:

1. A correct tunnel construction, starting with an 80% depth of corneal thickness at the location of radial incision;
2. One radial incision;

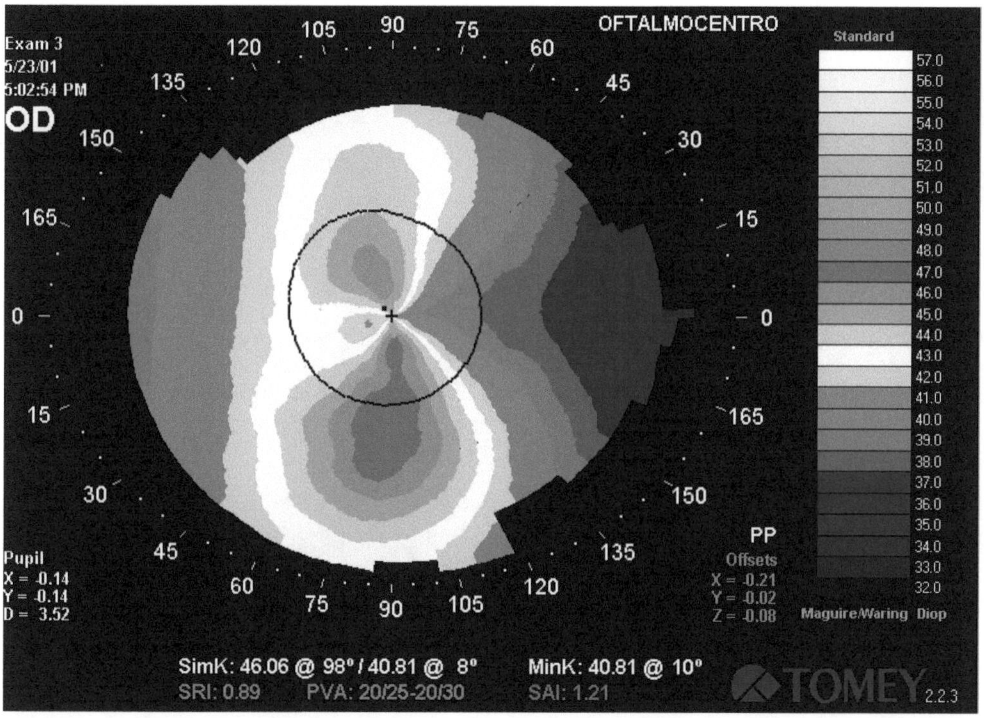

Fig. 12. Example of a preoperative topography of a keratoconus eye.

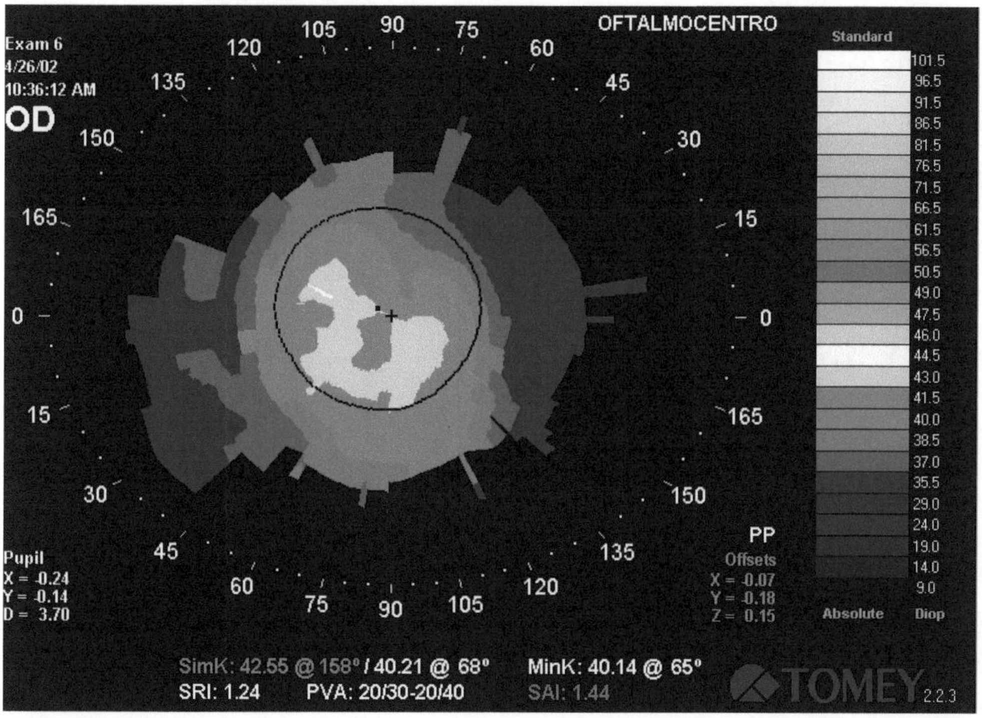

Fig. 13. Same eye of Fig. 10, after 11 mo of Ferrara's ring segments implantation under a corneal flap.

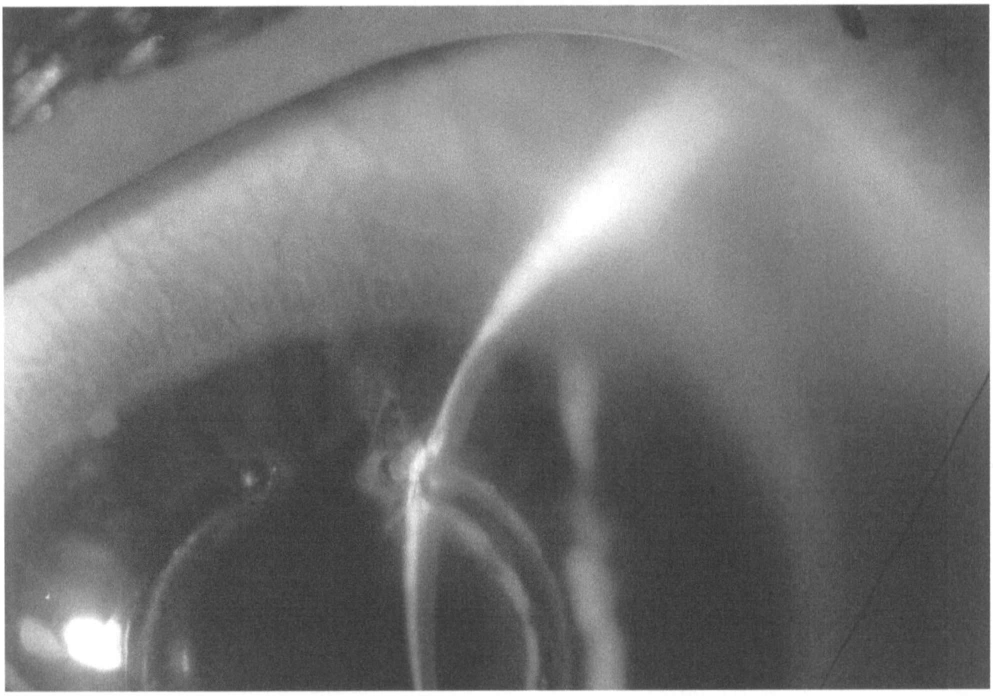

Fig. 14. Ring segment extrusion 5 mo after implantation.

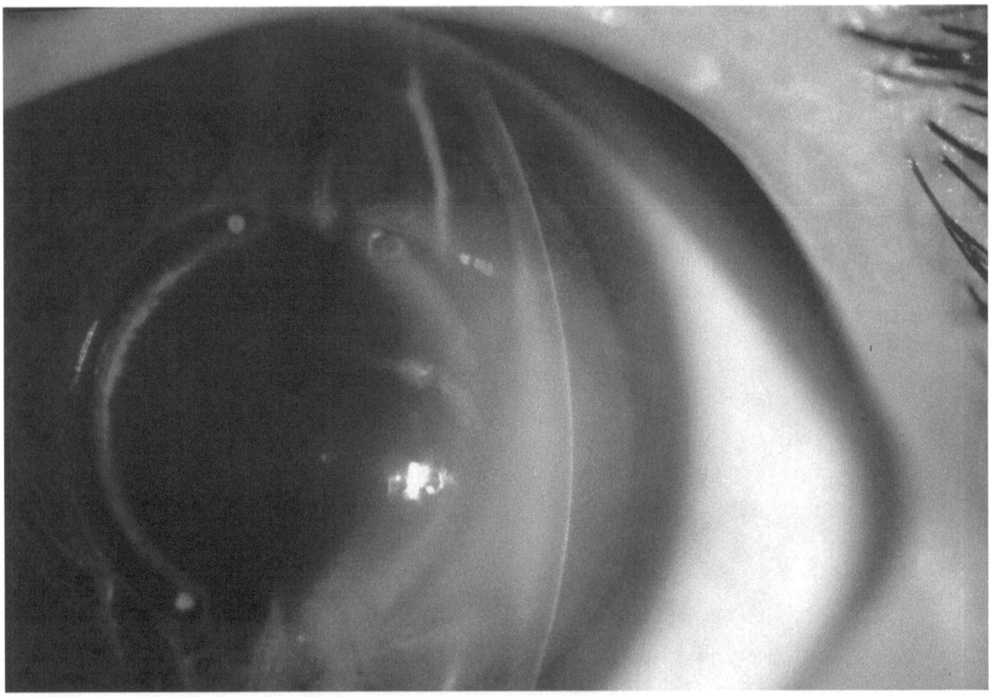

Fig. 15. Disciform keratitis adjacent to the segment, 7 mo after surgery.

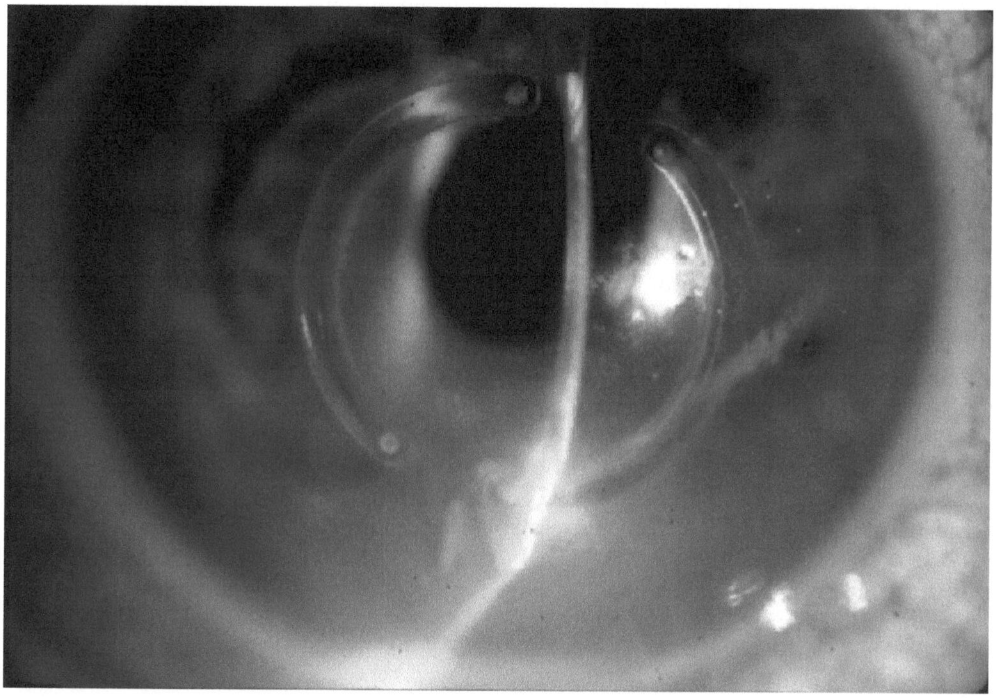

Fig. 16. Bacterial keratitis after ring extrusion.

3. Ring segment centration based on central corneal reflex;
4. Correct selection of ring segment position in the flat corneal meridian;
5. Selection of asymmetric rings based on corneal topography;
6. Simultaneous 160 μm corneal flap if corneal thickness is at least 500 μm in its thinnest point; and
7. Avoid implanting the ring segments in advanced keratoconus or if there is already a corneal opacity.

The experience suggests that FICRS have a definite place in the treatment of keratoconus, especially in those patients that are contact lens intolerant and are candidate for PK for visual improvement. This procedure has been proved to be successful for several keratoconus patients that are in the waiting list for a PK, being an extraocular fast surgery with topical anesthesia and having a very low ring rejection rate. Besides improving BCSVA, it will be possible to avoid or delay PK in many keratoconus eyes (76.5% of cases in author's personal experience).

In conclusion, FICRS implantation has the advantage of being reversible, adjustable, safe, and a low cost surgery, with reasonable predictable results that can recover BCSVA in several cases of keratoconus patients. Another advantage is that it does not interfere with a future PK if necessary, because it is placed inside the diameter of a PK. However, the long-term stability of these results is not known. Further clinical and experimental studies with more patients and longer follow-up are needed to improve accuracy and stability of FICRS results.

REFERENCES

1. Belin MW. Optical and surgical correction of keratoconus. In: Focal Points: Clinical Modules for Ophthalmologists, vol. 6, module 11. San Francisco, Am Acad Ophthalmol, 1988.
2. Bechara SJ, Kara-José N. Ceratocone. In: Belfort Jr R, Kara-José N, ed. Córnea clínica e cirúrgica, 1st ed., Ed. Roca, 1997;359–366.
3. Barraquer JI. Modification of refraction by means of intracorneal inclusions. Int Ophthalmol Clin 1966;6:53–78.
4. Bach RH, Maumenee AE. Corneal fluid metabolism. Arch Ophthalmol 1953;50:282–291.
5. Belau PG, Dyer JA, Ogle KN, Henderson JW. Correction of ametropia with intracorneal lenses: an experimental study. Arch Ophthalmol 1964;72:541–548.
6. Dohlman CM, Brown S. Treatment of corneal edema with a buried implant. Trans Amer Acad Ophthalmol Vis Sci 1981;21:107–115.
7. Nosé W, Neves RA, Buris TE, et al. Intrastromal corneal ring: 12 month sighted myopic eyes. J Refract Surg 1996;12:20–28.
8. Fleming JF, Wan WL, Schanzlin DJ. The theory of corneal curvature change with the ICR. CLAO J 1989;2:146–150.
9. Baikoff G, Maia N, Poulhalec D, et al. Diurnal variations in keratometry and refraction with intracorneal ring segments. J Cataract Refract Surg 1999;25:1056–1061.
10. Burris TE. Intrastromal corneal ring technology: results and indications. Curr Opin Ophthalmol 1998;9:9–14.
11. Cochener B, Le Floch G, Colin J. Intra-corneal rings for the correction of weak myopias. J Fr Ophthalmol 1998;21:191–208.
12. Asbell PA, Uçakhan OO, Durrie DS, et al. Adjustability of refractive effect for corneal ring segments. J Refract Surg 1999;15:627–631.
13. Ruckhofer J, Stoiber J, Alzner E, et al. Intrastromal corneal ring segments (ICRS, KeraVision Ring, INTACS): clinical outcome after 2 years. Klin Monatsbl Augenheidlkd 2000;216:133–142.
14. Burris TE, Baker PC, Ayer CT, et al. Flattening of central corneal curvature with intrastromal corneal rings of increasing thickness: an eye-bank eye study. J Cataract Refract Surg 1993;19 Suppl :182–187.
15. Burris TE, Holmes Higgin DK, Silvestrini TA, et al. Corneal asphericity in eye bank eyes implanted with the intrastromal corneal ring. J Refract Surg 1997;13:556–567.
16. Cunha PF. Técnica cirúrgica para correção de miopia. Anel corneano intraestromal. Rev Bras Oftalmol 1995;58:19–30.
17. Colin J, Cochener B, Savary G, Malet F, Holmes-Higgin D. INTACS inserts for treating keratoconus: one-year results. Ophthalmology 2001;108(8):1409–1414.
18. Colin J, Cochener B, Savary G, Malet F. Correcting keratoconus with intracorneal rings. J Catarct Refract Surg 2000;26:1117–1122.
19. Amsler M. Keratoconus: early diagnosis and microsymptoms. Ophthalmologica 1965; 149:438–446.
20. Schanzlin DJ, Asbell PA, Burris TE, et al. The intrastromal corneal ring segments: phase II results for the correction of myopia. Ophthalmology 1997;104:1067–1078.
21. Pisella PJ, Albou-Ganem C, Bourges JL, et al. Evaluation of anterior chamber inflammation after corneal refractive surgery. Cornea 1999;18:302–305.
22. D'Hermies, Hartmann C, von Eye F, et al. Biocompatibility of a refractive intracorneal PMMA ring. Fortschr Ophthalmol 1991;88:790–803.
23. Holmes-Higgin DK, Burris TE, Asbell PA, et al. Topographic predicted corneal acuity with intrastromal corneal ring segments. J Refract Surg 1999;15:324–330.
24. Kwitko S, Severo N. Ferrara intra-corneal ring segments for keratoconus. Cataract and Refract Surg J 2004;30:812–820.

25. Oliveira C, Moreira H, Wahab S, et al. Analysis of new technique for Ferrara ring implantation in keratoconus. Arq Bras Oftalmol 2004;67:509–517.
26. Cunha PF. Asymmetric ring segment implantation for keratoconus. VII International Congress on Cataract and Refractive surgery. São Paulo, Brazil, 2002.
27. Fabri PP. Ferrara ring under a corneal flap. I Brazilian Congress of Cataract and Refractive Surgery. Goiânia, Brazil, 2001.
28. Kwitko S. Ferrara ring under a corneal flap. Oftamologia em foco 2004;90:33–34.
29. Hofling-Lima AL, Castelo Branco B, Romano A, et al. Corneal infections after implantation of intracorneal ring sements. Cornea 2004;23:547–549.

Index

MIX
Papier aus verantwortungsvollen Quellen
Paper from responsible sources
FSC® C105338

If you have any concerns about our products,
you can contact us on
ProductSafety@springernature.com

In case Publisher is established outside the EU,
the EU authorized representative is:
Springer Nature Customer Service Center GmbH
Europaplatz 3, 69115 Heidelberg, Germany

Printed by Libri Plureos GmbH
in Hamburg, Germany